T1

A Survival Guide

Matthew S. Gast

Beijing · Cambridge · Farnham · Köln · Paris · Sebastopol · Taipei · Tokyo

T1: A Survival Guide
by Matthew S. Gast

Published by O'Reilly & Associates, Inc., 101 Morris Street, Sebastopol, CA 95472.

Editor: Mike Loukides

Production Editor: Linley Dolby

Cover Designer: Ellie Volckhausen

Printing History:

August 2001: First Edition.

Library of Congress Cataloging-in-Publication Data

Gast, Matthew.
T1 : a survival guide /Matthew Gast.
p. cm.
ISBN 0-596-00127-4
1. T1 (Telecommunication line)--United States. I. Title.

TK5103.15 .G37 2001
621.387'84--dc21 2001032861

[DS]

Table of Contents

Preface ... *vii*

1. History of the U.S. Telephone Network ... *1*
- 1876–1950: Analog Beginnings ... *2*
- 1951–1970: The Birth of T-carrier ... *4*
- 1970–Present: The Modern Telephone Network ... *5*

2. T1 Architectural Overview ... *8*
- Telecommunications Puzzle Pieces ... *8*

3. Basic Digital Transmission on Telephone Networks ... *17*
- Introduction to DS0 ... *18*
- Alternate Mark Inversion ... *18*
- B8ZS and Clear Channel Capability ... *21*

4. Multiplexing and the T-carrier Hierarchy ... *22*
- Building the T-carrier Hierarchy with Multiplexing ... *22*
- The Original Superframe ... *26*
- The Extended Superframe (ESF) ... *27*
- Telephone Signaling on T1 Links ... *30*

5. Timing, Clocking, and Synchronization in the T-carrier System ... *32*
- A Timing Taxonomy ... *32*
- T1 Circuit Timing ... *35*
- Slips: When Timing Goes Bad ... *41*

6. *Mysteries of the CSU/DSU* ... 44
Line Build Out: Moving Between Theory and Practice ... 44
T1 CSU/DSUs ... 48
CSU/DSU Configuration ... 51
Summary of Settings ... 53

7. *Connecting the Umbilicus: Getting T1 Connectivity* ... 56
Ordering ... 56
T1 Installation and Termination ... 57
Pre-Connection Tasks ... 60
Trading Packets ... 61
Post-Connection ... 62

8. *High-Level Data Link Control Protocol (HDLC)* ... 63
Introduction to HDLC ... 64
HDLC Framing ... 65
Cisco HDLC ... 70

9. *PPP* ... 76
Introduction to PPP ... 77
PPP Logical Link States and State Machines ... 78
PPP Encapsulation and Framing ... 83
Link Control Protocol (LCP) ... 85
PPP Network Control and the IP Control Protocol ... 109
Configuring PPP ... 112

10. *Frame Relay* ... 115
Frame Relay Network Overview ... 116
The Frame Relay Link Layer ... 118
Multiprotocol Encapsulation with RFC 1490 ... 120
The Local Management Interface ... 122
Configuring Frame Relay ... 130

11. *T1 Troubleshooting* ... 132
Basic Troubleshooting Tools and Techniques ... 132
Troubleshooting Outline ... 138
Physical Layer Problems ... 139
Link Layer Problems ... 145

Preface

In the past, T1 links formed the backbone of the telephone network in the U.S.; they were later used as the data circuits for the Internet backbone. Now, T1 is the standard for medium-speed access to the Internet. Widespread adoption is not to be confused with technical excellence, though. T1 is not elegant, efficient, easy to use, or even all that well suited to the needs of data transmission, largely because its development is intimately tied to the U.S. telephone system architecture of the late 1960s.

In spite of T1's limitations, no competing technology has presented a significant challenge. T1 is proven, reliable technology that currently fills the need for medium-speed Internet access for institutions of many sizes, especially those not in metropolitan areas. Newer technologies require widespread availability of fiber and extensive regional networks, and the population density in rural areas is not well suited to the expense of building large fiber networks. For Internet access off the beaten path, traditional leased lines are the only solution.

In spite of its age, T1 technology is often learned by apprenticeship, filing away little troubleshooting tidbits until they may be needed again. For technology older than many of its users, this is an inexcusable state of affairs. As more networks use T1s, more engineers will require practical information on how to set up, test, and troubleshoot them. This book is intended to be a practical, applied reference for networkers interested in using T1 as a data transport. It covers the following broad topics:

- How the history of the telephone network led to the development of T1 and how T1 was adapted to the needs of data transport
- What components are needed to build a T1 line, and how those components interact to transmit data effectively

- How to adapt the T1 physical layer to work with data networks by the use of standardized link layer protocols
- How to troubleshoot problems that arise, including how to work with all the vendors: the telephone company, equipment manufacturers, and Internet service providers

Audience

This book maintains a blatant, and sometimes overwhelming, focus on the U.S. In many ways, this concentration is a result of the fact that the T-carrier system was developed at Bell Labs and deployed by AT&T. European digital carrier systems took a different direction (and came out with a better result, in some respects). Nevertheless, this book is not likely to be useful for engineers who do not deal with American digital carrier systems.

This book approaches T1 from the fairly narrow perspective of data networking. Although T1 can be used for voice or video transport, or as a flexible network that can respond to changing demands, this book treats it as simply another type of network link. Many of the examples deal with Internet access explicitly, although several of these could easily be applied to private leased-line networks. Network engineers must frequently deal with T1 for a variety of reasons: it can be used to provide a single pipe for Internet access or 24 voice circuits for modem aggregation, or even as the distribution layer or core of a far-flung WAN.

Coordinating T1 networks involves a large cast. Telco employees are responsible for providing the T1s. ISPs may order T1s on behalf of their customers and support them directly with backup by the telco. Small businesses with a need for medium-speed Internet access now frequently turn to T1 technology to fill that need. In recognition of the popularity of T1 as a data transport, most network equipment vendors now offer interface cards with internal T1 CSU/DSUs. Engineers supporting T1 interfaces must be familiar with details as well as the "big picture" so that they can guide new T1 users through the steps of troubleshooting and contacting carriers and service providers to bring up T1s.

T1 may not be glamorous, but it fills a need for medium-speed, high-reliability Internet connectivity, and nothing is likely to displace it from that role in the near future.

Overture for Book in Black and White, Opus 3

Most of this book is theoretical. In parallel with describing T1 engineering, it discusses how engineering features apply to T1 systems as they are built and deployed in the real world. All the theory comes together in the final chapter on troubleshooting, which applies the theory to solving problems. Here is a breakdown of the chapters.

Chapter 1, *History of the U.S. Telephone Network*, traces the history of the telephone network in the U.S. from Alexander Graham Bell's first patent to the modern digital network that carries a great deal of Internet data.

Chapter 2, *T1 Architectural Overview*, provides a short overview of all the components that must work together to get a T1 running.

Chapter 3, *Basic Digital Transmission on Telephone Networks*, describes the basic building block of the T-carrier system in the U.S., the single voice channel.

Chapter 4, *Multiplexing and the T-carrier Hierarchy*, explains how the building blocks are put together to form a T1.

Chapter 5, *Timing, Clocking, and Synchronization in the T-carrier System*, presents the need for tightly controlled timing relationships between T1 components and explains several common timing architectures.

Chapter 6, *Mysteries of the CSU/DSU*, clarifies the role of the CSU/DSU, one of the most important, yet least understood, pieces of equipment in the T1 system.

Chapter 7, *Connecting the Umbilicus: Getting T1 Connectivity*, walks you through the process for turning up a T1.

Chapter 8, *High-Level Data Link Control Protocol (HDLC)*, describes one of the oldest link layer framing protocols ever standardized, the High-level Data Link Control protocol. With this chapter, the book shifts from the hardware realm of the physical layer to the software realm of packets, addressing, and framing.

Chapter 9, *PPP*, describes the Point-to-Point Protocol standard. PPP was designed with great flexibility and can run on links from the lowliest analog dial-up line to the SONET links at the Internet core. The purpose of this chapter is to introduce PPP and how it relates to data communications on T1 links.

Chapter 10, *Frame Relay*, introduces another link layer protocol derived from HDLC. Frame relay is a set of specifications that allow transmission capacity to be shared by multiple subscribers, and it is often the most cost-effective way to connect to larger networks.

Chapter 11, *T1 Troubleshooting*, has several charts and tables that lay out my accumulated T1 debugging wisdom and experience, plus the vicarious experience I have gathered by speaking with even older wizards.

Appendix A, *Access Aggregation with cT1 and ISDN PRI*, delves into the details of T1 as the telephone company sees it. High-speed modems rely on having only one conversion between digital and analog transmission, so they depend on service providers to deploy digital telephone technology. In the U.S., service providers can choose between channelized T1 and primary rate ISDN. Both can deliver up to 24 telephone calls digitally to an ISP.

Appendix B, *Multilink PPP*, builds on Chapter 9 by describing how to use multilink PPP to aggregate several small links into a larger logical link. Multilink PPP can be used with multiple T1s for increased speed, but it is more frequently used by subscribers to logically tie multiple modem or ISDN connections together.

Appendix C, *T1 Performance Monitoring*, explains the architecture used to collect T1 link performance statistics. In addition to "soft" factors such as response time and guaranteed repair time, telco service level agreements may include "hard" factors such as upper bounds on bit error rates.

Appendix D, *SNMP*, describes the T1-related MIBs. T1 administrators can use the DS1 MIB for T1-related statistics. Depending on the link layer in use, it may also make sense to monitor statistics from the frame relay DTE MIB and the PPP MIBs.

Appendix E, *Cable Pinouts and Serial Information*, has pinouts for the serial connectors and jacks that are commonly used in T1 systems.

Appendix F, *Reference*, is an annotated bibliography of standards documents relating to T1.

Finally, the Glossary lists terms and acronyms that are often used in T1 networking and defines common terms used by telephone companies. (When in Rome, it helps to speak Latin.)

Assumptions This Book Makes

When writing about telephony, a recurring challenge for authors is deciding what to leave out. No matter how detailed the text, some low-level details must be omitted to keep from inundating the reader with potentially interesting, but pointless, detail. I have left out a great deal of detail from the formal standards, published by ANSI. Appendix F lists a wide variety of relevant documents for interested readers.

This book is about using T1 to transmit data from one point to another. It does not address using T1 as a voice transport, except insofar as to explain why some parts of the T-carrier system evolved the way they did. No material on channel banks is in this book, except in some places to add a historical perspective. Finally, this book steers completely clear of pricing—it is about deploying and technically managing T1 services, not about using T1 for any particular economic reason.

Familiarity with the seven-layer OSI model is assumed. With rare exceptions, this book stays at the first layer (physical) and the second layer (data link). Basic understanding of electronics and the physics of wave propagation is helpful, but not required. Complete descriptions for several of the protocols would take up entire books. SNMP, HDLC, PPP, and frame relay are all discussed as they relate to T1; a relatively solid background is assumed.

Conventions Used in This Book

Italic is used throughout the book to indicate Internet addresses, such as domain names and URLs, and new terms where they are defined.

I use the common industry term *telco* when referring to telecommunications service providers. I use it interchangeably with the term *carrier*.

Telco references to *kilo-* and *mega-* are used as power-of-ten modifiers. That is, a kilobit per second is 1,000 bits per second. For clarity, I abbreviate a *1,000 kilo* (or "telephone company kilo") as *k* (e.g., 64 kbps for 64,000 bits per second) and a conventional computer science power-of-two kilo as *K* (e.g., 4 KB for 4,096 bytes).

Standards documents often use the word *octet* for a sequence of 8 bits. I prefer the term *byte*, though it is not technically accurate in all cases. Now that most machines with bytes of non-8-bit lengths have been retired from service, I feel comfortable doing so.

I use the standard C convention of prepending *0x* to a number to denote that it is hexadecimal. Unless otherwise noted, numbers in the text are decimal numbers.

ISO and ITU-T diagrams generally present bits in the order of transmission, which is usually least-significant first. In IETF diagrams, the fields are shown with the most-significant bytes first. Furthermore, IETF diagrams begin counting bits with zero, while other standards organizations begin with one. I shift between these two forms throughout the book. All diagrams use the bit order favored by the relevant standards organization, which means that ISO/ITU specifications use a different order than the IETF standards.

This icon indicates a tip, suggestion, or general note.

This icon indicates a warning or caution.

How to Contact Us

Please address comments and questions concerning this book to the publisher:

O'Reilly & Associates, Inc.
101 Morris Street
Sebastopol, CA 95472
(800) 998-9938 (in the U.S. or Canada)
(707) 829-0515 (international/local)
(707) 829-0104 (fax)

We have a web site for the book, where we'll list examples, errata, or any additional information. You can access this page at:

http://www.oreilly.com/catalog/t1survival/

To comment or ask technical questions about this book, send email to:

bookquestions@oreilly.com

For more information about our books, conferences, software, Resource Centers, and the O'Reilly Network, see our web site at:

http://www.oreilly.com

Acknowledgments

Any comprehensibility in the language I use is due to a long line of stellar teachers and professors who helped me to constantly improve my writing. I was fortunate to have attended an institution that prides itself on teaching all students to write, even those of us who majored in physics. Eliza Willis, my tutorial professor and initial academic advisor, had one of the strongest influences on my writing in my college years. One of her evaluations expressed the hope that I would continue to develop my writing, though I hope no readers hold my style against her. As my first academic advisor, much of her effort was spent ensuring that I did not take too many classes in the science building. My curriculum should be taken as definitive proof of her diplomacy and persuasiveness, as well as evidence that I even followed most of her advice. During my senior year, she received tenure. Granting her tenure should have been one of the easiest decisions any institution has ever had to make, and it guarantees that her influence on the world is only just beginning.

Several of my coworkers and friends provided encouragement and support along the way. In particular, Brian McMahon was invaluable. Brian is one of the extremely talented old-guard networkers from whom we have much to learn. I was fortunate to meet him in 1995, when the history of the Internet was shorter and simpler and protocols were fewer. (Some might even say that all the computers connected to it were above-average, too.) Brian has a deep understanding of T1s gained from two years with Cisco, and he graciously let me draw on his knowledge to provide a practical underpinning for a great deal of the material in this book. Most of his help with information on channelized T1 should be evident in Appendix A, especially any portions related to signaling. He also suggested the title for the book and is indirectly responsible for the choice of animal on the cover. I am also grateful for Brian's detailed technical review of the book, which helped me clear up a number of misconceptions that the initial draft may have inspired in readers.

My efforts to include photographs resulted from the aid of two colleagues in particular. Jerald Josephs, a colleague on the sales engineering team at Nokia, lent me his digital camera and took time out of his busy schedule to teach me how to use it. Giao Le let me into the telephone room at Nokia's Silicon Valley installation to photograph the smart jacks shown in Chapter 2.

Before I started writing, I had no idea how large a cast of characters went into book production. A great number of talented people have teamed up to provide me with a set of broad shoulders on which to stand. My family and friends, the group of people most familiar with my artistic (in)abilities, will easily see the tremendous work done by Jessamyn Read in the O'Reilly Illustration department. To fully appreciate her talents, you would have to see the obscure chicken-scratchings and mysterious hieroglyphics that metamorphosed into a stunning set of figures by her hand. Mike Loukides once again kept me on track and focused on the task, which I especially appreciated when writing about telecommunications. Standards are dense and multitudinous, and I could have easily turned this project into my life's work as I chased down "just one more standard." Mike's assistance was most welcome in balancing the goals of making the book complete and getting the book finished—and it is certainly better for his balance.

One of the major difficulties in writing a technical book is the desire to ensure accuracy. In telecommunications, more so than other fields, it is difficult to keep the writing interesting yet technically accurate. My excellent review team helped keep me honest and caught several mistakes that I found tremendously embarrassing. Steve Pinkston of ADC brought a formidable degree of expertise to the review, and his comments led to significant improvements in nearly every chapter. Thanks also go to Dianna Mullet of L3 Communications and Kevin Mullet of the University of Texas at Dallas. As authors themselves (of O'Reilly's *Managing IMAP*), they were able to provide subtle guidance on numerous aspects of the HDLC and PPP chapters. Dianna's notes helped me to pass through the minefield that is PPP without too much permanent damage. Kevin's observations led to a series of improvements in several parts of the book. One of Kevin's suggestions put me on the right path to improving the multilink PPP information provided in Appendix B, and a few of his notes wound up in the reference appendix (Appendix F).

1

History of the U.S. Telephone Network

All is flux, nothing is stationary.
—Heraclitus

We live in a digital world.

By digital, I don't mean the overused "digital economy" or anything remotely related to punditry. In the beginning, there was analog. Transmitted signals were subject to fading, interference, and noise. All that changed with digital encoding and transmission. Digital signals offer superior properties for information. Perfect regeneration is possible with sophisticated error detection and correction methods. Signals can be sent without loss or distortion over arbitrary distances. Think of all the digital systems people use on a daily basis: CDs are digitized audio, many cable systems are now digital, and the Internet cannot be overlooked. Then there is the telephone, which was one of the earliest experiments in digital technology.

Zeros and ones are not just the building blocks of the future—they are also the building blocks of the present.

It was not always so, of course. Telephony was initially all analog, and signal processing was not even a discipline. Social changes and increased mobility in the 1950s threatened to break the telephone system by swamping its capacity. Rescue came by way of digital technology and its increased capacity and quality. Telephone companies invested in the new digital technology to transmit digitized voices over long distances and, in doing so, laid the foundation for the long-distance data transmission so critical to the Internet. T1, though now used overwhelmingly for data transmission, found its roots in the U.S. telephone network in the digital upheaval of the 1960s.

1876–1950: Analog Beginnings

When Alexander Graham Bell invented the telephone in 1876, he followed a course of action familiar to many inventors today: he filed patent applications. His patent applications were approved in 1876–77, and he promptly sought financial backing to build a company to collect patent license fees. In 1877, he formed the Bell Telephone Company with two investors.

The year 1878 saw the construction of the first telephone exchange in New Haven, Connecticut. In the first exchanges, telephone calls were switched manually by operators using plugboards—a far cry from the modern central offices with computerized switching equipment. In the three years that followed, Bell Telephone licensed many local operators and constructed exchanges quickly in major cities.

Western Union attempted to enter the infant telephony market in 1861, after securing some patent ammunition of its own with the construction of the first transcontinental telegraph line.* Bell Telephone, by this point renamed American Bell, fought Western Union's entry tooth and nail in court until, eventually, Western Union agreed to stay out of telephony, provided that Bell did not compete in the telegraphy market. With the collapse of Western Union's ambitions in the telephone market, its major supplier, the Western Electric Manufacturing Company, was financially hobbled. Western Electric was an engineering firm with a record of successfully manufacturing novel inventions (including the incandescent light bulb), but was forced to sell a controlling stake to American Bell in 1881 to survive. Western Electric subsequently became the manufacturing arm of American Bell. In 1883, Western Electric's Mechanical department was founded. Ultimately, the Mechanical department grew into Bell Telephone Laboratories, a name interwoven with much of the technological developments later in this story.

In 1885, American Telephone and Telegraph (AT&T), founded to build and manage the long-distance network, started operations in New York; the long-distance network reached Chicago by 1892. Because electrical impulses degrade with distance, newer technologies were needed in order to reach farther. Loading coils, which reduce the decay of signal strength in the voice band (frequencies less than 3.5 KHz), enabled the long-distance network to reach Denver. Transcontinental service needed one more major enhancement, the addition of practical electronic amplifiers, which allowed AT&T to extend the long-distance network to San Francisco by 1915. In the midst of all this, AT&T became the parent company of American Bell in 1899.

---

* Some technology industry observers are amused at the way technology goes in cycles. It is an old phenomenon: the telegraph, a digital network (dash or dot), was replaced by an analog network (the early Bell System), which was then replaced by a digital network (the digitalized Bell System).

After Bell's patents expired in 1894, the company faced fierce competition from many upstart telephone companies. Theodore Vail, who had led American Bell's patent fight against Western Union, took the helm of AT&T in 1907 and guided the company to the dominant position it would hold for much of the 20th century. Under Vail, AT&T acquired many upstart telecommunications companies and became a regulated monopoly under the theory that telecommunications would work best when operating as a regulated monopoly dedicated to universal service.

To a certain extent, Vail had a point. One of the major problems faced by the telephone network without the uniformity of Bell's patents was interoperability. Different phone companies could choose different equipment manufacturers, and equipment from different vendors did not necessarily work together. With interoperability difficulties, interconnection was spotty, so it was not guaranteed that two telephone subscribers could call each other unless they were customers of the same telephone company!

The U.S. government accepted AT&T's monopoly status, but required AT&T to allow all local telephone companies to connect to its long-distance network. Interoperability was easy—everybody bought Western Electric equipment and, presumably, liked it. By 1939, AT&T served 83% of U.S. telephones, manufactured 90% of U.S. telephones, and owned 98% of long-distance connections.

In this period of growing dominance, AT&T was able to devote substantial resources to research and development. In addition to the famed Nobel Prize–winning basic research carried out by Bell Labs, AT&T made a steady stream of improvements to the telephone network, many of which were direct results of Bell Labs research. AT&T began experimenting with radio as a means of enabling transatlantic communication in 1915. They made commercial radio service available in 1927, but for a hefty price: calls cost $75 for 5 minutes.* During this time, Bell Labs researchers realized the potential of electromechanical relays for storing and processing information. In 1937, relays were first operated in a manner we would recognize today as being analogous to computer memory. In 1949, the U.S. government brought an antitrust suit against AT&T in an attempt to divest Western Electric. Seven years later, a consent decree restricted AT&T to running the nation's telephone network.

* To see the plummeting cost of telephony, it is interesting to adjust the 1927 rate for inflation. As measured by the consumer price index (CPI), the 1927 transatlantic telephone call would cost slightly more than $153 per minute in 2001 dollars. Thanks to the march of technology, today calls cost a mere fraction of that, even without signing up for a discount calling plan!

1951–1970: The Birth of T-carrier

It may not have been obvious at the time, but the telephone network was on the cusp of a fundamental change driven by several factors. Early long-distance connections, especially those across an ocean, were full of static and could fade quite abruptly. Development of appropriate cabling would eliminate the problems caused by the use of radio waves, but noise and distortion of the analog signal across such great distances posed a large problem. Postwar prosperity in the U.S. dramatically increased the demand for telephone service, bringing with it millions of new telephone users and requiring ever more operators to connect calls. Better means of providing service had to be found. Fortunately, two new technologies came to the rescue.

The Transistor and Computerization

The transistor, invented in 1947 at Bell Labs, made it possible to construct sophisticated electronics. Today's semiconductor and computer industries owe their existences to the transistor. Transistors also made AT&T's "Electronic Switching System" possible. In the early 1960s, AT&T built and field-tested specialized, programmable devices for controlling the telephone network. These devices, of course, were computers, but the terms of the 1956 consent decree prohibited AT&T from manufacturing computers. AT&T followed the time-honored tradition of redefining terms in the argument. AT&T deployed the first such device, the No. 1 ESS, in New Jersey in 1962 following a field trial.* Electronic switching made it possible to run the telephone network without having to dramatically increase the number of operators because the previous mundane tasks of taking numbers and connecting calls could be relegated to the phone switches. Electronic switching also made it feasible to build a telephone network with enough call-routing capability to eliminate operator intervention in connecting long-distance calls; by 1965, 90% of telephones in the U.S. were directly dialing long-distance numbers.

Digitalization

In parallel with computerization, Bell Labs researchers were investigating digital voice transmission. Previous efforts had begun before World War II and resumed in earnest in the 1950s. Prior to digital telephone networks, the trunk lines running between switching offices transmitted analog phone calls using *frequency*

* Many enhanced services that we take for granted today, such as touch-tone dialing, call waiting, and call forwarding, were made possible by the 1ESS. AT&T originally unveiled the ESS-enhanced telephone network in 1962 at the World's Fair in Seattle.

division multiplexing (FDM). FDM divides up the allocated resources into frequency bands and transmits each conversation continuously in its assigned frequency band, as shown in Figure 1-1.

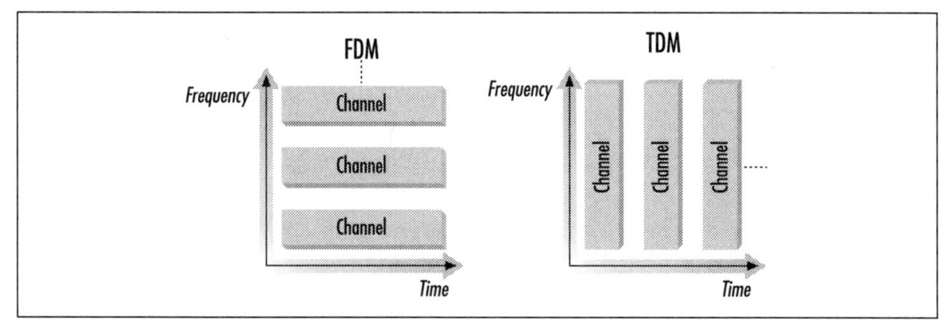

Figure 1-1. FDM versus TDM

Digital networks, by contrast, use a technique called *time division multiplexing* (TDM). FDM is a natural fit for analog signals, and TDM has a similar relationship to digital signals. In practice, nothing prohibits the use of FDM gear for digital networks or TDM gear for analog networks—those combinations are just more expensive than the natural fits. TDM divides the resources into time slots and transmits the time slots in turn, using the full capacity of the channel. Increased capacity was a major motivation for migrating the phone system to digital transmission. Each analog trunk could handle 12 calls, but digitized trunks, which used the same transmission facilities (wires), could handle 24 calls. Adding trunk lines took time and money—and lots of both. To add a trunk line, AT&T needed to evaluate potential routes, survey the chosen route, and acquire the right of way.

For overworked planners coping with mushrooming demand for telephone services, digitized trunks were perfect. In the early 1960s, AT&T began moving the U.S. phone network to digital carrier systems. In this context, *carrier* is the term telephone companies use to refer to trunk systems that can accommodate multiple calls. The U.S. digital hierarchy became known as *T-carrier* when Bell Labs engineers used the letter *T* for the digital carrier systems they were developing. There is no official significance to the letter T, though some sources suggest it is short for *terrestrial*. (Other systems in development in that era were the N- and L-carrier systems.) In the T-carrier hierarchy, the best-known members are T1 and T3.

1970-Present: The Modern Telephone Network

Digitalization of the telephone network sowed the seeds for a second government action against AT&T. The appearance of standardized digital carrier systems

Synchronous and Asynchronous Communications

One item of note about the T-carrier system is that it is *synchronous*. In synchronous communications, the data is sent as a stream. In contrast, *asynchronous* communications, such as dial-up modems, must have a method of informing the remote end when data will arrive and when the data transmission has finished.

Asynchronous communications are easier to implement in hardware because all that is necessary is to add start and stop sequences to the transmission and reception units. Synchronous communications are far more complex, as they transmit data as a continuous stream. For the remote end to pick out the data bits, a *clocking* signal is required. Furthermore, the two clocks must agree on when to pull a data bit out of the stream. Obtaining accurate clocking signals at high data rates and long distances, a major challenge for the T-carrier system, is explored in later chapters.

Like all technology decisions, choosing synchronous or asynchronous timing involves a trade-off. Synchronous communications have less overhead on the wire and can run at higher speeds, but require the maintenance of a clock signal and the associated electronics to recover the signal and extract the bit stream from the incoming transmission.

in the 1960s had blown apart Vail's justification of a natural monopoly based on incompatibility.

Round two of the antitrust wrangling with AT&T commenced in 1974. Although distracted by legal battles with the Department of Justice, AT&T remained the dominant telecommunications carrier. In the 1970s, AT&T pioneered fiber-optic transmission, beginning with an experimental system in Chicago that could carry 672 voice channels on a single strand of glass.*

Legal struggles concluded in 1982. AT&T retained businesses that operated in competitive marketplaces, namely long distance, Bell Labs and other R&D centers, and Western Electric. The *Regional Bell Operating Companies* (RBOCs), which were formed from AT&T's local exchange assets, continued to regulate local telephone service as a natural monopoly. After a long preparation period, AT&T split off from the RBOCs on January 1, 1984, in a momentous event known as *divestiture*. T1 had been used as a trunk line in the AT&T network for many years, but it was only after divestiture that T1 became a tariffed service that could be ordered by customers.

* This capacity is the equivalent of a T3, but is not a T3. No method for sending T3 signals over fiber was ever standardized.

The Problems with T-carrier

Problems with the T-carrier hierarchy stem from one simple fact: T-carrier systems were designed for AT&T's voice network. They can be used to transport data, but the adaptation to this use is not always a clean one.

Voice systems do not have the same requirements for operations, administration, maintenance, and provisioning (OAM&P) as data applications do, and the T-carrier hierarchy does not have enough overhead for even simple management tasks. For signaling purposes, the T-carrier system initially depended on robbed-bit signaling, which significantly diminishes data throughput, as we will see later.

Because T-carrier was developed by AT&T at Bell Labs, it is not surprising to note that Western Electric was the dominant manufacturer of T-carrier gear. While this may have been fine in the single-vendor/single-provider world that existed prior to divestiture, it was not adequate for the post-divestiture world. After divestiture, it was common for the ends of a T-carrier link to be under the control of different organizations, each of which had preferred vendors. AT&T had a simple answer: everybody could continue to purchase Western Electric equipment.* Research into solutions to the T-carrier's problems led to SONET, ATM, and the dreams of broadband ISDN.

The Rise of Data and the Internet

At this point, the story passes from the realm of history to a period too recent to have been judged by history. One signal event did recently occur: the early 1990s saw an explosion of data traffic with the rise of the Internet. In 1996, AT&T estimated that the telephone network carried more data traffic than voice traffic! Demand for data pushed the standard business-class connection from 56 kbps to T1.

T1 circuits are expensive, and the ordering and installation process can be ponderous. Many of the alternatives require either dense population concentrations or nearby fiber-optic facilities. While this is the case in major metropolitan areas, most of the country lags behind. In remote areas, T1 is the next available speed up from ISDN. T1 is also flexible enough to provide dial-in service for ISPs. Large ISP POPs use ISDN primary rate circuits to deliver the necessary density of incoming dial-in circuits. For many of these reasons, T1 circuits will be with us for quite some time to come.

* I doubt this was well received by anybody, except perhaps AT&T shareholders.

2

T1 Architectural Overview

The world hates change, yet it is the only thing that has brought progress.
—C. F. Kettering

One of the complaints that many data-networking veterans have when venturing into the telecommunications world is the bewildering number of acronyms and strange terminology that await them. Before diving into small details about different components of a T1, some background with the technology is essential. This chapter introduces the terms and basic structure of a T1 circuit so that successive chapters can delve into detail on the most important components.

Telecommunications Puzzle Pieces

Figure 2-1 shows a high-level diagram of the link between an Internet service provider (ISP) and a customer, delivered over an archetypal T1 circuit. Due to the number of components required to build a T1, Figure 2-1 is divided into four parts. In the classic case, two wire pairs, one each for reception and transmission, combine to form a T1. Devices called repeaters are used to allow for transmission of the signal over long distances. Signals degrade as they travel along the path, but repeaters recover the digital input data and retransmit the digital data at full strength. In theory, this allows for perfect transmission of data because a digital signal can be perfectly recovered and an exact bit-for-bit copy sent on to the destination. Repeaters divide the circuit into a series of *spans*, or paths between repeaters.

As technology marched on, the technology used to deliver T1 circuits changed. In most locations, only the end span connecting subscribers to the central office (CO) is delivered over the four-wire copper interface. Fiber-optic transport systems such as SONET are usually used to move data between the two COs at each end of the circuit. Among its many benefits, SONET provides *protection switching*. If a link in a SONET network fails, protection switching diverts connections onto a backup path within milliseconds.

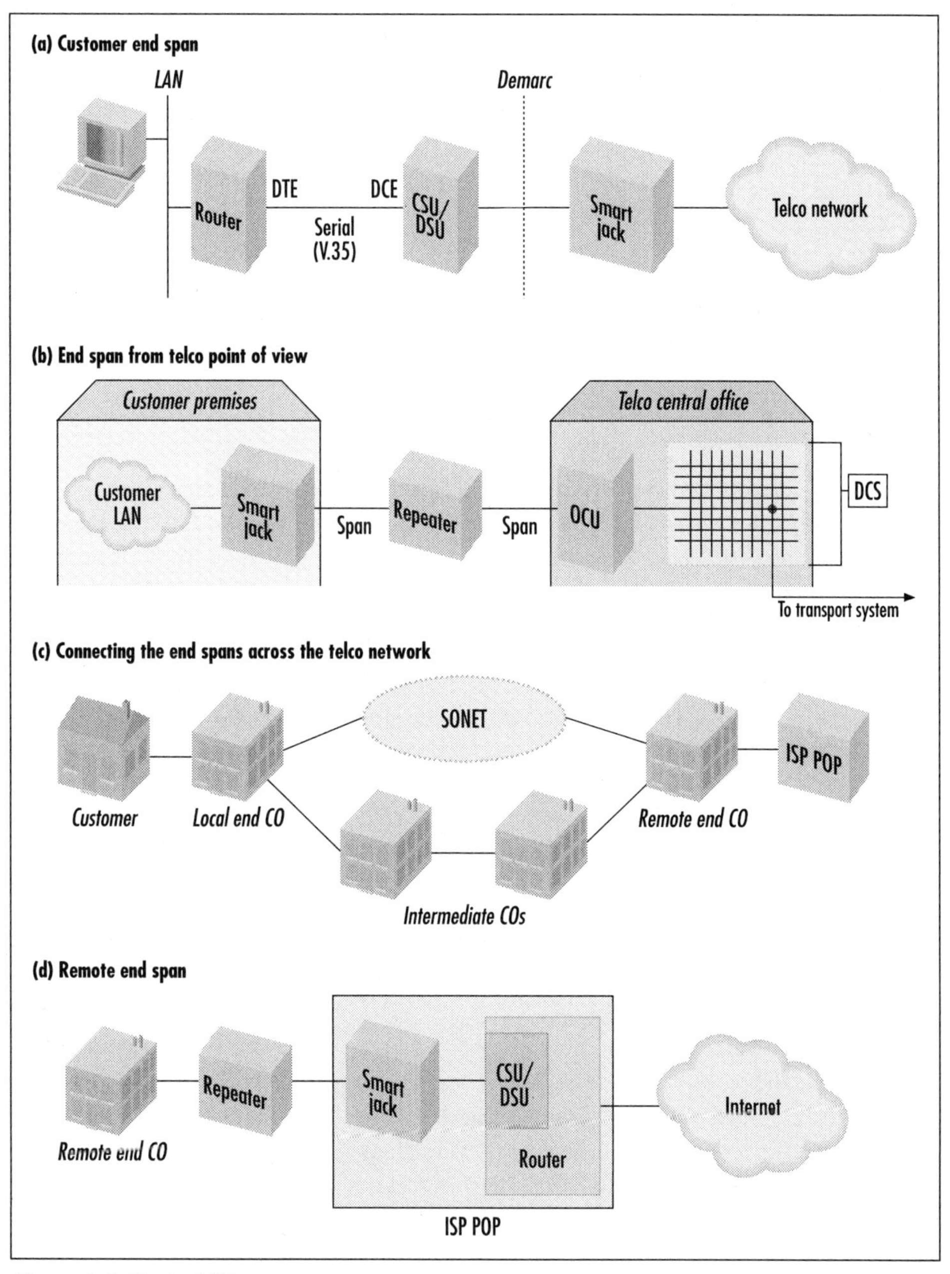

Figure 2-1. Typical T1 span

Once the span reaches the near-end CO, it is routed over the telco's network to the far-end CO. Between all the COs, redundancy and spare capacity is provided by the telco through SONET protection switching. Network outages due to telco equipment failure are far more common on the repeatered copper spans than on the SONET rings. It is not uncommon for these spans to be interrupted by construction equipment. When construction activity breaks up the end spans, protection switching is not available because only one path exists from the customer location to the CO. This type of interruption is frequently called *backhoe fade* by network engineers to distinguish it from other types of signal fading on the T1 line.

Figure 2-1a shows the end span on the local end, which is the most familiar part to subscribers. The computer, LAN, and probably even the router are nothing new and exciting—these are components that you probably have looked at many times before. Most likely, some form of Ethernet is used as the LAN medium, and the path from a computer to the router is a familiar one with well-known concepts such as MAC addresses, ARP, and so forth. In simple situations like that in Figure 2-1a, routers are not complex beasts, either. Any packets addressed to distant parts will be relayed to their destinations over the T1.

Past the router, the territory gets decidedly less familiar, even for those of us who deal regularly with telecommunications. Many network engineers have nearly endless opportunities to break Ethernet networks, but WAN lines are much more mysterious, so they tend to be left alone.

From the Router to the CSU/DSU with V.35

A serial connection joins a router to a T1. Serial communications make sense, after all: the T1 takes a stream of bits from Point A and moves them (in order, we hope) to Point B. T1s run quite fast, especially when compared with pedestrian serial standards such as the ubiquitous RS-232 on personal computers. Typically, the router-to-T1 connection is made using the *V.35* serial standard. For the purposes of establishing the V.35 serial circuit, the router is considered the *data terminal equipment* (DTE) and the CSU/DSU is considered the *data communications equipment* (DCE). As with modems, DTE simply means "generator of data," and DCE means "responsible for sending data somewhere else."

V.35 is an electrical specification, so it specifies voltages and currents, but does not specify what the connector looks like. The most common connector is the *Winchester block*, which may be simply called a *Winchester* connector, an *M-block* connector, or an *MRAC-34* connector. (Several people I know think that the Winchester connector looks like an instrument of torture. Judge for yourself: Figure 2-2 shows one.) Nothing prevents the V.35 electrical signal from being transmitted over a different type of connector, though. Some equipment uses 25-pin D-sub cables, which are extremely common connectors in the personal computer world. Proprietary connectors are also not unheard of.

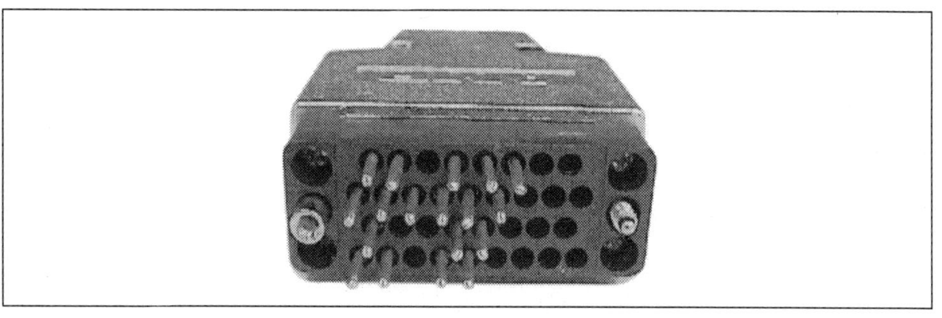

Figure 2-2. V.35 Winchester connector: instrument of torture, or just a bunch of pins?

The International Telecommunications Union

Interoperability has been a thorn in the side of the communications industry ever since long-distance electronic communications have existed. The ITU (*http://www.itu.int*) was founded to provide a forum for developing standards to assure uniform telegraph interfaces and has assumed an increasingly important role as telecommunications has become more involved. Prior to the early 1990s, a United Nations body called the Consultative Committee on International Telephone and Telegraph (CCITT) carried out standardization efforts. As part of a reorganization in the early 1990s, the CCITT was renamed and became the ITU-T, or the ITU, telecommunications sector. (Because this book deals only with telecommunications, I'll drop the "-T" and refer simply to the "ITU.")

The ITU divides standardization work into several series and labels the resulting standards with the series letter and a document number. ITU Series V standards are collectively labeled "data communication over the telephone network." Modem users are undoubtedly familiar with ITU standard V.90 for 56 Kbps downstream data transmission with a modem. (Depending on how long you have used a modem, you may also remember V.34, V.32 bis, V.32, and so on.)

V.35 is a standard for synchronous serial communications. In 1988, the ITU deprecated V.35 in favor of the V.10 and V.11 standards, but the market for V.35 equipment is still quite healthy, especially now that nearly every T1 CSU/DSU has a V.35 connection. More details on V.35, including a pinout, can be found in Appendix E.

Details of V.35 are really beyond the scope of this book. Most T1 problems are caused by other pieces of the puzzle, and it is quite rare to do anything other than swap out a bad V.35 cable from the router to the CSU/DSU.

The CSU/DSU: Your Network to the Telco Network

To connect a data network to the telephone network, you need a device called a *CSU/DSU.* CSU/DSU stands for *Channel Service Unit/Data Service Unit.** As the name implies, the CSU/DSU includes two separate functions. The DSU is responsible for handling the V.35 interface with the router. The CSU handles connection to the telco network; the reason for the name "channel" in CSU will become apparent later.

In terms of the OSI model, CSU/DSUs are layer 1 devices that translate the V.35 signals into telco network–compatible signals. When these signals are sent on the telco network, some degree of framing across the network is needed, but the details are not relevant to this chapter's high-level overview.

The End Span: From the CSU/DSU to the CO

In addition to a V.35 port, the CSU/DSU has a *network interface* (NI) to hook up to the telco network. A variety of AT&T technical reports first specified characteristics of the network interface. To create a true multivendor interface, the American National Standards Institute (ANSI) later undertook standardization efforts and published them as ANSI T1.403. The T1 interface uses four wires: one pair for transmit and a second pair to receive. Figure 2-3 shows CSU/DSU network interfaces.

Ensuring cable compatibility is easy. Standard jacks are registered with the FCC and given RJ (registered jack) numbers. T1 network interfaces follow the RJ-48X standard. RJ-48X jacks contain shorting bars, which automatically loop the line back if the cable is unplugged from the jack.

CSU/DSU equipment uses one of two methods to connect to the T1 span. Some equipment uses a second RJ-48 jack and a cable with two modular connectors. Other equipment uses a DA-15 jack at the CSU/DSU end and a modular connector at the line end.

Cable runs from the CSU network interface to the *demarc*, a term short for *demarcation point.* You are responsible for repairing problems on your side of the demarc, and the telco is responsible for anything beyond the demarc. Frequently, the demarc is the telco's Network Interface Unit (NIU), nicknamed a *smart jack.* Smart jacks perform a variety of network maintenance functions and make it much easier for the telco to manage the network remotely. Figure 2-4 shows a smart jack.

* The CSU and DSU functions are distinct. (Prior to divestiture separate boxes provided each, but the CSU and DSU are always integrated into a single box when T1 is used for data communications.) Because both functions are required to connect data communications equipment to telco networks, they are almost always combined into one box. Even though the functions are equally important and are both needed to connect to the telco network, it is common to hear people refer to the entire unit as either a "CSU" or a "DSU."

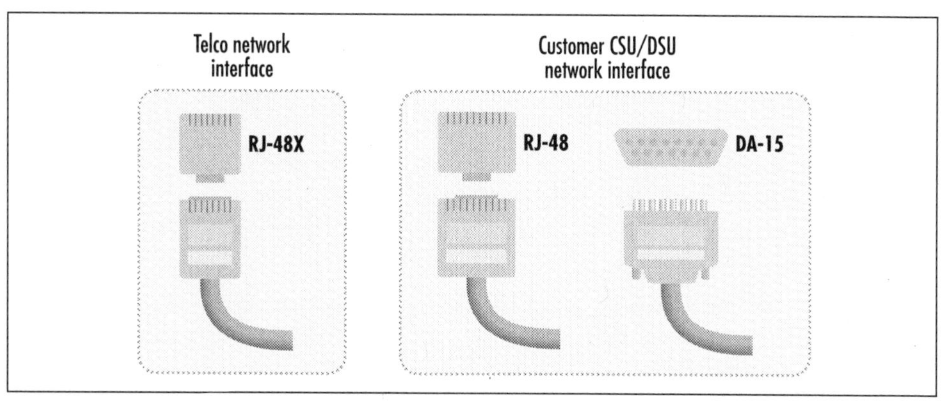

Figure 2-3. CSU/DSU network interfaces

Figure 2-4. A smart jack

The smart jack is located with other telephone equipment at your building. It is placed where it is convenient for the architect (and maybe the telephone company), but those considerations do not necessarily have anything to do with the network. In situations where the smart jack is far away from the CSU/DSU location in a machine room or network operations center (NOC), wiring must be run from the demarc at the smart jack to the CSU/DSU location. This is called an *extended demarc*. Wiring is extended from the smart jack to a more convenient location, where it is terminated by what looks like an overgrown telephone jack. (I once worked at a company that rented space in a multitenant office building. The telco closet was used for the entire building, which meant that anybody in the building who ordered a T1 needed to install an extended demarc.)

Because of their small, squarish appearance and off-white color, these jacks are frequently referred to as *biscuits*. Extended demarcs can be tricky to get right and are a major source of problems with T1 circuits. Wiring an extended demarc is not as simple a task as installing an Ethernet drop. In many cases, shielded cable must be used to preserve signal quality. Additionally, the pinout for the smart jack interface is different from other data-communications pinouts, so unless the installer is familiar with the pinout, the line will pick up outside interference. For a fee, the telco will typically install the extended demarc for you. Unless you have extensive experience installing extended demarcs, pay somebody else to do it right. In the long run, it will save time and money. Figure 2-5 illustrates an extended demarc.

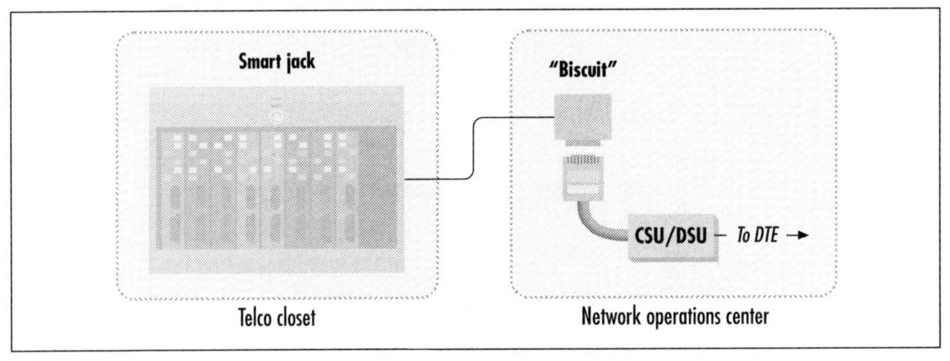

Figure 2-5. An extended demarc

The End Span and Entry into the Telecommunications World

From the smart jack, the connection runs to the nearest telco central office. Depending on the facilities the telco uses to provide the wiring, the end span may be over a four-wire repeatered line or it may be partially run over a metropolitan fiber network. Figure 2-1b shows a T1 using repeaters the entire way to the CO. At the CO, the line terminates at the CO-side equivalent of a CSU, called an *Office Channel Unit* (OCU).

T1 specifications allow each span to have up to –30 dB loss before the signal is unintelligible. (Actually, the specifications allow 33 dB, but the last 3 dB serve as a safety factor.) To keep signal loss at or below this level, repeaters are required at approximately 6,000-foot intervals. (One unsubstantiated legend suggests that the repeater interval is where the "1" in "T1" comes from—repeaters are required at one-mile intervals.) In the purely copper scenario in Figure 2-1b, the parts of the line that are provided using a four-wire interface must be composed of carefully selected pairs in a conduit to avoid crosstalk and distortion. Additionally, the line must be properly conditioned for T1 service, which means removing a variety of devices such as bridge taps and load coils that improve the quality of analog voice service.

What Is a Decibel (dB)?

Decibels are simply a way to express large ratios in a more compact form. When ratios are large, they can become unwieldy. Decibels are calculated according to the following formula, which compares a reference level (x_{ref}) to a measured quantity (x), as in:

$$dB = \log (x/x_{ref}) \times 10$$

x_{ref} is a reference level chosen for the task at hand. For sound volume, it is chosen as the threshold of human hearing. For electrical components, it is chosen as the initial level of a signal. When $x=x_{ref}$ the ratio is 1, which is 0 dB.

Amplifiers make electrical signals larger. The input power of an amplifier is the reference level, and the measured quantity is the output power. The gain of an amplifier measures the ratio between the output power and the input power, which is always a positive number.

As signals travel along a path, there is some level of resistive loss because real-world systems do not use the ideal loss-free wire with which students of physics are accustomed to dealing. To quantify the loss, the reference level is taken as the strength of the input signal, and the measured quantity is the size of the signal at some point down the line. In this case, the ratio is described by a negative decibel number because the measured quantity is smaller than the reference quantity.

Extensive line conditioning, cable selection, and repeater installation are all expenses that increase as the length of the span increases. As a result, T1 pricing typically depends on the distance of the span, and installation time may be long.

Even if most of the T1 is run over fiber, the fiber-optic network multiplexors must be configured to create a 1.544 Mbps digital transmission channel through the fiber-optic network. Provisioning a new stream through a fiber-optic network is also a time-consuming affair, though not as bad as eliminating crosstalk between T1 repeaters in the field by careful wire selection in a conduit.

At the CO, the physical line from the customer must be connected to the ISP (if both are served by the same CO) or to the telco's transport network (which moves the bits from the local CO to the remote CO that serves the ISP). In the very distant past, these connections were physical—wires were spliced together to create a long wire loop between the two locations. Today, logical connections are made with the help of a *digital cross-connect system* (DCS), which is sometimes called a *digital access and cross-connect system* (DACS). DCS hardware is essentially a switch. All the lines coming into a CO are connected to the DCS so the incoming digital streams can be logically connected to outgoing streams. DCSs also perform multiplexing functions, so several incoming T1s can be combined and mapped onto a single higher-speed output stream.

Once the input data stream connects to the transport data stream, it can be moved across the telco network to the remote CO, as shown in Figure 2-1c. Depending on the telco's physical infrastructure, this may happen in a variety of ways. In densely populated metropolitan areas, SONET rings may be used to ship the data directly from the local CO to the remote CO. SONET rings can move hundreds of megabits per second or even gigabits per second, but their chief advantage is that they allow for equipment to easily "peel off" the data destined for the local end without having to demultiplex the entire signal. Areas without extensive SONET infrastructures may use a variety of techniques, including simple repeatered copper pairs through intermediate COs. Both approaches are shown in Figure 2-1c.

At the remote end, the CO connects the incoming signal from its DCS, to the output path for the T1, to the ISP POP. The remote end of the circuit, shown in Figure 2-1d, is quite similar to Figure 2-1a. ISPs are responsible for supplying their own CSU/DSUs and routers. Because ISPs may terminate huge numbers of customer-access lines at a particular point of presence (POP), the ISP router will frequently have an integrated CSU/DSU to terminate the line to save rack space.

3

Basic Digital Transmission on Telephone Networks

The dead govern the living.
—Auguste Comte

As mentioned in Chapter 1, T1 is a time-division multiplexed stream of 24 telephone calls. Each call is carried by a 64-kbps digital stream called a DS0. Several meanings are ascribed to the acronym DS; you may hear any combination of the words data, digital, service, stream, speed, and signal. DS0 is the bottom rung of the T-carrier hierarchy. Higher levels of the hierarchy are built by multiplexing lower levels together. Understanding the T-carrier hierarchy starts with understanding DS0 transmissions.

AT&T's initial digital leased-line offering was called the Digital Dataphone Service (DDS). DDS was offered at several different speeds, ranging from 2,400 bps to 56 kbps. Service initially topped out at 56 kbps because a portion of the signal is required for timing overhead. DDS circuits formed the Internet backbone in December 1969. Traffic growth eventually overwhelmed the limited-circuit capacity, and the Internet backbone was upgraded to T1 circuits in the late 1980s. Mushrooming traffic led to a further network upgrade to T3 in the early 1990s. Eventually, economics conspired to kill DDS as a standalone service. T1 is not much more expensive, so companies that required more throughput, higher reliability, and guaranteed service levels shifted to T1, while budget-conscious users migrated to cheaper technologies such as DSL.*

* Any institution that received even a moderate news feed felt the pressure early. Leased 56k connections running full time at 100% utilization can transfer only slightly more than 575 megabytes in a day.

Introduction to DS0

The telephone network is a *circuit-switched* network. Each telephone call is assigned to a dedicated path through the network for its duration. Telephone calls require a 64-kbps path through the network. At each hop between switching offices, the trunk lines are divided into 64-kbps channels, which are called DS0s. These individual 64-kbps channels are the building blocks of the telephone network because a DS0 has sufficient capacity for a one-voice telephone call.

To the phone company, 64 kbps means 64,000 bits per second, not the 65,536 bits per second a computer engineer might expect. However, this basic quantity is used by all telephone companies throughout the world. Bundling is different: in the U.S., the T1 standard bundles 24 DS0 channels, plus framing overhead, into a single T1. In Europe, an E1 circuit is composed of 30 DS0 channels, plus framing and signaling overhead. Given the differences in telecommunications between the U.S. and the rest of the world, it should be obvious that since we all use a given standard, there must be something special about 64,000 bits per second.*

Transmitting understandable human speech requires using frequencies up to 4,000 kHz. To adequately represent an analog signal in digital format, the analog signal must be sampled at twice the maximum frequency in the signal, a result known as the *Nyquist criterion* or *sampling theorem*. To adequately digitize voice, therefore, requires a sampling rate of 8 kHz. Each sample is represented by 8 bits using a technique called Pulse Code Modulation (PCM), where the value of each sample is transmitted as an 8-bit code. In one second, 8,000 samples are transmitted. With a sample size of 8 bits, the resulting data rate is 64,000 bits per second.

DS0s do not impose any higher-level structure on the data stream—they are simply an unframed, raw sequence of bits. The telephone company takes the bits from one location and moves them to another location; to the telephone company's equipment, it does not matter whether the bits are a voice telephone call or a data circuit, as long as the stream of bits obeys the rules.

Alternate Mark Inversion

In common encoding schemes, ones are represented by voltage pulses and zeros are represented by the lack of a voltage pulse. Each pulse is approximately 3 volts in amplitude and has a 50% duty cycle, meaning it takes up half of the time slot for pulse transmission. Pulses have a tendency to spread out in the time domain as

* Technically, the U.S. and the rest of the world do not quite use the same standard. U.S. digital-carrier systems are based on a digital encoding technique called the *mu-law*, while most of the rest of the world uses a different method called the *A-law*.

they travel down a line, as illustrated in Figure 3-1. Restricting the initial transmission to occupy half the time slot helps the repeaters, and the receiving end, find the middle of each time slot and stay synchronized.

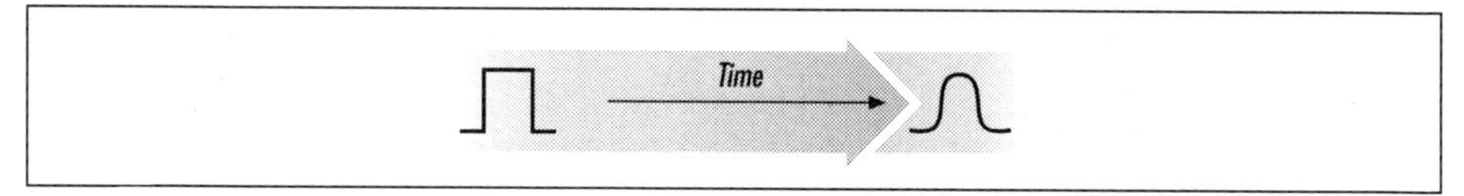

Figure 3-1. Time domain spreading

Commonly, a scheme called *bipolar return to zero* or *Alternate Mark Inversion* (AMI) is used. One and zero are sometimes referred to as *mark* and *space*, respectively, in communications jargon. AMI gets its name from the fact that only ones, or marks, result in pulses on the line. Successive pulses are encoded as positive and negative voltages, as shown in Figure 3-2.

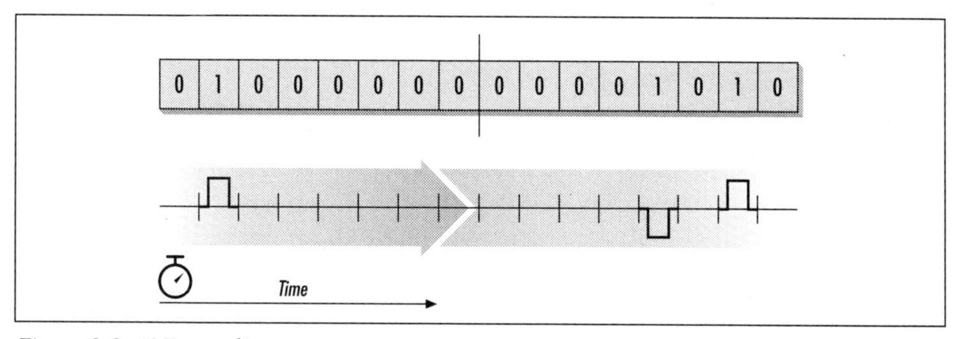

Figure 3-2. AMI encoding

Alternating pulse polarity enables a quick-and-dirty form of error detection. AMI specifies that polarity must alternate, so two successive pulses of the same polarity, such as the sequence in Figure 3-3, might be an error. (Bipolar schemes can catch many errors, but not all of them. Errors in voice transmission result in odd sounds and can be shrugged off. Data transmission is far less forgiving, which is why far better error-detection techniques were developed.)

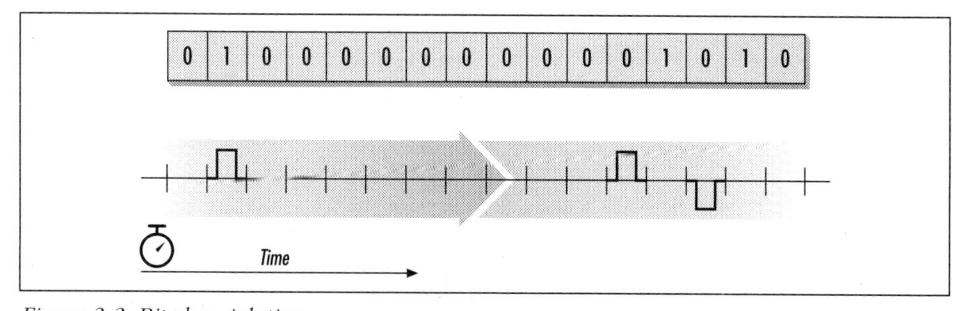

Figure 3-3. Bipolar violation

Errors on Leased-Line Circuits

The T1 standards describe a variety of error-detection features. Nothing in the voluminous standards is about error *correction*, however. Various T1 features offer different degrees of error detection. Error information is reported to higher-layer protocols, and those protocols must decide what to do with potentially altered bits.

In the voice-networking realm, getting information to the receiver quickly is often more important than getting information to the receiver in perfect sequence with no errors. Altered bits may lead to small auditory anomalies, but the next bits of data are likely to be correct. This trade-off is not appropriate for the data world, where the difference between a one and a zero can mean a great deal. Reporting errors to higher-layer protocols allows for those protocols to take the appropriate corrective action (usually retransmission).

Timing and Synchronization

In a sidebar in Chapter 1, I made a brief reference to synchronous and asynchronous communications. This book is about synchronous communications: both sides of links must agree on when to look for a bit. Long strings of zero bits result in no activity on the wire. If a string of zeros were to persist for a long period of time, the two ends of the link might disagree on its length and be forever out of sync. To prevent two sides from disagreeing on when to look for a pulse, the standards specify a maximum number of successive zeros and a minimum pulse density. Both constraints enable the electronics in the CSU/DSU to lock on to pulses and synchronize their clocks.*

The simple-minded solution is to require transmission of pulses at regular intervals to prevent long zero strings—a simple matter of injecting one bits. In the initial schemes, every eighth bit was stolen for timing purposes and transmitted as a one. Only seven data bits were available in every PCM sample slot, which limited effective throughput to 56,000 bits per second. Injected ones were data on the line, even if the data-communications equipment at both ends ignored them; as a result of this background activity, the transmit and receive lights glowed faintly even when no user data was being transferred.

Blindly stuffing a one into the eighth bit time of each sample period restricted transmission to seven bits. When data services were used to transmit ASCII text,

* Requiring synchronization between the two ends is a slight fudge on my part. Strictly speaking, the requirement is that each side maintains a fixed offset relative to the other side's clock; that is, the two clocks must remain *phase locked*.

throughput did not suffer because ASCII characters are only seven bits. With the evolution of data services beyond text, bit stuffing removed one-eighth of the throughput—a condition that became more and more unacceptable with increasing bandwidth demands.

B8ZS and Clear Channel Capability

Throughput suffers when the eighth bit is blindly stuffed with a one. Maintaining synchronization requires only that enough pulses are sent down the line. An alternative to straight AMI encoding is to use a scheme based on code word substitution. A second encoding method, called *Bipolar with Eight Zero Substitution* (B8ZS), is able to transmit an arbitrary bit sequence. When eight consecutive zero bits are scheduled for transmission, a B8ZS transmitter replaces the eight-zero sequence with a code word that contains intentional bipolar violations (BPVs), as shown in Figure 3-4. The code word takes the form of 000VP0VP, where V is a pulse of the same polarity as the previously transmitted pulse and P is a pulse with the opposite polarity as the previous pulse.

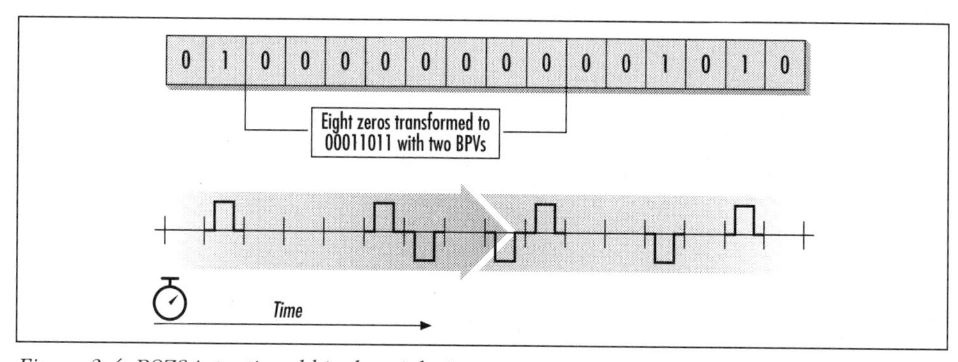

Figure 3-4. B8ZS intentional bipolar violation

When B8ZS-capable CSU/DSUs receive the B8ZS code word containing bipolar violations, the code word is replaced with eight consecutive zeros before passing the data on to the user. Transmission at the full line rate is possible with B8ZS because it does not require a portion of the circuit capacity to be used for synchronization. Because the full line rate is available, the telco may refer to the circuit as one that has *Clear Channel Capability* (CCC), or some similar term. Widespread use of B8ZS is a relatively recent development. Most early digital CO equipment was designed to catch and flag bipolar violations. Moving to B8ZS required massive upgrades to remove all of these "helpful" pieces of equipment. In the late 1980s, less than 1% of telco equipment was B8ZS-capable. Extensive equipment upgrades in the past 10 years have made widespread use of B8ZS possible, so virtually all new T1s are deployed using B8ZS.

4

Multiplexing and the T-carrier Hierarchy

The tyranny of the multitude
is a multiplied tyranny.
—Edmund Burke

To move beyond the DS0 into higher-bandwidth realms, additional layers of multiplexing are needed. This chapter describes how DS0s are bundled into DS1. DS1 refers to a digital signal operating at 1.544 Mbps; T1 refers specifically to a DS1 delivered over a four-wire interface. Most people simply use the term T1 to refer to any digital signal at that speed and, to avoid breaking the common convention, so does this book.

Higher levels of multiplexing are used to generate further levels of the T-carrier hierarchy, such as DS3. DS3 is different from T1, though, because the much higher speed requires different encoding methods, far more precise timing, and new network-to-router interfaces. To avoid getting lost in DS3 details, therefore, this chapter details only the DS0 to DS1 multiplexing process.

Building the T-carrier Hierarchy with Multiplexing

Assembling higher-speed links in the T-carrier system is conceptually easy. Take a collection of lower-speed links and bundle them together as *channels* in a TDM framework. When 24 DS0 streams are bundled together, the result is a higher-level digital stream: DS1. Multiple DS1s are bundled together to form DS2s, and DS2s are tied together into DS3s. Table 4-1 shows the standardized data rates in the T-carrier system. While the data rate for DS4 was standardized, most of the network interface details were not.

Table 4-1. T-carrier comparison

Stream type	Speed (Mbps)	Equivalent T1s	Equivalent voice channels
DS0	0.064	1/24	1
DS1	1.544	1	24
DS1C	3.152	2	48
DS2	6.176	4	96
DS3	44.736	28	672
DS3C	89.472	56	1344
DS4	274.176	168	4032

Due to differences in control bits and framing bits in different levels of the hierarchy, discrepancies may appear to exist. For example, DS1 is composed of 24 DS0 channels, each operating at 64 kbps, but 24 × 64 kbps = 1.536 Mbps. DS1 adds extra framing and control information that pushes the line rate to 1.544 Mbps, even though only a maximum of 1.536 Mbps is available for user data.

Within the T-carrier hierarchy, DS1C and DS3C never existed as services that could be ordered. The data rates corresponding to DS1C and DS3C are intermediate multiplexing rates that could exist in a digital cross-connect. DS2 was never widely deployed, and DS4 was never thoroughly standardized.

Multiplexing to Form the T1

At the receiving end of a T1 bit stream, some method is required for distinguishing where one channel stops and another channel starts. Each channel could be individually framed with a unique header, but such an approach would add a great number of header bits and use too much of the transmission capacity of the carrier simply for control information. Instead, T1 transmits each channel in turn and adds a single framing bit at the beginning, as shown in Figure 4-1. Framing bits are used to synchronize clocks and for rudimentary error indication.

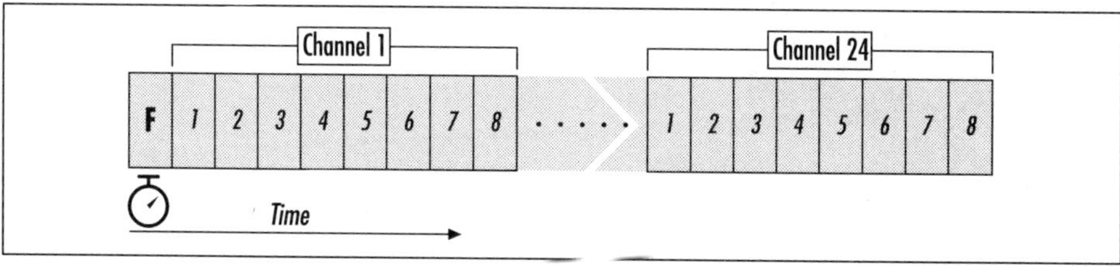

Figure 4-1. Generic frame format

Multiple frames can be aggregated together into *superframes*. A complete header is constructed by using the framing bit from each frame and concatenating bits over several frames.

Use of the terms *frame* and *superframe* is different here than in the data world. I think of a frame as an entity that has a complete header, not a twelfth or a twenty-fourth of the header. If it had been up to me, I would call a succession of 192 data bits plus a framing bit a *subframe*, and assemble subframes into frames. Thus, my subframe is what the telco would call a frame, and my frame is what the telco calls a superframe. I obviously did not have any influence over the terms, though, so we are stuck with frame and superframe; I use the industry standard terms because the usages are entrenched.

When you first plug in a T1 cable, an alarm indicator will still complain about the lack of framing. Loss-of-framing warnings will persist for a short while as the CSU/DSU searches for the framing bits, locks on, and synchronizes its clock. Synchronization is a pretty amazing process: the CSU/DSU must identify the 193rd bit position in each frame and then identify the start of the framing sequence. After synchronization is complete, bits can be transmitted to the remote end, which locks on to the transmitted frames and clears any remote alarm conditions.

Fractional T1 (FT1) Service

For financial reasons, fractional T1 service is popular in many areas. It allows you to purchase multiple DS0s with an easy upgrade path to higher speeds. Fractional T1 equipment is the same equipment used to provide full T1 service, so there is no additional equipment cost to upgrade from fractional to full T1 services. (Of course, the full T1 CSU/DSU is much more expensive than a 56k CSU/DSU.)

Fractional T1 is provided using the same facilities as full T1. The telco providing fractional T1 will transmit data only on some of the 24 time slots in the T1 frame. For example, a 256 kbps FT1 is a T1 with only 4 of the 24 time slots active. All CSU/DSUs allow only certain time slots in the frame to be active, which makes provisioning FT1 easy.

Ones Density and the T1

Because clock synchronization is maintained by monitoring pulse times, the T1 specifications mandate a certain pulse density so that both ends stay synchronized. Two minimums are imposed by the standards:

1. Overall, a 12.5% ones density.
2. A maximum of 15 consecutive time slots without a pulse. Modern digital repeaters can handle much longer strings of zeros, but this requirement was instituted well before such equipment was available. There is no easy way of knowing the capability of repeaters on any stretch of cable, so all commercial equipment is still built to the older specification.

When a T1 is installed, the telco technician may hook a handheld device with lots of buttons and blinking lights on it up to the new T1 jack. The testing device can perform stress tests on the new line to measure its quality and clarity. After looping back the T1, the testing device injects a specific bit stream. Returning bits are compared against the original sequence to determine the bit error rate (BER). Several common tests are used:

QRS tests
: These tests use a *quasi random signal* specified in ANSI T1.403. The QRS bit sequence is not guaranteed to meet pulse-density requirements. On B8ZS-encoded links, CSU/DSUs will perform zero substitution and the transmitted signal will meet pulse-density requirements without a problem. AMI-encoded links should not alter the signal to meet minimum pulse-density requirements, but misconfigured equipment on the line may be configured to do so. Errors observed in the QRS test indicate one of two things: a line that is bad, or a line that has misconfigured equipment. One common source of misconfiguration is related to pulse stuffing on the CSU/DSUs or other line components. Some components can be configured to "stuff" pulses to ensure minimum pulse density; the stuffed pulses alter digital data and are not acceptable. Poor QRS test performance may occur because pulse stuffing is mistakenly enabled on the CSU/DSU or other line equipment.

3-in-24 test
: When used on AMI-encoded links, the 3-in-24 test sends the framed pattern 010001000000000000000100, which meets both the minimum pulse density and maximum zero length, and stresses the link by sending the lowest density signal allowed by the specification.

1-in-8 test
: B8ZS-encoded links use this test, which transmits a framed sequence of bytes in which the second bit is a one (01000000). 1-in-8 is a stressful test for B8ZS links because it contains repeated strings of seven consecutive zeros, which is the longest duration of inactivity on a line that is allowed by the B8ZS line code. B8ZS zero substitution may occasionally occur with a 1-in-8 test pattern.

The last channel transmitted in a frame has six consecutive zeros, and the first data bit in the first channel of the next frame is a zero. If the intervening frame bit is a zero, zero substitution will occur.

All-zeros test

B8ZS links may use this test, which transmits a framed sequence of zeros. When used on a B8ZS link, the all-zeros test should result in continuous zero substitution. On lines configured for AMI encoding, this test will report large numbers of bipolar violations.

All-ones test

Both B8ZS- and AMI-framed links may use this test, which transmits a continuously framed sequence of ones. This test checks for *ringing* and *crosstalk*. Ringing occurs when transmitted signals reflect off boundaries and the reflections bounce back and interfere with later transmitted signals. Crosstalk occurs when the cable pairs are not separated correctly, so that the transmitted signal induces a similar signal in the receive pair.

The Original Superframe

In the beginning, T1s were deployed to users as high-capacity voice trunks and were terminated by devices called *channel banks*. Western Electric called the first channel bank, which was first placed in service in 1962, the D1. Later channel banks received successively higher numbers: the D2 channel bank was introduced in 1969, the D3 was completed in 1972, and the D4 made its debut in 1976. Each new product introduced new features. With the D2 channel bank, Western Electric introduced the *superframe* (SF) standard. Some manuals may refer to the SF format as the D4 format, which is somewhat erroneous—the D2 channel bank was sold in the late 1960s, long before the D4 channel bank became commonplace nearly twenty years later. Superframe formatting groups 12 channels together into a single structure called a superframe, as shown in Figure 4-2.

24 DS0 channels are bundled into frames. Twelve frames are then put together to form one superframe. The framing bits from the 12 constituent frames form a frame bit sequence, which is always 100011011100, as Figure 4-2 shows. Odd bit positions in the superframe are *terminal framing bits*, abbreviated F_t(101010), and even bit positions are *signaling framing bits*, abbreviated F_s(001110).

Alarms and the Superframe

AT&T's introduction of the D2 channel bank brought link monitoring into existence—a major step forward. Alarm signals indicated severe error conditions. Table 4-2 summarizes the alarm conditions available on superframe links. Alarms initially had color names because of the corresponding color of the indicator lights on channel banks.

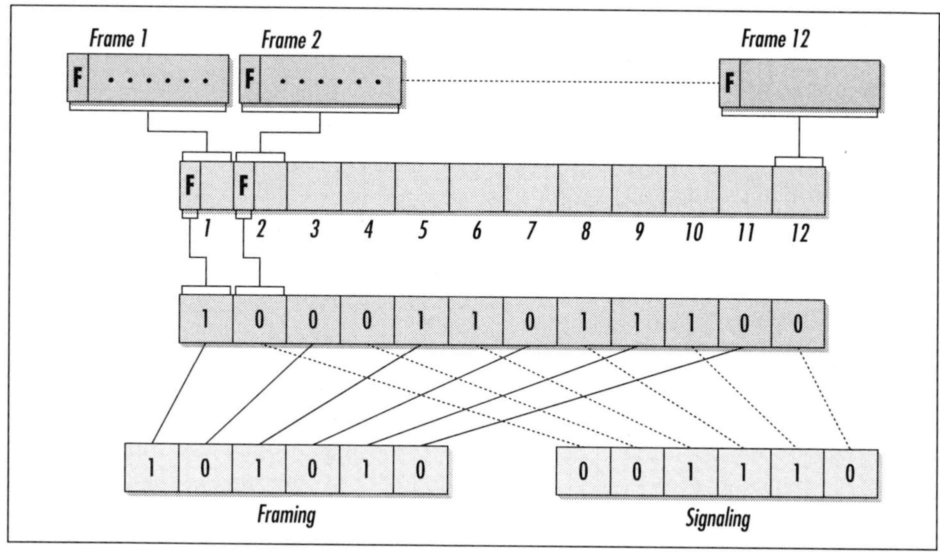

Figure 4-2. The superframe

Table 4-2. Superframe alarm conditions

Name	Cause
Red alarm Receive alarm	This is declared when the local CSU/DSU is not able to detect an incoming framing signal.
Yellow alarm[a] Remote alarm	This is sent as the outbound signal when a CSU/DSU enters the red alarm state to inform the remote end of a potential failure. The most important thing about the yellow alarm is that, in most cases, the span from the remote end to the local end must be functional to send the yellow alarm signal. On SF-framed links, a yellow alarm is indicated by setting the second bit of each channel to zero.
Blue alarm Alarm indication signal (AIS)	When no incoming signal is detected, a CSU/DSU transmits an unframed all-ones pattern. AIS signals serve two purposes: the continuous transitions may help resynchronize the network, and they indicate a problem to the network.

[a] Some equipment may refer to the yellow alarm as the *remote alarm indicator* (RAI), which is the ITU name for a yellow alarm.

The Extended Superframe (ESF)

SF-framed links have two notable drawbacks. First, a yellow alarm is transmitted by setting the second bit to zero in all of the time slots in a frame. When the yellow alarm is present, no data is received. Unfortunately, setting the second bit position to zero is something that can happen frequently in user data. Altering the bits is not acceptable with data transmission, so the only solution is to use one time slot for yellow-alarm prevention and to set the second bit to one. Although this prevents a false yellow alarm, it sacrifices one DS0 worth of bandwidth. Secondly, the error-detection mechanism with SF links is quite limited. Bipolar violations are a *line error check*, which means that they can flag potential problems in the local copper portion of the T1 span. Errors may be introduced anywhere along the span, however. Any corruption introduced at the central office, or in the high-speed optical components, cannot be detected by T1 equipment and must be detected by higher-layer protocols. What is needed is a *path error check*, which verifies data integrity across its entire path from one end to another, no matter what type of transport is used.

In response to these limitations, AT&T developed the *extended superframe* (ESF), which was introduced on the D5 channel bank in 1982. Advances in electronics made it possible to use a smaller proportion of the frame bit sequence for synchronization and devote it to solving the problems of the SF framing format. Figure 4-3 shows the ESF superframe. As with the SF superframe, it begins with frames, each of which is made up of 24 8-bit time slots with a single frame bit at the beginning. The 24 frames are put together into a single superframe.

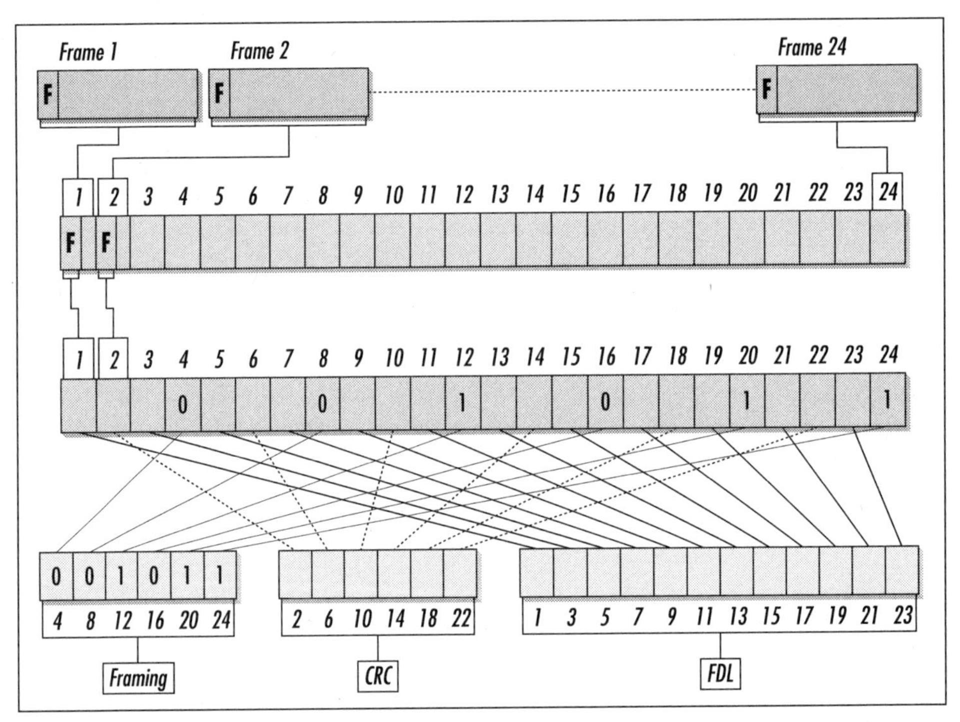

Figure 4-3. The ESF superframe

Of the 24 framing bits, only six are needed for synchronization. Every fourth frame contributes a bit to the synchronization pattern, 001011. CSU/DSUs can easily identify the synchronization pattern because it cannot shift onto itself. Synchronization requires 2 kbps from the aggregate T1 capacity, which is only half of the bandwidth required by SF framing.

To provide a path error check, six bits in the frame bit sequence are used for a *cyclic redundancy check* (CRC). Like the frame-synchronization pattern, the CRC requires another 2 kbps. The CRC is a value that is calculated by using the data payload of the superframe together with the frame bit sequence as an input. After sending a superframe, the sender calculates the CRC and places that value in the six bits in the framing bit sequence of the next superframe. Upon reception, the receiver calculates the CRC and compares it to the CRC received in the subsequent superframe. A difference indicates potential corruption somewhere in the previous superframe.

ESF Signaling

Making the T1 transparent to arbitrary user data required changing the signaling method from altering bits of user data to using a separate signaling channel. The remaining 12 bits of the framing bit sequence are used to create a 4 kbps channel called the *facilities data link* (FDL). Alarms and performance data are reported over the FDL. Telcos commonly use it to collect circuit performance data. The performance report message format is fully specified in T1.403. The FDL is not limited to performance monitoring, however. It may be used as a DSU-to-DSU communication channel, or adopted for proprietary purposes.

Code words: ESF alarms and instructions

Urgent messages, such as alarm conditions, are reported over the FDL by sending code words. Carriers may also use the FDL to send commands to the customer's CSU/DSU to activate loopbacks.* When idle, the pattern 01111110 is sent continously. Table 4-3 shows common ESF code words along with their message types. Priority messages override all other code words and LAPD messages.

Table 4-3. ESF code words

Code word	Type	Description
01111110	N/A	Idle code transmitted when no code word or LAPD message is present; also used as LAPD demarcation flag
11111111 00000000	Priority	Yellow alarm/RAI
11111111 01010100	Priority	Loopback retention
11111111 01110000	Command	Line loopback activation

* Bellcore specified an unused command (00010010 11111111) to put the remote smart jack into a loopback state.

Table 4-3. ESF code words (continued)

Code word	Type	Description
11111111 00011100	Command	Line loopback deactivation
11111111 00101000	Command	Payload loopback activation
11111111 01001100	Command	Payload loopback deactivation
11111111 00100100	Command	Universal loopback deactivation

Alarm conditions are transmitted for at least one second and until the condition causing the alarm is repaired. A 1-second interval is required between successive alarm signals.

Command/response code words are transmitted ten times. When the carrier initiates a line loopback condition for maintenance and testing purposes, the FDL transmits the *loopback retention* signal to avoid sending any loopback control codes or performance reports back to the carrier.

Other FDL code words can be used to trigger protection switching, which is a controlled switchover to backup transmission facilities that occurs when the primary path fails. FDL code words may also indicate how accurate the clock is on one end of a span.

ESF performance reporting

Non-alarm messages are transmitted using a message-based protocol that resembles the *Link Access Procedure, D Channel* (LAPD), an ITU protocol originally designed for ISDN signaling messages. Use of the FDL for message-based performance reporting is described in detail in Appendix C.

Telephone Signaling on T1 Links

On circuit-switched networks, nodes must occasionally send administrative messages to each other to establish, reroute, and tear down circuits. Call-control signaling is incorporated into T1 links by a technique known as *bit robbing*, illustrated in Figure 4-4. Every sixth frame steals the least-significant bit from each channel to create an auxiliary channel for network messages. When the least-significant bit is altered on a PCM word, the auditory effect on a modulated voice signal is negligible.

Changing bits of data, though, is disastrous. When using robbed-bit signaling on a T1, data communications equipment is usually set to ignore all least-significant bits in case a signaling bit has clobbered a data bit. Ignoring the eighth bit limits the throughput of each channel to 56 kbps. Robbing is an appropriate term indeed, given that available capacity falls from 1.536 Mbps to 1.344 Mbps!

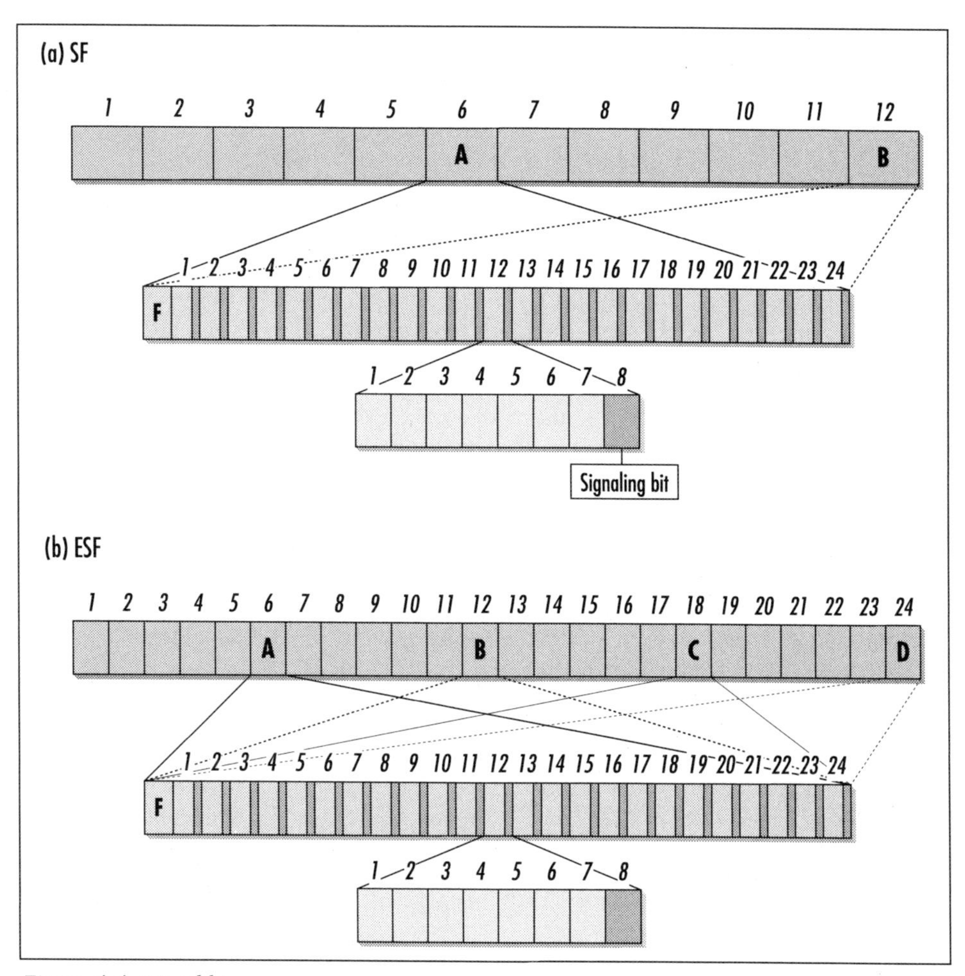

Figure 4-4. Bit robbing

Each robbed bit in a frame is used to create a signaling channel. SF-framed links have two channels, which are called the *A channel* and the *B channel*. ESF-framed links have four signaling channels: the *A channel*, the *B channel*, the *C channel*, and the *D channel*. (The last channel should not be confused with the D channel in ISDN.) Use of the robbed-bit signaling channels is discussed in Appendix A.

5

Timing, Clocking, and Synchronization in the T-carrier System

Time is the extension of motion.
—Zeno

Faster networks depend on accurate timing. As the number of bits per second increases, the time in which to look for any particular bit decreases. Getting both sides to agree on timing becomes more difficult at higher speeds. Synchronous networking is largely about distribution of accurate timing relationships.

A Timing Taxonomy

Synchronous communications do not depend on start and stop flags to mark the beginning and end of meaningful data. Instead, the network constantly transmits data and uses a separate *clock* signal to determine when to examine the incoming stream to extract a bit. Distributing clock information to network nodes is one of the major challenges for synchronous network designers. Three major types of timing are used on networks: asynchronous, synchronous, and plesiochronous. All three terms derive from the Greek word *kronos*, meaning time. The three differ in how they distribute timing information through the network.

Asynchronous systems do not share or exchange timing information. Each network element is timed from its own free-running clock. Analog modems are asynchronous because timing is derived from start and stop bits in the data stream. Free-running clocks are adequate for dial-up communications because the time slots are much longer than on higher-speed digital networks.

Synchronous systems distribute timing information from an extremely accurate primary system clock. Each network element inherits its timing from the primary clock and can trace its lineage to the common shared clock. When AT&T operated the U.S. telephone network, the system derived its timing from the *primary reference source* (PRS), a cluster of cesium clocks located in Hillsboro, Missouri.

Synchronous networks may have several layers of accuracy, but the important feature is that each clock can trace timing to a single reference source. In the case of the Bell system, the primary source was labeled Stratum 1. Less-accurate devices were in higher-numbered strata. Tandem offices, also called toll offices, serviced the long-haul portions of the telephone network and were located in Stratum 2. Local switching offices were located in Stratum 3, with end-user devices such as CSU/DSUs in Stratum 4.

Figure 5-1 sketches the basic system, along with its end goal: distribution of accurate timing to peers at each stratum level. The master PRS at the top of the picture is the source of all timing goodness throughout the network. The network distributes timing information from the primary reference to the toll offices and from the toll offices to local switching offices. Customers attach to the local switching offices. Because timing information even at the lowly customer-equipment stratum is derived from the master clock, two pieces of customer equipment at the end of a T1 link between different local offices can operate within the strict timing tolerances required with 648-nanosecond bit times.

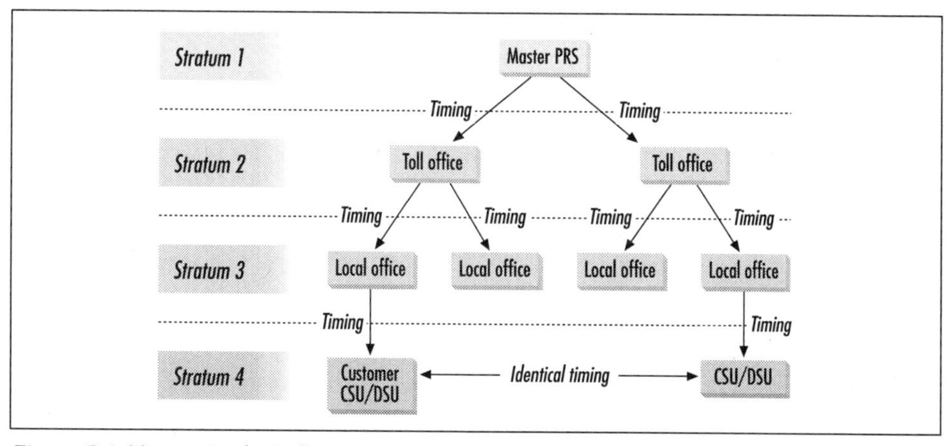

Figure 5-1. Timing in the Bell system

Maintaining a single network timing source is extremely expensive, and the timing distribution must be carefully engineered. Having only one timing source did not fit into the model of the post-divestiture telecommunications landscape in the U.S.

Plesiochronous networks are networks in which the elements are timed by separate clocks, which are very precise and operate within narrow tolerances. Within one telco's network, there may be multiple "primary" (Stratum 1) clocks; each telco maintains its own set of timing information.* For this reason, the U.S. telephone network is really a plesiochronous network. For contrast with a synchronous network, Figure 5-2 illustrates a plesiochronous network.

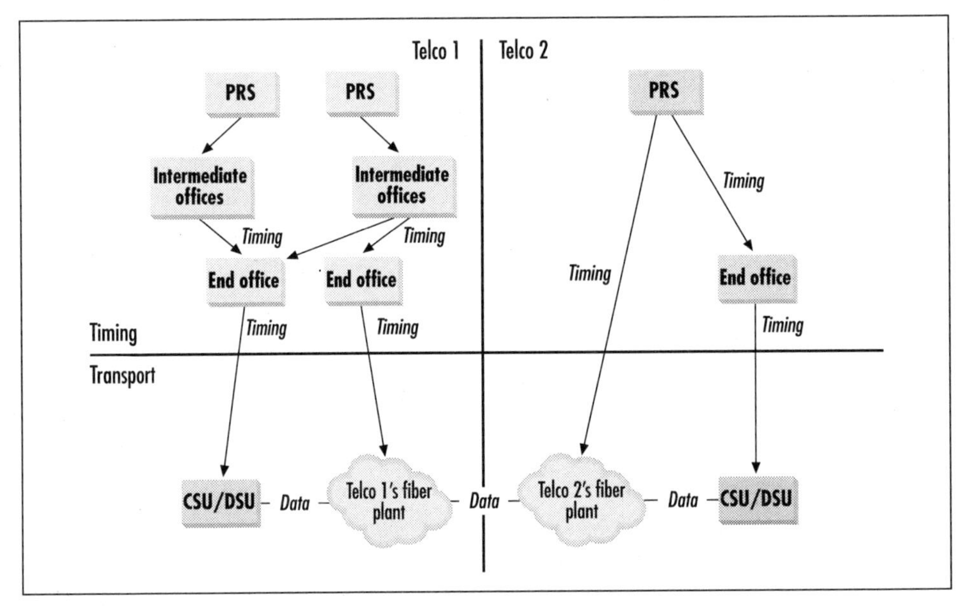

Figure 5-2. A plesiochronous network

Figure 5-2 divides the network into its timing components on top and the data-transport facilities on the bottom. It also shows the facilities of two different carriers, one on the left and one on the right. A T1 transports data between the customer CSU/DSU on the left and the CSU/DSU on the right. Both telcos maintain their own primary reference sources to feed timing information to switching offices and to the devices making up the transport facilities. Even though the T1 is provided by two different carriers, precise timing tolerances allow them to cooperate in providing the T1 without needing a single shared (and trusted) source of timing information.

T-carrier systems are technically known as the *plesiochronous digital hierarchy* (PDH). In practice, though, the distinction between plesiochronous and synchronous is a hair-splitting one. "Synchronous" has acquired a connotation of describing any system that depends on extremely accurate timing, and the T-carrier system is occasionally referred to as the *synchronous digital hierarchy* (SDH). In mafsny cases, the two are combined into one acronym: PDH/SDH.

* Telcos may now use multiple primary reference sources for their networks. Navigational systems like the Global Positioning System (GPS) and the Long Range Navigation (LORAN) network depend on having extremely accurate timing, and it is now quite inexpensive to build specialized receivers to extract timing information from the radio signals at each office, instead of building a transnational timing network.

T1 Circuit Timing

CSU/DSUs are like bridges. They have one interface in telco territory and one interface in data-communications territory. Both are serial interfaces that make use of tight timing tolerances. Appropriate configuration of the CSU/DSU to work within the timing straitjacket is essential.

Receive Clock Inference on the Network Interface

In the T1 world, clock signals are not transmitted separately from the data stream. Instead, receivers must extract the clock from the data signal based on the stream itself. Each bit time slot is 648 nanoseconds. Pulses are transmitted with a 50% duty cycle, meaning that for the middle half of the time slot, the voltage is at its peak. Based on these characteristics, the receiving CSU/DSU infers time slot boundaries from incoming pulses. Ideally, each pulse comes in the middle of a time slot, so finding time-slot boundaries is simply a matter of going 324 ns in each direction. Figure 5-3 illustrates clock inference from pulse reception.

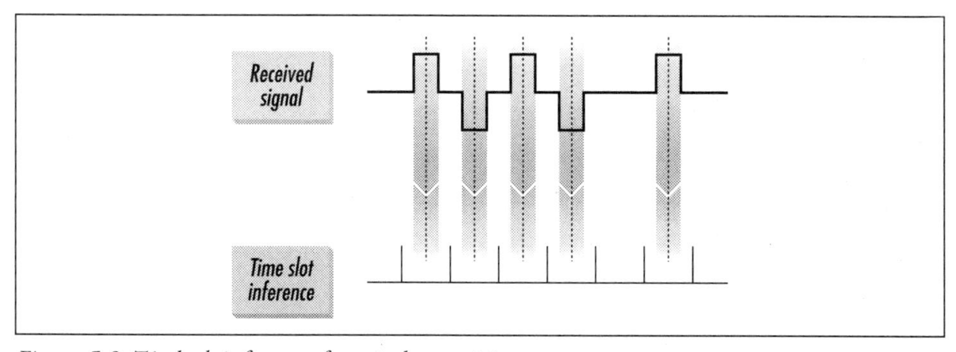

Figure 5-3. T1 clock inference from pulse reception

In practice, of course, things are never quite as simple, and CSU/DSUs must compensate for a variety of non-ideal conditions. Clock signals may exhibit both short-term and long-term irregularities in their timing intervals. Short-term deviation is called *jitter*, and long-term deviation is referred to as *wander*.

Timing on the T1 network interface from the telco is implicit and based on the content of the pulse stream. On the other hand, the serial circuit that connects the CSU/DSU to the router makes use of explicit timing. V.35, for example, includes two pairs (four leads) for sending timing signals and one pair (two leads) for receiving timing.

Transmit Clocking at the Network Interface

At the interface to the telco network, clocking on the received data is based on inferring where the bit times fall. The CSU/DSU does not send an explicit clock for use with transmitted data, but uses internal circuitry to determine when to send a pulse. The internal clocking circuitry can typically operate in one of three modes, which go by different names for different vendors. Descriptively speaking, the typical modes of operation are to derive the transmit clock from the telco, to use an internally generated timing signal, or to take the transmit clock from the attached DTE. Of the three options, the first two are by far the most common in data-transmission applications.

Master/slave timing (also called network or loop timing)

In master/slave timing, the CSU/DSU takes its timing from the telco network. The telco network maintains an extremely accurate timing source and uses that to send pulses to customer locations. At the customer CSU/DSU, the receive clock is extracted from the incoming pulses. In master/slave timing, the extracted receive clock is used for the transmit clock on the network interface, as shown in Figure 5-4.

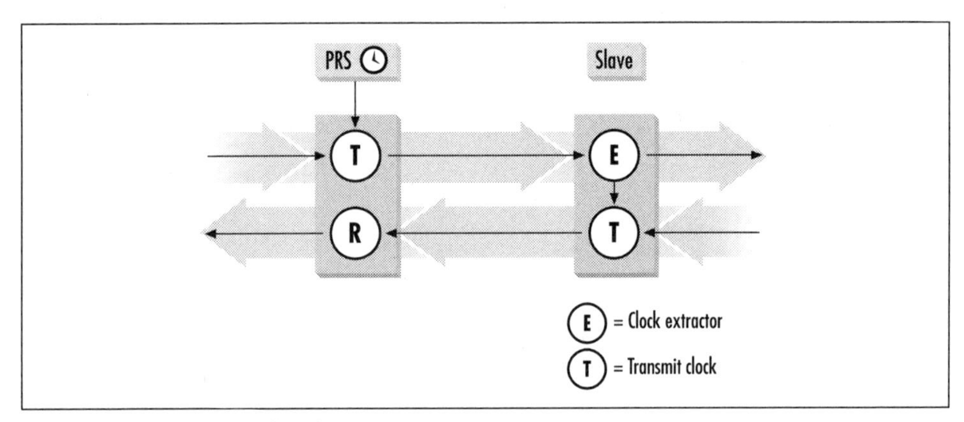

Figure 5-4. Master/slave (loop) timing

Master/slave timing ensures that the less-accurate clock in the customer premises equipment does not drift significantly, relative to the telco's accurate timing system. Several sources may drive the telco transmit clock. One common source is the *building-integrated timing supply* (BITS), which ensures that all the equipment in the CO is running from the same signal. BITS can be linked to an external clock, illustrated in Figure 5-4 as the PRS. At the customer side of the link, the CSU/DSU extracts the receive clock, rather than relying on an internal oscillator in the CSU/DSU. Master/slave timing is also called *loop timing* because the clock is extracted from signals on the digital loop, or *network timing* because the clock source is from the telco network interface.

Internal timing

Internal timing uses an internal oscillator in the CSU/DSU as the transmit clock source. No special measures are taken to ensure that the timing of transmitted pulses matches the timing of received pulses because the two operations are logically independent, as Figure 5-5 illustrates.

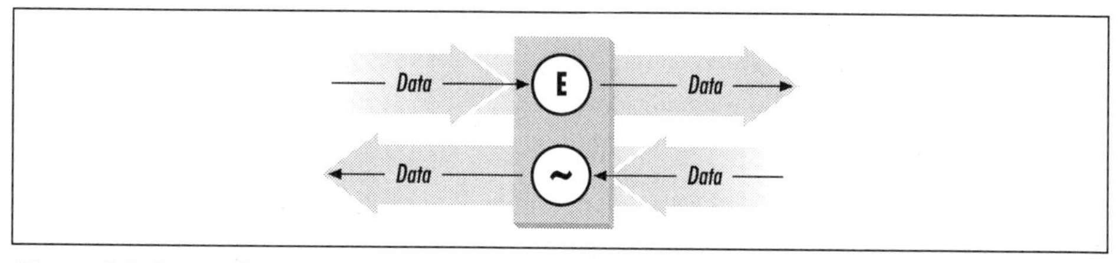

Figure 5-5. Internal timing

All circuits must have one timing source. In most cases, the telco will supply timing because the entire telco network must operate with unified timing to deliver the T1 circuit. In some cases, however, a simple copper wire pair can be leased from the telco. For spans with less than 30 dB attenuation, an unrepeatered copper pair can cost much less than a full-service line. In private-line applications, one end of the line must provide the clock, as Figure 5-6 shows. The remote end is set to loop timing, so the remote transmit clock is derived from the local transmit clock.

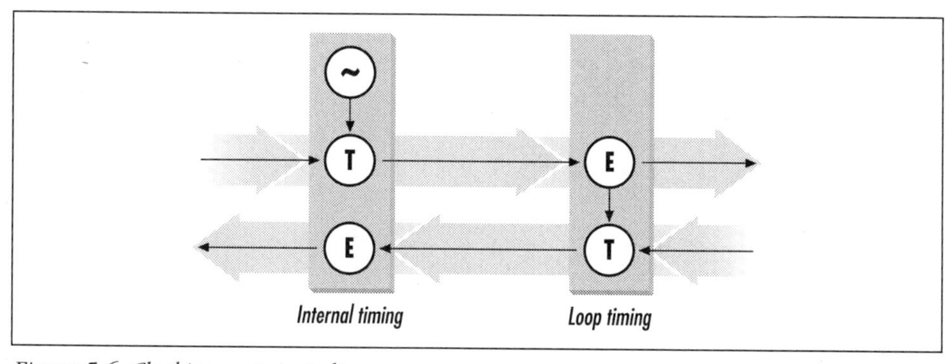

Figure 5-6. Clocking on private lines

Clocking at the Data Port

Data ports on CSU/DSUs are synchronous serial ports. CSU/DSUs transmit data as a varying voltage on the line, with a high voltage representing one and a zero voltage used for zero. A second signal, the clock signal, triggers a voltage measurement and extracts a bit from the voltage stream. Figure 5-7 illustrates the use of the explicit clock signal on a synchronous serial port.

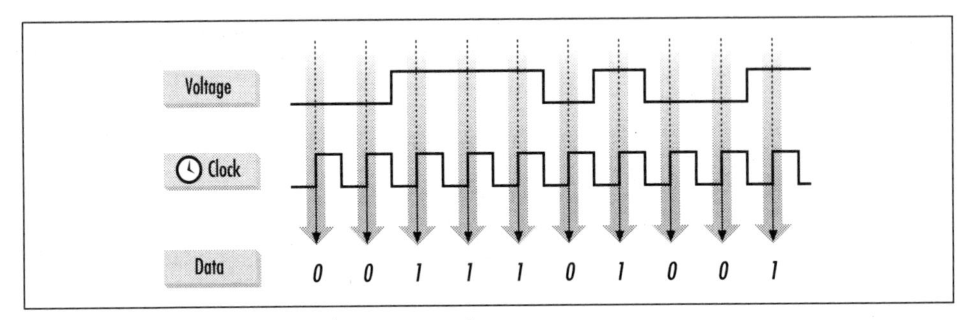

Figure 5-7. Clocking on a synchronous serial port

In Figure 5-7, the external clock signal triggers a measurement when the clock signal goes from a low voltage to a high voltage. Aligning the clock signal with the voltage plateaus is important. Ideally, the clock signal should trigger a voltage measurement at the middle of the bit time. If the clock signal falls too close to a voltage transition, the reading will be unreliable.

Receive clock timing

The clocking on the serial circuit from the CSU/DSU to the router is a synchronous serial line that uses explicit timing. Clock signals for the received data are extracted from the incoming pulse train at the carrier network interface. The extracted data is then transmitted with the extracted clock signal out the data port. No configuration is necessary on T1 equipment to configure the clock signal for received data. Figure 5-8 demonstrates receive clock timing.

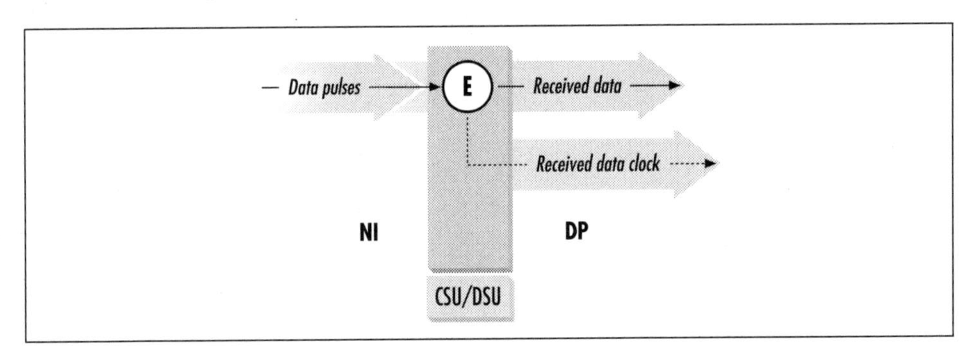

Figure 5-8. Receive Clock timing for the data port

Internal data port clocking

Forwarding the received data out the data port and on to the router is only half of what a T1 does. Transmitting data from the data port out to the telco network successfully requires that the data be correctly received from the data port and processed accurately. The simplest clocking method is to allow the CSU/DSU to control the clocking of transmitted data, too. V.35 interfaces allow the CSU/DSU to supply timing to the DTE. Data arrives at the CSU/DSU and is then extracted using the transmit clock supplied to the DTE, as Figure 5-9 illustrates.

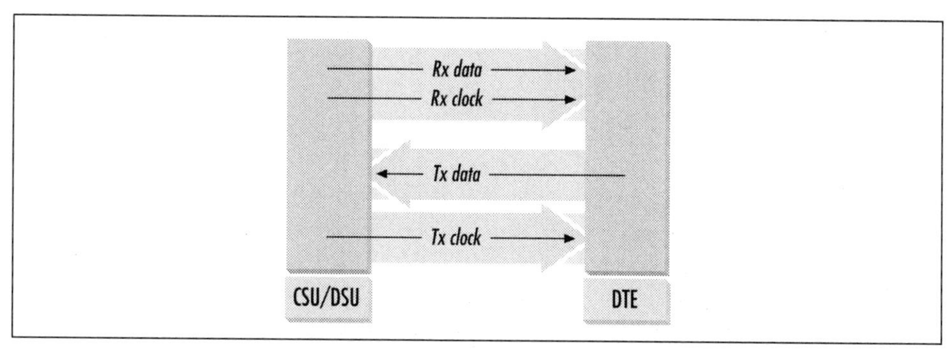

Figure 5-9. Internal data port clocking

In most applications, internal data port clocking provides acceptable performance. If the clock signal drifts out of phase with the data transmission, however, the clock will trigger measurements too close to the transition between the high voltage and the low voltage. Measurements to extract bits still take place, but those measurements may not reflect the data that was supposed to be transmitted. Figure 5-10 illustrates the problem.

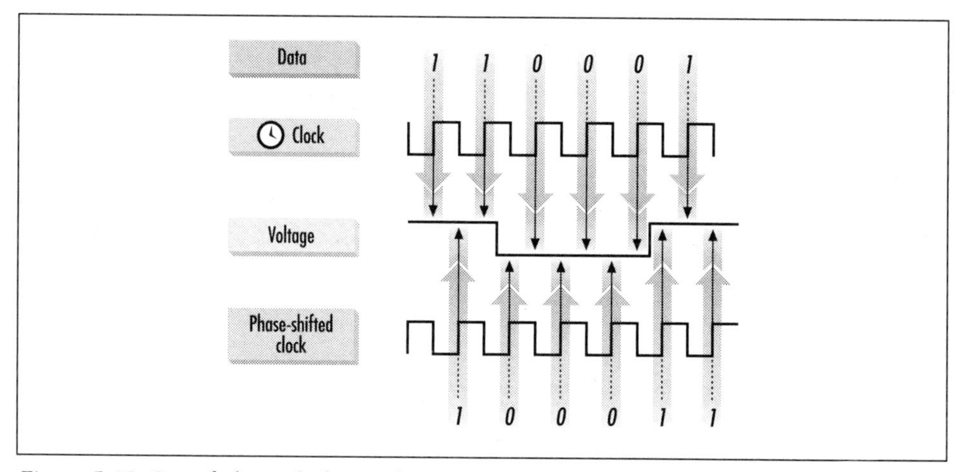

Figure 5-10. Out-of-phase clock signal

As in Figure 5-7, the clock signal triggers a measurement at the rising edge. The lower clock signal is delayed, or phase shifted, so measurements occur too late and extract incorrect data. One common reason for the delay is a long cable between the CSU/DSU and the router. The router synchronizes its transmissions with the transmit clock, but the data must travel from the router to the CSU/DSU. In the time it takes for the data to travel along the cable to the CSU/DSU, the clock signal has moved on and the measurements it takes are tardy. Problems may also occur at high-transmission rates, because the bit times are shorter, or when the router has a significant processing latency.

External data port clocking

One way to address the problem of a phase-shifted clock signal is to change the source of the transmit clock at the data port. Routers can be designed to accept the transmit clock from the CSU/DSU and use it to drive the external clock line. The external clock line is often labeled XCLK or SCTE (an abbreviation for serial clock timing external). The router assumes the responsibility of synchronizing the transmitted data with the external clock signal. With external clocking, the clock and data must take the same path and are subject to the same delays, so the data and its clock signal stay in phase over the cable. This is illustrated in Figure 5-11.

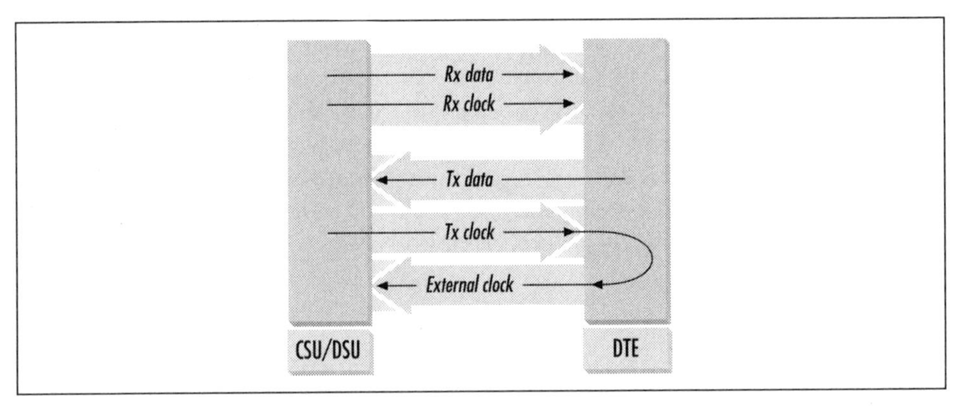

Figure 5-11. External data port clocking

Many routers, however, do not support looping the received clock signal back to the CSU/DSU. When these routers are used with a CSU/DSU that expects a transmit clock, nothing will be transmitted because no clock signal is transmitted to the CSU/DSU.

Using a router that does not supply a transmit clock with a CSU/DSU can be a particularly difficult problem to pinpoint without the right equipment. Protocol analyzers that tap into the V.35 connection between the CSU/DSU and the router may not see a problem because the V.35 connection is fine.

This often manifests itself as a router that insists that it is transmitting data even though nothing is received by the remote end. Because nothing can be transmitted, the link layer protocol cannot initialize the link. If you see this symptom, check with the router's vendor to ensure that they support sending a timing signal to the V.35 DCE.

Inverting the internal clock signal

A second method of addressing a phase shift between the transmitted data and its clock signal is to invert the internal clock, which has the effect of shifting the clock signal by a half-cycle. The goal of clocking on the serial port to the router is to make sure that the clock signals trigger bit extractions in the middle of the bit time. Problems occur when the clock fires at the edge of a bit time. Moving the clock trigger half a cycle returns the clock signal to the middle of the bit time.

Slips: When Timing Goes Bad

T1 equipment employs a variety of techniques to compensate for variations in timing signals. Intermediate network equipment may buffer the 192-bit frames to ensure that frames are complete before forwarding them on to their destinations. CSU/DSUs are equipped with *phase lock loop* (PLL) circuitry to track with the more accurate clocks at the local exchange office. Occasionally, though, these measures are not enough, and timing problems occur.

Imperfect timing conditions may force network equipment to replicate or delete data in a process called a *frame slip*. Slips are divided into two categories. *Controlled slips* replicate or delete a complete 192-bit frame of data, but do not cause any problems with the T1 path. *Uncontrolled slips*, which are also called *change of frame alignment* (COFA) events, are much more severe because they disrupt the framing pattern. Controlled slips are the more benign of the two because the path remains available. Uncontrolled slips indicate more severe problems with the circuit.

Controlled slips always involve complete frames, and can be the result of either a buffer overflow or underflow. Both conditions are illustrated in Figure 5-12. In the overflow case, the second frame is lost in time unit 1 because the buffer overflows and replaces it with the third frame. Both the second frame and its framing bit are lost. Receivers use the disruption in the framing bit sequence to detect controlled slips. Controlled slips may also occur because of a buffer underflow, which causes frames to be repeated. In Figure 5-12, the buffer underflow in time unit 1 means that no fresh data is available for transmission in the second time unit.

Uncontrolled slips are far more severe. If a buffer overflow or underflow causes a partial frame to be lost, an uncontrolled slip will occur. Framing bits shift within the bit stream, as illustrated in Figure 5-13.

In the overflow case, part of the second frame is lost. Following the partial frame 2, the third frame begins with its framing bit. The underflow case is similar in that the first frame is only partially replicated in the second frame slot. Immediately following the partial frame in the second frame slot is the framing bit for the third frame. Partial frames lead to a change in the frame alignment—instead of being

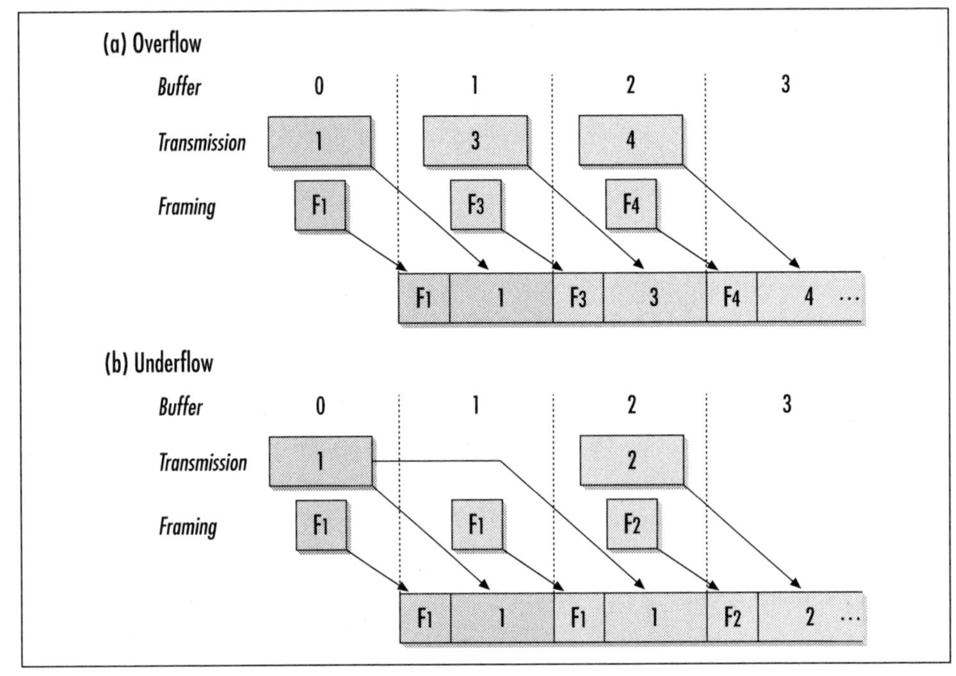

Figure 5-12. Controlled slip operations

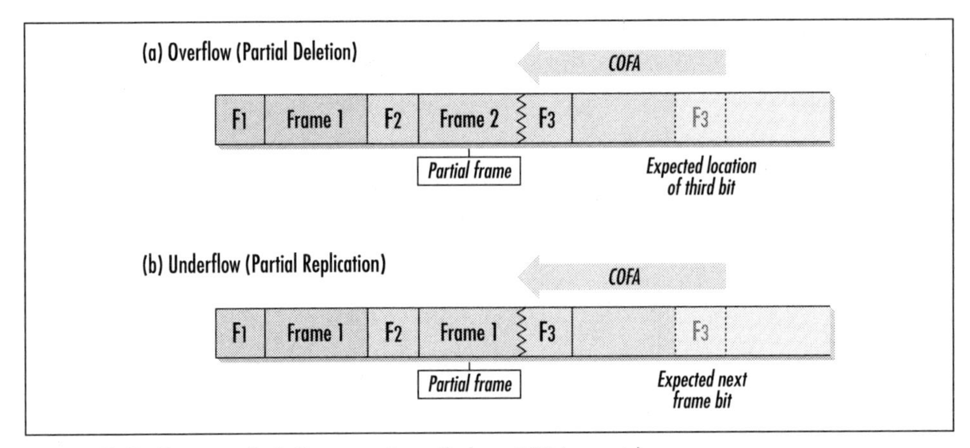

Figure 5-13. Uncontrolled slip operations (a.k.a. COFA events)

greater than 192 data bits, it will be fewer than 192. Because the framing bit is in an unexpected location, the receiving framing unit will need to examine the incoming bit pattern carefully to determine the new frame alignment and reframe appropriately.

Reframing can be a time-consuming process. Instead of a simple higher-level retransmission, the equipment at both ends of the T1 must resynchronize, which takes a significantly longer time to accomplish. Uncontrolled slips are usually the result of equipment that does not buffer full 192-bit frames. Full frame-size buffers are expensive because they require more memory and more complicated handling. Additionally, handling and checking full frames increases the latency of the frames as they pass through the device, which most equipment vendors prefer to avoid.

Avoiding Slips

Underlying timing problems on the circuit are what cause slips to occur. Two rules can prevent configuration errors with regard to the clock source:

1. There must be no more than one clock source. Synchronous networks depend on having one source of clocking truth, and free-running clocks at both ends of a circuit will corrupt data as the two clocks drift in and out of synchronization.
2. There must be no less than one clock source. If two CSU/DSUs are both operating in slave mode, each will look to the other as the source of timing information. Any changes in the timing of one CSU/DSU will affect the partner because both CSU/DSUs are attempting to lock on to the timing information supplied by the other.

Combining these two rules leads to the obvious conclusion that there must be only one clock source. Typically, the clock source is the telco and the CSU/DSUs at both ends are set for loop timing. On untimed circuits, one of the CSU/DSUs should generate the transmit clock, and the other should be set for loop timing. Network administrators should choose one clock source and configure other devices to accept that clock.

6

Mysteries of the CSU/DSU

A man is as old as his arteries.
—Pierre J. G. Cabanis
Epigrams

When bringing up a T1, a CSU/DSU is required by FCC part 68 to protect the telco network from your equipment. If the CSU/DSU does not function correctly, whether due to component failure or misconfiguration, the line will not come up. Unfortunately, producing clear, understandable printed documentation has never been a goal of data communications equipment vendors. CSU/DSU manuals have improved to the point of basic intelligibility, but even the best manuals require a solid understanding of telco networking.

Line Build Out: Moving Between Theory and Practice

In a perfect theoretical world, wires have no resistance and voltage pulses can travel forever. The real world, however, is rarely anything like the world of theory. (There is a reason that physicists joke about spherical frictionless cows, after all.) Repeaters and signal regenerators along T1 spans are the first admission of a practical world. *Line build out* (LBO) is the second concession to practicality.

Two different types of connections are used on T1 lines. A *long-haul* connection is made to the telco network. The first repeater may be up to 3,000 feet away from the end-user location. CSUs must be able to drive the pulse up to 3,000 feet, so the arriving pulse at the first repeater is still strong enough to be regenerated. Long-haul line build out is used to avoid problems that arise from having multiple T1 customers in an area. It works by reducing the signal to the level required by the telco network. *Short-haul* connections are digital cross-connect links up to 655 feet that are made between user-owned devices such as PBXs. Short-haul line build out is also called *line equalization*. Equalization increases the signal strength above the reference level. For historical reasons, a network interface is abbreviated *DS1*, and the local side is often labeled *DSX* or *DSX-1*. Figure 6-1 contrasts short- and long-haul line build out.

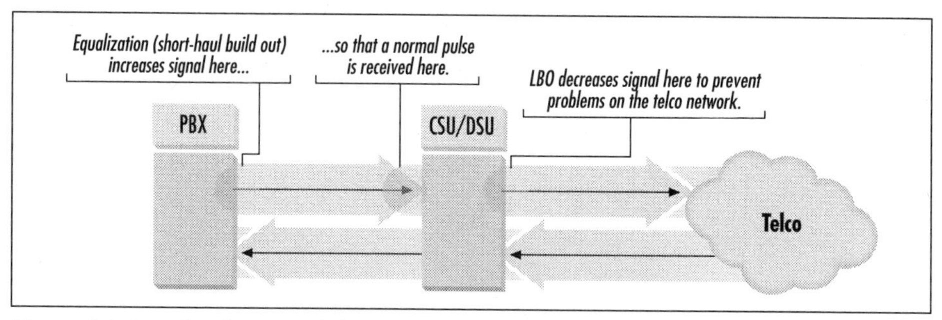

Figure 6-1. Long-haul LBO and short-haul LBO

Long-Haul Build Out

All T1 connections to a telco network must be made with long-haul settings. Two main reasons motivate the use of long-haul LBO. The first relates to T1 repeaters, which are designed to detect incoming pulses at a wide variety of voltages. Some repeaters, however, cannot sense pulses unless they exhibit a loss of at least 7.5 dB from the nominal transmission level. Such repeaters will not register one bits from pulses that are not sufficiently attenuated and will therefore reconstruct a stream of continuous zeros. To work with these older repeaters, some CSU/DSUs may apply a default LBO setting of –7.5 dB.

T1 also uses line build out to prevent crosstalk at "non-repeatered line junctions." In Figure 6-2, two T1 lines run in the same conduit before reaching the first repeater. One customer is farther away, so its signal in the shared conduit is much weaker than the signal from the nearer customer. If crosstalk exists between the two paths when the difference between the signals exceeds 7.5 dB, the resulting noise will disrupt the operation of the repeater and cause errors. To prevent this, T1 uses line build out to make sure that the signals from both customers are at approximately the same level at the route junction. Rather than adding extra cable to the nearby customer's path, the LBO setting is adjusted on the nearby customer's CSU/DSU to create signal loss.

Long-haul LBO settings insert a deliberate signal loss into the path. The loss that LBO inserts matches the loss from cabling, so LBO can be used to equalize the virtual path length even though the physical path length is different. Figure 6-3 illustrates this.

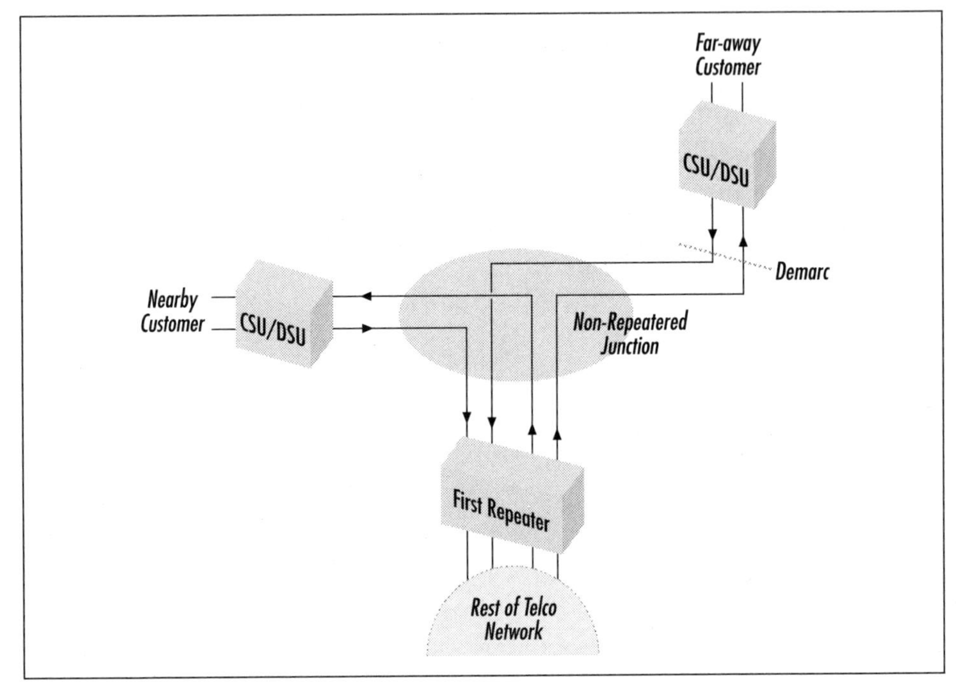

Figure 6-2. Non-repeatered line junction

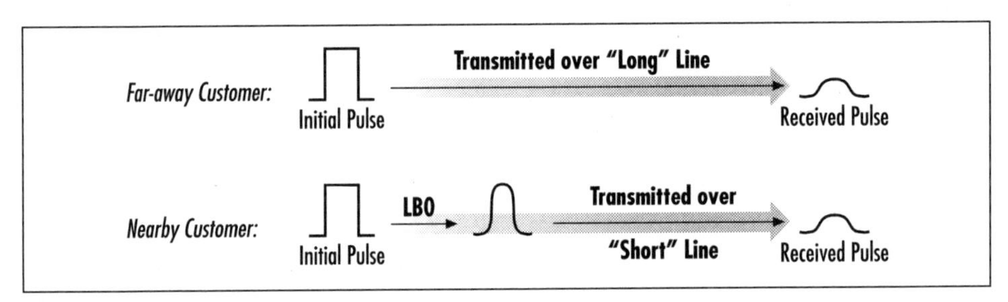

Figure 6-3. Long-haul LBO loss insertion

In Figure 6-3, the shorter line uses line build out circuitry to give it the same transmission loss as the longer line. LBO-induced loss must therefore mirror the loss induced by a cable. T1.403 lays out a reasonably complex mathematical transform to accomplish the goal of mimicking cable loss.

End spans in T1 lines typically have smaller loss budgets than the main interoffice sections. Inter-repeater sections may have losses up to 30 dB, but end spans are usually designed for a maximum loss of 16.5 dB, which corresponds to approximately 3,000 feet. Line build out is used to ensure that all signals arriving from customers have approximately the same level. Most customer equipment allows you to select between 0 dB and 22.5 dB loss in steps of 7.5 dB.

The telco supplying the line determines the appropriate LBO setting and includes this with the order. If the LBO setting is not specified, ask the installation technician from the phone company. (The installer won't want to make a trip back simply to change the LBO.) Most new installations use smart jacks, which generally use a full-strength input signal and thus a 0 dB LBO setting. Some vendors may ship equipment with a default LBO setting of –7.5 dB because there are repeaters that will not detect pulses unless a 7.5 dB attenuation has taken place.

Short-Haul Build Out (Line Equalization)

In the beginning, the demarc was the interface between the CSU and the DSU, which was known as a DSX-1. Although the demarc has now shifted to the network side of the CSU, companies continue to build some telephone equipment with a DSX-1 interface. The most common examples of this are phone switches (PBXs). Channel banks and multiplexers may also fall into this category. Connecting equipment with a DSX-1 interface to the telephone network requires a CSU. To simplify the electronics, a pulse received at the DSX-1 interface must have the same characteristics as a pulse transmitted out the CSU network interface.

Cable loss gets in the way of this ideal goal, so equipment that connects to CSUs must provide line equalization, which is sometimes called short-haul line build out. This feature increases the signal level so that it is received at the end of the cable matching the nominal pulse that the CSU transmits. The level is increased, and pulse edges are enhanced so that the cable loss wears them down to a "normal" shape.

Equalization is set in multiples of 133 feet and can range from 0 to 655 feet. Typical settings are 0–133 feet, 134–266 feet, 267–399 feet, 400–533 feet, and 534–655 feet. Naturally, many vendors have implemented different configuration methods.

Configuring LBO Correctly

When using T1 as a transport for external data links, a CSU/DSU is used to connect to the telco network. To connect to the telco network, follow the telco's rules and configure LBO to the setting that the telco requires. Most new T1 installations place the telco network interface unit (NIU) in the same room as the customer router. Smart jacks can accept a variety of pulse heights, but it is generally advisable to set the LBO to 0 dB. Many CSU/DSUs incorporate automatic LBO settings. A variety of approaches are used to select the correct setting, but a typical one is to monitor the amplitude of received voltage pulses and set the line build out accordingly. T1 uses line equalization (short-haul build out) only when two pieces of customer equipment are being connected. T1 also commonly uses line equalization when connecting a PBX to a CSU.

Table 6-1 lists common line build out settings.

Table 6-1. Common line build out settings

Short-haul options	Long-haul options[a]
0–133 feet (no boost)	0 dB (no loss)
134–266 feet	7.5 dB
267–399 feet	15.0 dB
400–533 feet	22.5 dB (maximum loss)
534–655 feet (maximum gain)	

[a] Vendors may use different signs on the long-haul build out. In spite of the confusion, long-haul build out settings always go in increments of 7.5 dB and are always a smaller output signal.

T1 CSU/DSUs

Figure 6-4 shows a typical CSU/DSU. It has a series of lights and a switch that is used to control the loopback test mode. The meanings of various common indicator lights are discussed in Table 6-2.

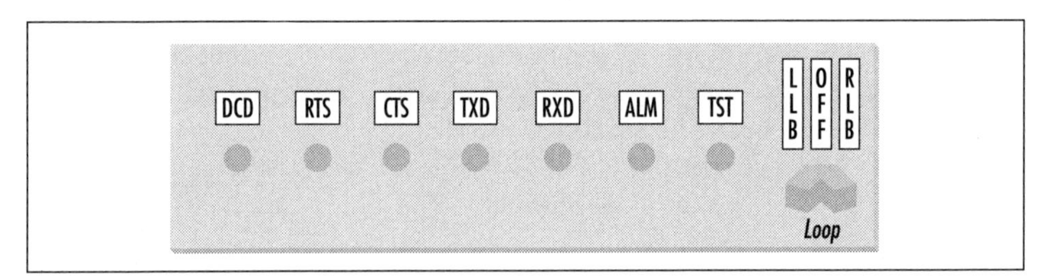

Figure 6-4. Typical CSU/DSU

Table 6-2. Common T1 CSU/DSU indicators and their meanings

Indicator	Common labels	Meaning
Clear to send	CTS, CS	The CSU/DSU is ready to receive data.
Request to send	RTS, RS	The DTE is ready to send data.
Carrier detect	DCD, CD	The CSU/DSU is generating carrier signal.
Send data	TX, TXD, SD	The CSU/DSU is transmitting pulses to the telco network.
Receive data	RX, RXD, RD	The CSU/DSU is receiving pulses from the telco network.
Loss of signal	LOS	When no pulses arrive within 100 to 250 bit times, LOS is declared. Under normal conditions, at least a few pulses would be present in that interval.
Out of frame	OOF	The CSU/DSU triggers this when framing bits are in error and clears it when frame synchronization is regained.

Table 6-2. Common T1 CSU/DSU indicators and their meanings (continued)

Indicator	Common labels	Meaning
Loss of frame (red alarm)	LOF	The CSU/DSU asserts this when OOF has persisted continuously for 2.5 seconds. This alarm is usually cleared when frame synchronization has been obtained for at least one second, but some CO hardware may not clear the alarm for much longer (15 seconds or more).
Keepalive transmission	KA	When framing is lost, the CSU/DSU transmits a keepalive signal to other network components. (See the following section.)
Remote alarm indication (yellow alarm)	RAI	The remote end is reporting a loss of signal. (See the following section.)

RAI/Yellow Alarm Signal

When the incoming signal is lost, a CSU/DSU transmits the RAI/yellow alarm signal in the opposite direction, as Figure 6-5 shows. If equipped, the local CSU/DSU may then light up the appropriate indicator to show that it is receiving the RAI/yellow alarm signal.

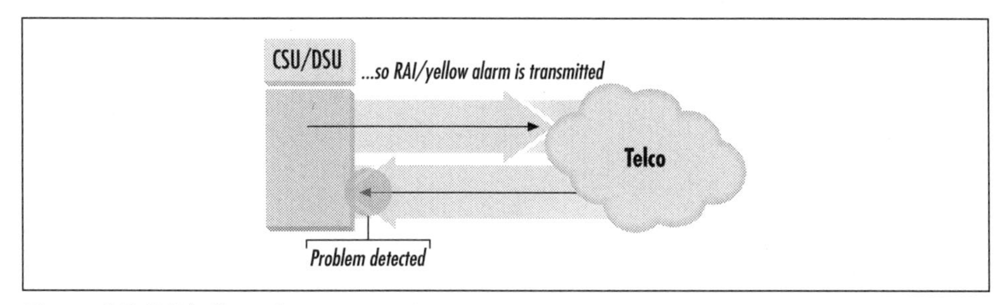

Figure 6-5. RAI/yellow alarm generation

Transmission of the RAI/yellow alarm signal depends on having a signal that is correctly in-frame in the outbound direction. To the receiver, RAI/yellow indicates a potential problem with the outgoing path, but its reception indicates that the incoming path is functioning normally. In simple terms, it is a report from the remote end that nothing appears to be transmitted on the circuit.

Keepalive/AIS

When some condition prevents a T1 component from transmitting data, it transmits an unframed all-ones keepalive (KA) signal, which is also called the alarm indication signal (AIS). Using all ones helps to keep the timing in the network synchronized.

T1 generates AISs for two main reasons. The first, which Figure 6-6a illustrates, is to inform other network equipment of the fault. One specific situation that arises with some regularity in the T1 world is shown in Figure 6-6b. If one of the repeaters detects a fault or is placed in loopback mode, it transmits a keepalive signal in the other direction. If the CSU/DSU is receiving a keepalive signal, call the telco to report the problem. When an AIS is coming in from the telco, it typically means that some piece of equipment along the circuit has not been configured, connected, or turned on. It may also mean the local smart jack is in loopback.

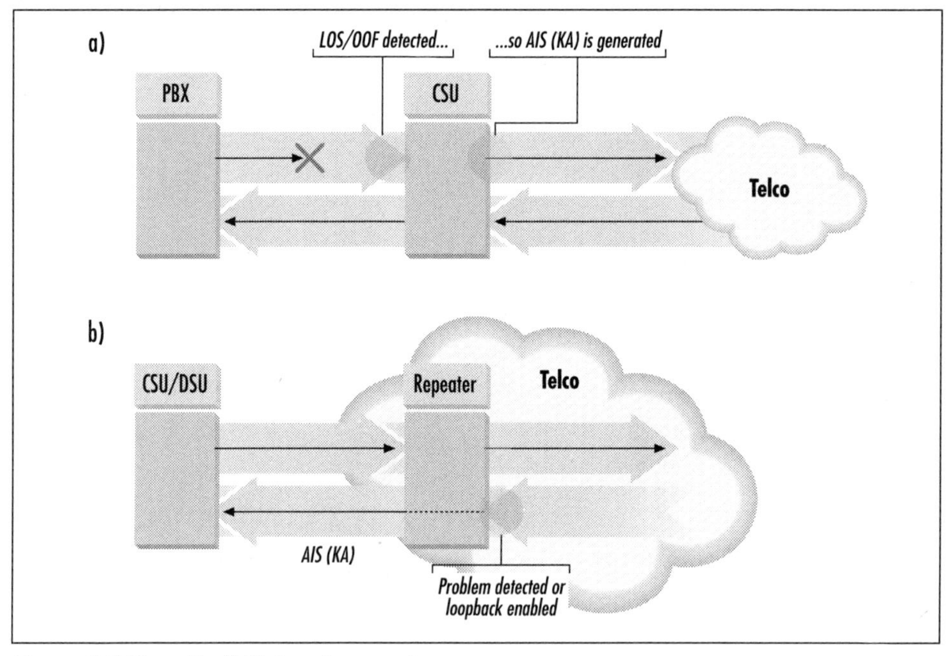

Figure 6-6. Keepalive/AIS signal generation

Loopback and Other Test Functions

All CSU/DSUs have loopback options to assist with testing. *Local loopback* sends the data from the DTE right back to the DTE, and *remote loopback* sends data received from the far end of the line right back out on the line. Both loopback options are illustrated in Figure 6-7.

More expensive CSU/DSUs may have other options that allow finer control over some of the serial control signals. Leave them set to the default unless they appear to cause problems. If problems arise, contact the DTE vendor. Fancier CSU/DSUs may also improve testing capabilities by incorporating functions that allow the generation of test patterns. Chapter 11 describes the use of loopback settings for testing purposes.

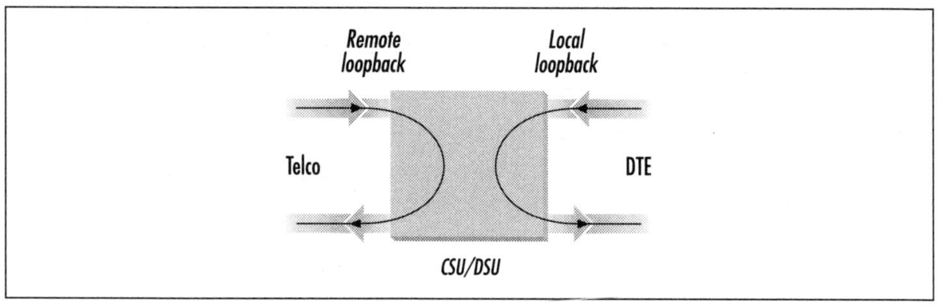

Figure 6-7. Local and remote loopback

CSU/DSU Configuration

CSU/DSU configuration can be a frustrating affair for many reasons. Each vendor has a slightly different way of describing features, and terms in the manual may differ from the terms your telco uses. In many cases, your service provider or telco will include a CSU/DSU in the startup charges for a new line and will even pre-configure the CSU/DSU to work with the new line. Depending on the model, configuration can be frustrating. Some CSU/DSUs are configured with jumpers or DIP switches, and the cheapest ones even require a power-cycle to reread the jumper or switch settings. Higher-end CSU/DSUs have menu-driven configuration and are far easier to use.

While the array of configuration options may be bewildering, the following options are critically important and account for most of the problems you may see:

Framing
: The choice is between SF and ESF, and the former is rare and getting rarer. This setting is supplied by the telco because it must match the telco equipment.

Line code/DS0 channel speed
: These two settings are related. AMI encoding is almost always associated with 56k component DS0 channels, and B8ZS is always associated with 64k component DS0s. Depending on your CSU/DSU, configuration may be required for both line code and DS0 speed or just one of the two. Like framing, these settings are supplied by the telco.

Line build out
: By the time a pulse reaches a T1 repeater, it is assumed that a certain amount of attenuation will have already occurred. T1 uses LBO to adjust the outgoing transmitted pulses so that they will be attenuated to the normal amount by the time the pulses arrive at the first repeater. Naturally, LBO depends on the distance from the first repeater. If the CSU/DSU is connected to a smart jack, the LBO can usually be set to 0 dB. Further distances from the first repeater

require higher build out levels. The telco must supply this setting. Some CSU/DSUs have an automatic LBO mode that sets the build out based on the attenuation of received pulses.

Clock source

Precise timing is the foundation of the modern digital telephone network. Like many other things in life, the key to timing success with T1 spans is to have exactly one source of timing. When connecting to the telco network, you should receive clock from the telco. Unless you are deploying extremely special (and expensive) gear, the telco clock will be more accurate than yours. Take your lead from them, and obtain the transmit clocking from the network. Consult the CSU/DSU documentation, because this setting may go under several different names: *external* (because the clock is external to the CSU/DSU), *loop timing* (because the clock comes from the digital loop), and *network* are all common terms for "take the clock from the telco." Internally generated clocking should be used at one end—and one end only—of a private line that does not go through the telephone network.

DTE clocking

Data that the CSU/DSU receives on its data port must also be clocked. The CSU/DSU may use its own clock or may take its clocking from the DTE. When clocking is done internally, the setting is typically called *internal* by the CSU/DSU vendor. If the CSU/DSU-to-DTE cable is long, the CSU/DSU clock may be out of phase with the DTE clock and look for bits at the wrong times. To address this, many CSU/DSUs incorporate an *internal/inverted* setting, which phase-shifts the clocking signal to the correct place. Alternatively, the DTE can supply clock to the CSU/DSU, which is usually called *external* timing. This setting is DTE-specific. Some vendors supply clock to the CSU/DSU, and others do not. If the CSU/DSU is expecting clock from the DTE and it is not supplied, no data will ever be clocked on to the line and nothing will be transmitted out the span. Some CSU/DSUs incorporate an automatic switching feature. Automatic data-port clocking uses the external clock if it is transmitted by the DTE. If no external clock is present, the CSU/DSU measures the latency to the router and back and selects internal or inverted clocking, depending on the delay.

Time slots

If you order a fractional T1 (FT1), only some of the DS0 slots will be available for data traffic. Make sure you and the telco agree on which slots those are. In most cases, the T1 uses time slots starting with the first one and continuing in sequential order until the number of time slots provides the fractional bandwidth ordered. Table 6-3 shows typical examples.

Table 6-3. FT1 configuration guidelines

Line speed	Time slots used
256k	1–4
512k	1–8
768k	1–12

For documentation purposes, it is an excellent idea to keep information on each circuit in use. At a minimum, collect the telco's (and perhaps the ISP's) circuit ID, along with the framing type, line coding, and line build out settings. DTE clocking settings are specific to each piece of DTE-vendor equipment, so you might also want to keep separate lists of the required CSU/DSU settings for each DTE vendor.

Embedded CSU/DSUs

Now that T1 is a common type of access link for remote sites, many vendors offer the option of a T1 card with an integrated CSU/DSU. Simply plug in the cable, and the physical layer is done.

This is an extremely positive trend. Vendors can preconfigure the CSU/DSU-to-DTE setting so that it will always work, which saves you time by leaving only the work of determining the appropriate CSU/DSU-to-telco settings. It eliminates finger-pointing between the CSU/DSU vendor and the DTE vendor and gives the DTE vendor's support staff a known platform for which engineering knowledge is directly available.

Summary of Settings

Table 6-4 summarizes the most common CSU/DSU configuration options, along with available settings, common industry defaults, and additional remarks.

Table 6-4. T1 CSU/DSU configuration options

Option name	Possible settings	Common CSU/DSU default	Who sets this option?	Remarks
Framing	Superframe (also SF or D4) Extended superframe (ESF)	Equipment defaults to SF; most lines are ESF Some equipment can autoselect	Telco	

Table 6-4. T1 CSU/DSU configuration options (continued)

Option name	Possible settings	Common CSU/ DSU default	Who sets this option?	Remarks
Line code	AMI B8ZS	Equipment defaults to AMI; lines are usually B8ZS	Telco	
Transmit clock source	Network (also looped timed): transmit clock derived from receive clock DTE Internal CSU/ DSU oscillator	Varies	DTE vendor; the network setting is generally safest	There is a wide variation in terminology. Consult the manual.
DTE clocking	Internal CSU/ DSU oscillator Inverted signal from oscillator Supplied by DTE (also called external)	Internal Sometimes can be automatically selected	DTE vendor	Generally internal. Long cables may require inverting the internal clock signal.
Line build out	0 dB 7.5 dB 15 dB 22.5 dB	0 dB Some equipment can autoselect	Telco	Usually 0 dB when connected to a smart jack, though it could be anything.
Transmit FDL performance reports	Transmit Do not transmit	Transmit	Telco	Performance reports exist only on ESF links.
Transmit yellow alarm	Transmit Do not transmit	Transmit	Telco/ISP	
Inband signaling	Yes No	No	Telco	Steal 8 kbps from channel 1 for signaling.
Channel data rate	64k 56k	56k	Telco	B8ZS must be used for 64k. Either B8ZS or AMI can be used with 56k, though AMI is more common.
Number of channels; channel map	1–24	Varies	Telco	Based on subscription level for fractional T1.

Table 6-4. T1 CSU/DSU configuration options (continued)

Option name	Possible settings	Common CSU/ DSU default	Who sets this option?	Remarks
Channel order	Continuous Alternating	Continuous	Telco	Alternating is used to ensure pulse density on non-B8ZS links. Half of the channels carry data, and the other half are used to maintain ones density. The bandwidth penalty is very high, so the alternating option is rarely used.
Data inversion	Do not invert Invert	Do not invert	Telco	Assures ones density on some types of HDLC lines; generally not needed.
Bit stuffing	Do not stuff Stuff	Disable	Telco	Required only when line does not ensure correct ones density; generally not used.

7

Connecting the Umbilicus: Getting T1 Connectivity

Solitude is impractible, and society fatal.
—Ralph Waldo Emerson
Society and Solitude

With a good deal of theory behind you, some practical information is in order. This chapter describes the procedure for hooking up a T1 circuit to an ISP.

Ordering

If you want people to do things for you, money is usually an excellent way to make it happen. To get a T1 installed, you need to place an order and promise to pay a bundle of money. There are two main routes to go here. One way is to do it yourself. On the other hand, many ISPs will happily sell you a startup package that includes any necessary equipment (router, CSU/DSU, cabling, etc.) and will place the circuit order with the telco for you. I strongly recommend the latter approach. Circuit ordering and installation often take longer than scheduled, and ISPs that deal with telcos on a regular basis are familiar with the issues involved. Many larger ISPs even have "Telco Liaison" positions staffed by people whose jobs are working with phone companies on WAN link issues. Large ISPs bring buying power to the table as well. It may not help your pricing, but it will help smooth over the installation process.*

* Increasingly, large ISPs are subsidiaries of major telecommunications firms and help route circuit orders to the parent telco umbrella.

T1 Installation and Termination

After you place an order for a T1, be prepared to wait. Avoiding crosstalk on T1 circuits requires that the telco carefully select the wire pairs in a conduit. Repeaters must be adequately separated so that they do not interfere with each other, but must be close enough together to regenerate and recondition the signal. Line cards may be needed at each CO involved. Delays of a month or more from the initial promised date are not uncommon with the installation of high-speed links.

All the waiting does offer you the opportunity to gather the equipment you need and nail down the configuration. When you've ordered the circuit, the telco will give you the basic configuration information (line coding, speed, and build out). T1s are priced based on distance. To save money, you could purchase a T1 that links to the nearest entry point to a public data network, such as the telco's regional frame relay net. This is illustrated in Figure 7-1. You'll need to find additional configuration information for connections that are not made directly to the ISP.

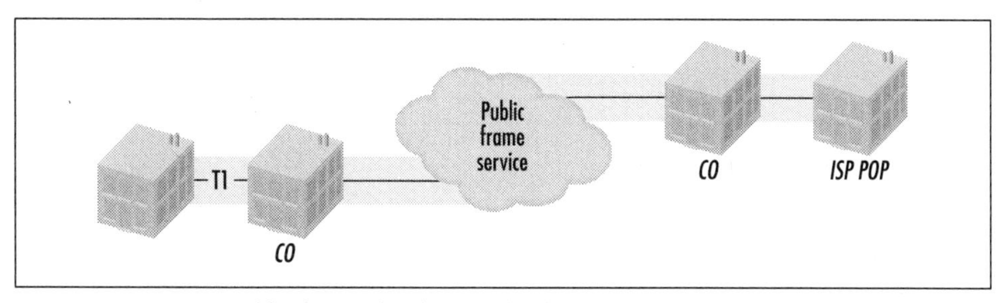

Figure 7-1. Using a public frame cloud to cut the distance

The Demarc

A telco technician will come to your site to connect the circuit. After the telco brings the line from the CO to the building, the technician installs a jack of some sort, which is often called the *demarc*. Anything beyond the demarc is up to you (and your ISP) to sort out, while anything up to the demarc is the telco's responsibility. As part of this process, the technician may need to speak with somebody at the CO. If possible, make sure there is a phone handy near the place where the demarc is to be installed.

If the telco entry point into the building is not convenient for you, an extended demarc is the solution. Convenience may at times be a legal matter, too. Office buildings may have only one telco closet, even if there are multiple businesses in different suites. In such situations, using an extended demarc is required to keep as much control as possible over your network infrastructure. The most likely problem with the wiring in an extended demarc is *near-end crosstalk* (NEXT), caused by the difference in signal levels between the transmit and receive pairs in the extended demarc. When the difference between the two is 7.5 dB or greater, inductive coupling of the transmit pair and the receive pair can cause a ghost of the stronger signal on the weaker pair. At the CSU/DSU, the transmission will be a full-strength, 0 dB transmission with a peak voltage difference of 6.0 volts, but the received signal will be attenuated by the extended demarc wiring. If the signal on the receive pair is –7.5 dB or lower, inductive coupling between the transmit pair and the receive pair may cause crosstalk and corrupt data.

You can combat crosstalk by using cable with appropriate shielding. Most modern LAN wiring uses Category 5 *unshielded twisted pair* (UTP) cable, which relies on tightly twisting the component wires of a pair together so that interference affects both component wires of a pair equally. In T1 applications, UTP is subject to interference if the difference in strength between the signals on adjacent pairs is 7.5 dB. Adding shielding dramatically improves the picture because it prevents signal leakage from pairs that carry a strong signal and helps resist inductive coupling in pairs that carry a weaker signal. *Individually shielded twisted pair* (ISTP) cable wraps each pair of wires in a foil shield, then adds *drain wires* to help terminate the shields. Figure 7-2 contrasts UTP and ISTP.

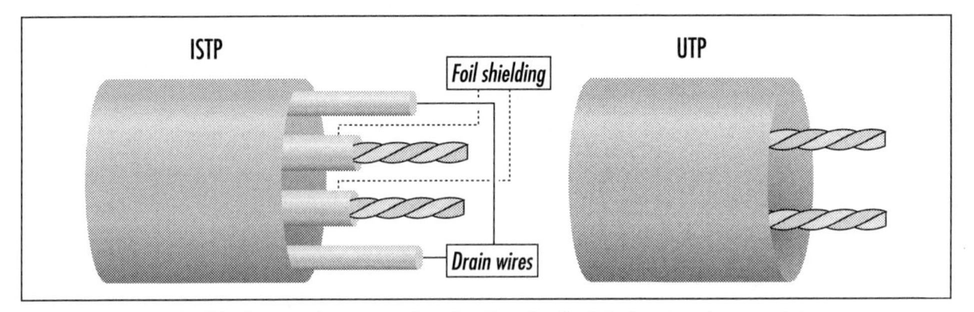

Figure 7-2. Unshielded twisted pair and individually shielded twisted pair wiring

Shielded cable prevents crosstalk in nearly all circumstances, but it costs more and is more complicated to work with because the shield foil must be peeled back and the drain wires terminated. Unshielded cable can be used when the signal differential is less than 7.5 dB. Most smart jacks will output a 0 dB signal, so unshielded cable is fine for extensions of up to the 1000–1200 foot range. Beyond those distances, shielded cable must be used. Digital cross-connects between devices such as a PBX and an add/drop CSU must be less than 655 feet, so unshielded cable can be used without a problem in most cases.

After the demarc is installed, the technician may hook up a small box to test the line. At the CO, the line is looped back to your location. Any bits sent down the line to the CO should bounce back without error. Test equipment sends standard patterns down the line and counts errors on the return path. Severe errors require that the technician speak with other telco personnel so the telco can correct any problems on the line.

If all goes well, the telco should eventually pronounce the line good and affix a sticker to the demarc with configuration information, such as the circuit ID. Depending on the telco, the sticker may also include an installation date and other information. Even though the end span is installed at your location, the entire line may not be up because the full circuit may need to wend its way through other COs and the ISP-side end span may need to be installed. Your end span is frequently the final piece, though, because the telco may take many T1 access lines from the same area and multiplex them into a higher-speed line, such as a T3, to the ISP's point of presence.

Circuit Identification Information

As part of the ordering process, you should be able to easily obtain the following information:

Line framing from the telco
: This provides enhanced monitoring and error checking; nearly all new lines use ESF.

Line code from the telco
: Most new T1s are used for data and the demand is for full data throughput, so nearly all new lines are B8ZS.

Line speed from the telco
: Line speed may depend on a few factors: the line code and whether it is capable of clear channel transmissions, and how many of the component DS0s are used for transmission.

Line build out from the telco
: In most areas, customers now connect CSU/DSUs to smart jacks and LBO is 0 dB. The telco can estimate this before the circuit turn-up, but you should double-check the value with the technician who installs your circuit.

IP addressing and routing information from the ISP
: When you order the circuit, your ISP will assign IP address space as well as any uplink port-configuration information. Most likely, the routing will be a simple static default to the other end of the serial link to hand traffic off to your ISP.

Configuration information from the public data network provider
: If you are connecting to your ISP through another public data network, such as the telco's regional frame cloud, you will need this information. In the case of a frame relay network, you will need the DLCI and LMI types.

Pre-Connection Tasks

When the circuit is up and the ISP is ready, the ISP will call to complete the circuit turn-up. When everything goes according to plan, this usually happens a day or two after the telco installs the circuit. If the telco circuit is delayed, the ISP may be ready whenever the telco is finished and may even await your call. During this time, there are a few things you can do to make sure the turn-up is smooth:

1. Check that you have the proper cables to connect your CSU/DSU and router. You will probably need to obtain a V.35 cable from your router vendor; most vendors use proprietary serial-port pinouts and require you to buy special cables. If you are ordering new equipment, such as a router with a V.35 port, make sure that you ask the vendor if you'll need a new cable.
2. To connect the CSU/DSU to the telco network, you'll need a second cable—most likely one with an RJ-48 jack (also sometimes called an RJ-48C or RJ-48X; see Appendix E for details). RJ-48 uses wires 1 and 2 as a pair and wires 4 and 5 as a pair. To prevent crosstalk, make sure that the component wires of each pair are tightly twisted together: the wires leading to pins 1 and 2 should be twisted together, and the wires leading to pins 4 and 5 should be twisted together. Twisting the appropriate pairs together is only the minimum step. Long runs should use shielded wiring.
3. Configure the CSU/DSU. If the ISP included the CSU/DSU, they may have shipped it preconfigured for installing the line. Nevertheless, it doesn't hurt to check the configuration. Verify that the CSU/DSU matches the circuit requirements and that the "big five" (framing, line code, build out, clock source, and DTE clocking) are set correctly. If you have any questions, refer to Table 6-4 and contact the appropriate party.
4. Configure the terminating DTE on your side. "Terminating DTE" is really a fancy term for router, which is what you will be hooking up to the CSU/DSU. Even though the connection may not be live, you can configure IP addresses and the default route. Depending on the vendor, make sure you can hook up your internal LAN to the router. Some routers have only AUI ports. If yours is AUI-only, purchase and install the appropriate transceiver.
5. If there is any easy way, get the IP addresses of "important" hosts at your ISP. By important, I mean likely to be up and running. Good hosts are the FTP, web, or DNS servers. If you have the IP addresses you can test basic connectivity after the circuit is turned up, even if DNS is not yet configured correctly.
6. Bring power close to the demarc so you can turn everything on easily. As the day approaches, you can wire everything together and turn it on in anticipation. Connecting the CSU/DSU may save time, if the ISP can see the circuit is operating end-to-end before the turn-up call.

Be careful about connecting the CSU/DSU without the telco's permission. Technically, Part 68 of the FCC rules requires that you obtain permission from the telco before hooking up a CSU/DSU—many CSU/DSU vendors include affidavits for the telco at the end of their instruction manuals. The ISP may file these on your behalf.

Some telcos may not rigidly enforce the notification requirement. In times past, using a T1 to transport data was an exotic application (hence the requirement). Now that most T1s are used for data, many telcos do not strictly enforce the requirement.

Trading Packets

In many cases, the ISP network engineers assigned to completing circuit turn-up are the best engineers you will have an opportunity to work with directly. (More talented engineers do exist, but they're at work designing the ISP's next-generation network architecture or maintaining the important peering connections in the current network.)

When the call comes, the ISP's engineer usually starts off by discussing the configuration to make sure both of you are reading from the same sheet of music. Do not be surprised if he asks some simple questions—remember that turn-up staff deals with anybody who orders a new circuit, not just those who are fully prepared.

Both of you will bring up the link layer software on your respective routers, and it will attempt to negotiate. Assuming the link layer comes up, you should be able to send traffic out to the world. Do not be surprised if traffic does not come back at first. Many of the larger ISPs have such large networks that they must extend their routing protocol timers to much longer intervals than the defaults. It may take a few minutes to advertise and establish routes to your circuit. When the ISP engineer advises you that everything should be working, send a few ping packets—first out to destinations within the ISP's network, then out to destinations on other ISP networks. Unless you are using Network Address Translation (NAT), initiate these connections from behind the router so the source IP addresses are in the address space the ISP delegated to you.

Finally, be prepared to be billed. When the circuit is up and running, the turn-up staff will undoubtedly send a note to the billing department so you start getting billed. If you want to take advantage of any other services offered by the ISP, such as web site or name server hosting, check with the turn-up staff to figure out with whom you should speak.

Troubleshooting

If trouble arises, stay calm. Turn-up staff members are familiar with the problems that can occur when lines are activated. They will likely ask you to put your CSU/DSU into loopback mode for testing. For background on what they may do, see Chapter 11 in this book.

Post-Connection

Gather all the information about your new line so that if problems occur, you will have all the information you need in one place. Here's a list of what I keep handy about circuits for which I am responsible:

- Line framing
- Line code
- Line speed (channel speed and number of DS0s)
- Line build out
- Transmit clocking and DTE clocking settings, along with the make and model of the DTE
- Link layer information, such as frame relay DLCI or ATM virtual path and virtual circuit IDs
- IP addressing and routing information from the ISP
- Telco circuit ID number and contact information
- ISP account ID number and contact information

8

High-Level Data Link Control Protocol (HDLC)

A precedent embalms a principle.
—Benjamin Disraeli

Up until now, this book has focused on strictly the physical layer of the OSI model: volts, pulses, timing, plugs, and so on. Stretching it, we can see as far as bits, flowing together in an unending stream of zeros and ones. Nothing, however, has touched upon creating packets, and without packets, there is no data network.

Packets are lumps of data with addresses that are independent of any physical medium. Putting packets on the wire (or on the fiber, or on the air) requires a lower-level construct. Below the medium-independent addressing at layer 3 of the OSI model is the frame, which lives down at layer 2. Frames are a network-dependent wrapper for packets. Because physical networks may be used for several different network protocols, distinguishing between them is the most important function of the link layer. To accomplish this, frame headers include tags for network protocols. Each individual link layer decides how its frames are tagged.

LAN technology presents a relatively homogenous interface to network protocols, in part due to the dominance of IEEE 802 methods. WAN technologies, however, present a wide field of diverse requirements. As a result, several link layer protocols are in widespread use on serial links. Riding at the link layer, framing protocols are well positioned to monitor link quality, report problems to higher-layer protocols, and discard frames that have been mangled due to line noise. Most importantly, though, the use of link layer protocols smooths over idiosyncrasies of the WAN physical layer.

T1 lines use three main link layers, all of which are derived from one of the most successful protocols ever specified—the ISO's High-level Data Link Control protocol (HDLC). In addition to HDLC, Point-to-Point Protocol (PPP) and frame relay are commonly used. To understand any of these link layer protocols you should first understand HDLC, its framing, and the set of extensions developed by Cisco that are used by most of the networking industry to adapt HDLC to data-communication environments.

Introduction to HDLC

HDLC is one of the hidden success stories in data communications. HDLC underpins most serial communications, whether explicitly or through its progeny. Its wide reach is partially due to its age—IBM developed its predecessor, the Synchronous Data Link Control (SDLC) protocol, in the mid-1970s for mainframe communications. One of the most common applications of SDLC was to link 3,270 terminals to mainframe frontend processors.

Because SDLC was designed for use in mainframe communications, it has a centralized mindset. One end of the link, connected to the computing resource, was identified as the primary end; the terminal (secondary) end of the link had only diminutive computing power. Primary stations controlled all communication, and secondary stations could not initiate communication except at the order of the primary. SDLC allowed several types of physical topologies. In addition to the common point-to-point links, multidrop links could carry SDLC with one primary station and multiple secondaries, as well as loop and hub topologies. While this model was well suited to communications in which one end of the link possessed all the computing horsepower, times were changing and processing power was becoming more distributed. HDLC was enhanced to add the following transfer modes to SDLC's lone mode:

Normal Response Mode (NRM)
: This is the essential SDLC model. Secondary stations must obtain clearance from the primary to transmit.

Asynchronous Response Mode (ARM)
: This lifts the restriction that secondary stations must obtain permission to transmit, but secondary stations are still of subordinate importance.*

Asynchronous Balanced Mode (ABM)
: This is the most common transmission mode today, largely because several of the HDLC derivatives specify ABM-style communication. Nodes in ABM environments are referred to as *combined* nodes, which means that they have "situational dominance" (any combined node may initiate a conversation without first obtaining permission from other nodes).

* The asynchronous in ARM and ABM refers to the link-control method, not the type of data link. Stations can transmit without clearance, so the transmissions are asynchronous. Asynchronous response modes can be used on any type of link, including synchronous links.

ISO first published the HDLC specification in 1979. The ITU subsequently adopted HDLC as the basis for the Link Access Procedure (LAP). Several varieties of LAP have been developed for different purposes.* Ethernet's 802.2 Logical Link Control (LLC) is also derived from HDLC.

HDLC Framing

HDLC's frame format is specified in ISO/IEC 3309 and shown in Figure 8-1. In the figure, the control field is ordered with the most-significant bit first, though when frames are transmitted on the network, the least-significant bits are transmitted first.

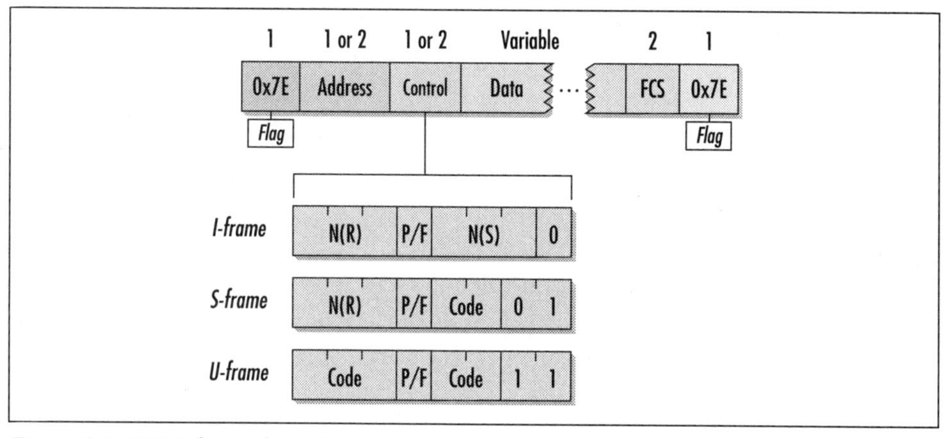

Figure 8-1. HDLC frame format

A note on both byte and bit order for this chapter: ISO and ITU diagrams typically show bits in the order they are transmitted on the wire, or least-significant bit first. IETF diagrams, on the other hand, typically display most-significant bits first. Except where noted, this chapter uses the most-significant bit first notation.

HDLC frames open and close with the frame flag (0111 1110; 0x7E). Flags assist devices at either end of the link by bounding frames for error-checking procedures. Idle HDLC links may still be physically active. Two idle states are defined by ISO. Two frames may also share the flag sequence. When sharing the flag between successive frames, the closing flag in the first frame is also treated as the opening flag of the next frame.

* The most common strains of LAP are LAPD (Link Access Procedure on the D channel) for ISDN signaling, LAPB (Link Access Procedure, Balanced) for X.25 networks, LAPM (Link Access Procedure for Modem) in the V.42 specification for error detection, and LAPF (Link Access Procedure for Frame). All LAP frames share common HDLC characteristics, such as opening and closing flags.

After the flag, HDLC transmits an address byte. HDLC was designed to support multidrop links, with one primary station and many secondaries all connected to the same circuit. Stations discard frames addressed to other stations. Use of the address byte depends on the adaptation of HDLC used on the link.

After the address byte, HDLC transmits the control byte. HDLC defines three frame types, each with their own control-byte format, as shown in Figure 8-1. Control information has one or two bits to determine the frame type, a *poll/final* (P/F) bit, plus sequence numbers and coding information. The poll/final bit is derived from SDLC. Primary stations use it to demand an immediate response from secondaries, and secondaries use it to signal the end of data transmission to the primary.

Following the control data, a variable-length data field contains the upper-layer protocol packet. After the data field is a 2-byte *frame check sequence* (FCS), which ensures data integrity, and a closing flag to enable HDLC framers to find the end of the frame. To prevent data or the FCS from being confused with the flag, any sequence of more than five ones in the data field is altered by the injection of a zero bit. Receivers discard zero bits that follow five consecutive ones. The later section, "HDLC Transparency and Bit Stuffing," further discusses this procedure.

In addition to the 1-byte control and address fields, extensions exist that allow both 2-byte address fields and 2-byte control fields. Extended addressing and control allow for longer sequence numbers, but there is no fundamental difference between the two.

Information Frames (I-Frames)

Information frames carry higher-layer protocol packets in the data field. HDLC assigns sequence numbers to frames and can be configured to provide positive acknowledgment. Acknowledgments may be piggybacked on to I-frames or carried in S-frames.

HDLC specifies a window, much like the sliding windows in TCP. HDLC can transmit several frames without an acknowledgment, and acknowledgments can report the reception of multiple I-frames. Not all vendors implement the HDLC window. Cisco's adaptation of HDLC explicitly does not use the window.

Figure 8-2 compares the basic (1-byte) and extended (2-byte) control information used with I-frames. HDLC implementations recognize I-frames by picking out the zero in the least-significant bit of the control information. Basic control information is also called the *mod-8* form because three sequence number bits lead to a sequence space of eight frames. Senders and receivers keep track of the total number of frames and use the remainder, after dividing by eight, as the sequence number on the wire. Likewise, extended control information is often referred to as the *mod-128* form for a similar reason.

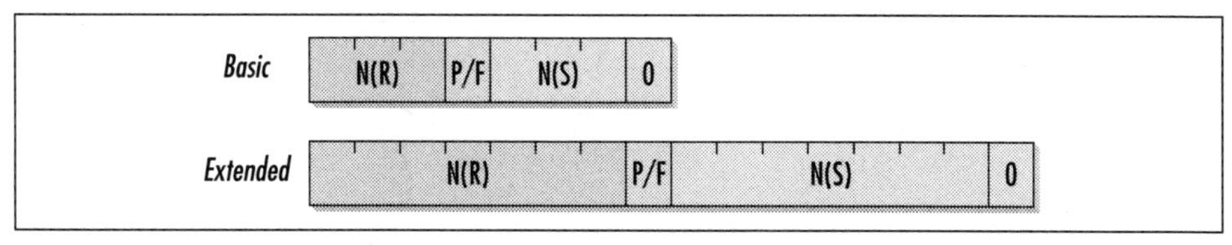

Figure 8-2. Basic and extended control information for I-frames

Supervisory Frames (S-Frames)

HDLC uses supervisory frames for flow control. Four types of S-frames are defined:

Receiver Ready (RR)
: This frame acknowledges receipt of all data up through the included sequence number.

Receive Not Ready (RNR)
: This frame acknowledges receipt of all data up through the included sequence number, but also requests that the sender stop sending. It is the HDLC analog of a TCP acknowledgment with a zero window.

Reject (REJ)
: This is a null acknowledgment frame. An included sequence number reports the last sequence number that was received without problems; this frame type requests the retransmission of all frames after the specified point.

Selective Reject (SREJ)
: When a single frame is missing, this frame requests retransmission of the missing frame only. An included sequence number identifies the frame requiring retransmission. Balanced HDLC modes allow only one selective retransmission request on the wire at any given time.

Figure 8-3 shows the use of the S-frame control information in both the basic and extended formats. Like Figure 8-2, it is ordered most-significant byte first and most-significant bit first. HDLC implementations recognize S-frames by setting the first two transmitted bits of control information to 10. The two forms of control information share S-frame codes.

In the extended control information for S-frames, the high-order four bits in the least-significant byte are reserved and set to zero.

Unnumbered Frames (U-Frames)

Unnumbered frames are used for link initialization, disconnection, and miscellaneous operations. The U-frame's common operations include setting the link for ABM using the Set Asynchronous Balanced Mode (SABM) command and transferring supervisory information with the Unnumbered Information (UI) type.

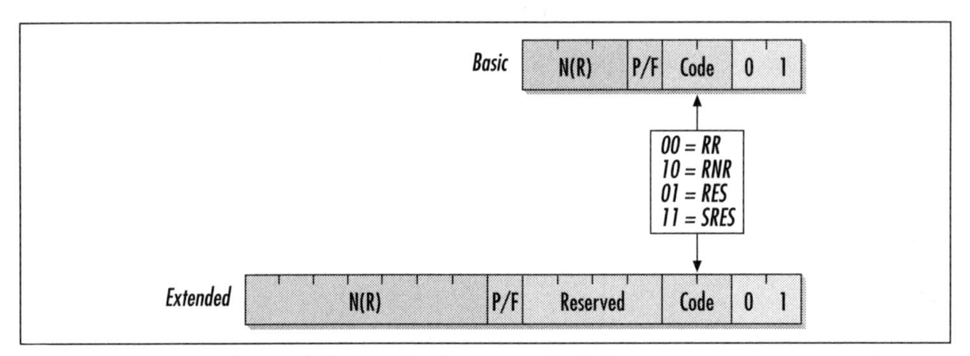

Figure 8-3. Basic and extended control information for S-frames

Figure 8-4 shows the basic and extended U-frame formats. Bit positions containing code bits are labeled with the ISO-specified bit position numbers. (The figure, like all others in this section, is shown most-significant byte first and most-significant bit within each byte first.) The five labeled bits are used to create a code.

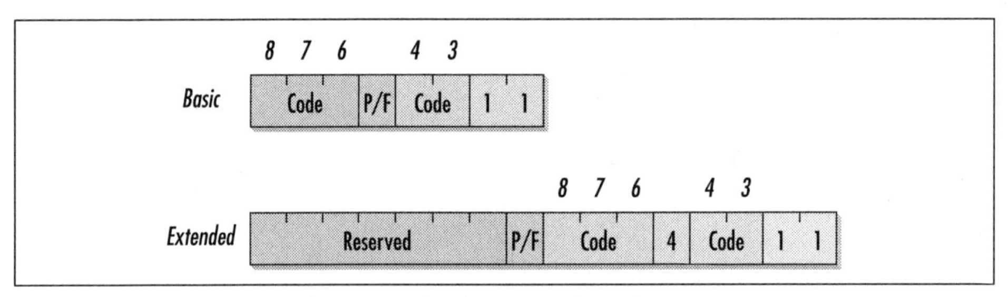

Figure 8-4. Basic and extended control information for U-frames

When U-frames use extended control information, bit 5 in the control information is unspecified, and the high-order seven bits in the most-significant byte are reserved and set to zero. Codes used for U-frames are described in Table 8-1.

The Set...Extended commands are used to select extended (2-byte) control information. On T1 links, the important commands are SABM, which ensures that the routers at both ends of the link are able to send frames, and UI, which is used by HDLC-derived protocols to transfer information.

Table 8-1. U-frame codes

Code[a]	Name	Command/ response[b]	Description
00-001	SNRM	C	Set Normal Response Mode. Designates the receiver as the secondary station in an NRM configuration.
11-000	SARM	C	Set Asynchronous Response Mode. Designates the receiver as a secondary station, but ARM allows the secondary to transmit without permission.

Table 8-1. U-frame codes (continued)

Code[a]	Name	Command/ response[b]	Description
11-100	SABM	C	Set Asynchronous Balanced Mode. Sets the link to ABM mode.
00-010	DISC	C	Disconnect. Places the secondary station into disconnected mode.
11-011	SNRME	C	Set Normal Response Mode Extended. Same as SNRM, but uses 2-byte control information.
11-010	SARME	C	Set Asynchronous Response Mode Extended. Same as SARM, but uses 2-byte control information.
11-110	SABME	C	Set Asynchronous Balanced Mode Extended. Same as SABM, but uses 2-byte control information.
10-000	SIM	C	Set Initialization Mode. Orders the secondary station to perform initialization procedures.
00-100	UP	C	Unnumbered Poll. The "reach out and touch someone" frame to see if the receiver is alive.
00-000	UI	C/R	Unnumbered Information. Used to send information to the secondary station.
11-101	XID	C/R	Exchange Identification. Sent by the primary station to the secondary to request information about the secondary.
11-001	RSET	C	Reset. Resets the state information, including sequence numbers, in the receiver.
00-111	TEST	C/R	Test. TEST frames are echo requests. Receivers respond to a TEST frame with a TEST frame as soon as possible.
11-000	DM	R	Disconnected Mode. Sent by secondary stations when they are in the nonoperational disconnected mode.
00-110	UA	R	Unnumbered Acknowledgment. Sent in response to SNRM, SARM, SABM, SNRME, SARME, SABME, RSET, SIM, and DISC commands.
10-001	FRMR	R	Frame Reject. Sent when a frame is valid, but was received incorrectly.
00-010	RD	R	Request Disconnect. Sent by the secondary station when it wishes to enter disconnected mode.
10-000	RIM	R	Request Initialization Mode. Sent by secondary stations to request initialization information from the primary station.

[a] Codes are displayed in bit order: 34-678.
[b] Codes may overlap because HDLC distinguishes between commands and responses.

HDLC Transparency and Bit Stuffing

Early SDLC and HDLC networks often used Non-Return to Zero Inverted (NRZI) line coding. To transmit a zero, NRZI changes the voltage. Ones bits are transmitted by sending the same voltage as in the previous bit time. As a result, long sequences of ones appear as a flat voltage lines with no state changes. Bit times are not readily apparent from the flat line, and the ends of the connection may fall out of synchronization. To avoid losing synchronization, HDLC stuffs in a zero after five consecutive ones between the framing flags. HDLC receivers then monitor for inserted zeros following five ones and remove the stuffed zeros before passing data up to higher layers. Figure 8-5 illustrates this.

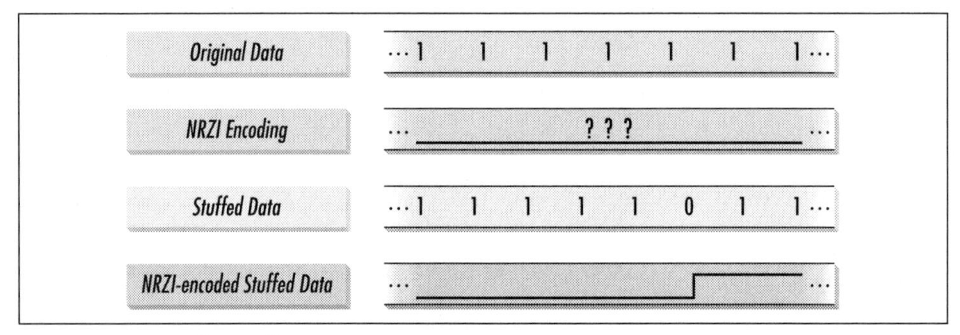

Figure 8-5. HDLC bit stuffing

Figure 8-5 shows how HDLC's bit stuffing helps to maintain synchronized timing. In the original data, a long string of ones is present. When that is encoded with NRZI, it becomes a flat line with no voltage changes, and a timing unit may find it difficult to maintain accurate timing throughout that absence of state changes. On synchronous links, HDLC stuffs in a zero after the first five ones, which leads to the generation of a voltage transition. The bit stuffing rule also guarantees that the flags create a unique pattern for framers to search for—any occurrence of 01111110 in the frame header or data field will be stuffed to 011111010.

ISO 3309 also includes procedures for data transparency on 7-bit links as well as procedures to escape control characters. Transparency allows special control characters such as XON, XOFF, and DELETE to pass through asynchronous links without affecting intermediate communications equipment.

Cisco HDLC

As initially specified, HDLC did not include multiprotocol support or link quality monitoring and was not useful for anything other than frame description. These shortcomings were addressed by Cisco's extensions to HDLC (cHDLC).* cHDLC is an HDLC frame with an Ethernet type code to identify the network protocol carried in the frame. The basic frame structure is shown in Figure 8-6. Framing components are identical to standard HDLC, with the opening and closing flags and FCS.

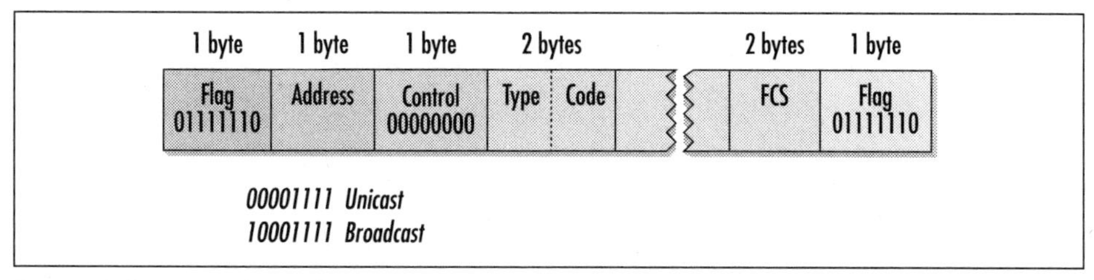

Figure 8-6. cHDLC frame format

Addressing in cHDLC is simple. The address is either 0x0F for unicast packets or 0x8F for broadcast packets. Labeling a frame as broadcast or multicast has very little significance because IOS does not support multidrop links. Broadcast is simply a reflection of higher-level protocols designating a frame as a broadcast. Control information is set to zero; cHDLC does not support the HDLC window. To distinguish between packets, the third and fourth bytes contain a type code, which is usually the Ethernet type code for the higher-layer protocol. Table 8-2 shows common type-code assignments. After the type code, the higher-level packet is encapsulated as it would be on an Ethernet network. After the data, a frame check sequence and flag close out the frame.

Table 8-2. cHDLC protocol codes

Protocol	cHDLC protocol code (hexadecimal)
IP	0x0800
Cisco SLARP (not RARP!)	0x8035
EtherTalk	0x809b
AARP	0x80f3
IPX	0x8137
DECnet phase IV	0x6003

SLARP

Frames using the Reverse ARP Ethertype are used for Cisco's Serial Line ARP (SLARP). SLARP is a router-to-router protocol that provides a method of dynamic address assignment as well as link-integrity verification. To distinguish between

* Some works describe these extensions as "proprietary." While the Cisco extensions were proprietary in that Cisco developed them, the details were widely distributed and not kept secret. "Proprietary" is often (incorrectly) used interchangeably with "secret" or "single-vendor" in the industry today, and the Cisco HDLC encapsulation does not fit with the connotation.

the two operations, SLARP frames begin with a 4-byte type code. Currently, only three codes are defined. Address requests use a code of zero, and replies use a code of one. Keepalive frames use a code of two.

Address request and response packets are similar in structure; both frames are shown in Figure 8-7. In address request packets, the address and mask fields are zero, and the last two bytes are not defined. Replies include the address and mask of the network. Usually, point-to-point serial links use a 30-bit mask.

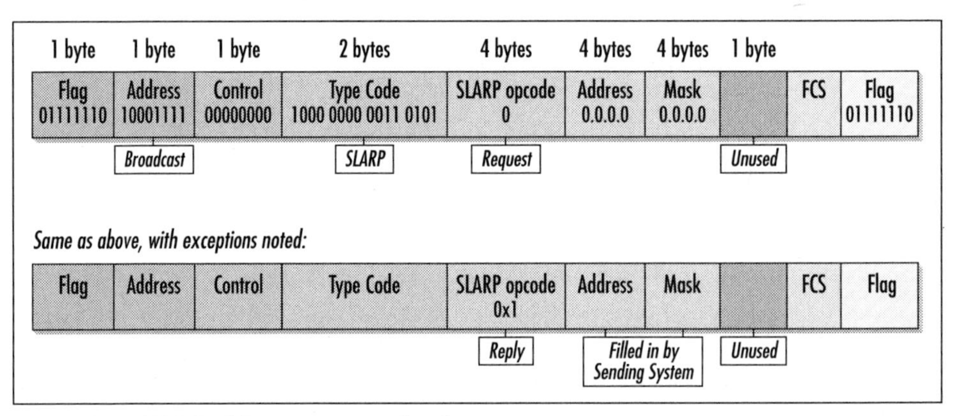

Figure 8-7. cHDLC address request and reply

SLARP does not support complex or subtle configurations. It assumes that both ends of the link are on the same IP subnet and that one system is host 1 and the other system is host 2. A logical IP network with a 30-bit subnet mask (255.255.255.252) has four IP numbers. The first IP number has a host field of all zeros and is commonly used to refer to the network itself. The second and third IP numbers are used for two hosts on the network, and the fourth IP number is the all-ones broadcast.

SLARP assumes a 30-bit network and deduces the host number from the advertisement. If SLARP receives a response, and the host number on the logical IP network is 1, the requesting system deduces it is host 2. Conversely, if the response indicates that it is from host 2, the requesting system assumes it is host 1. Figure 8-8 shows a request-response cycle.

To ensure that links with physical layer connectivity can be used for data transport, SLARP includes a basic keepalive mechanism. Both ends of the link send out polling messages and monitor received polls to keep track of the link state. Figure 8-9 shows the basic format of a SLARP keepalive. Following the type code of 2, two 4-byte sequence numbers track line quality, and a 2-byte reliability field is set to all ones.

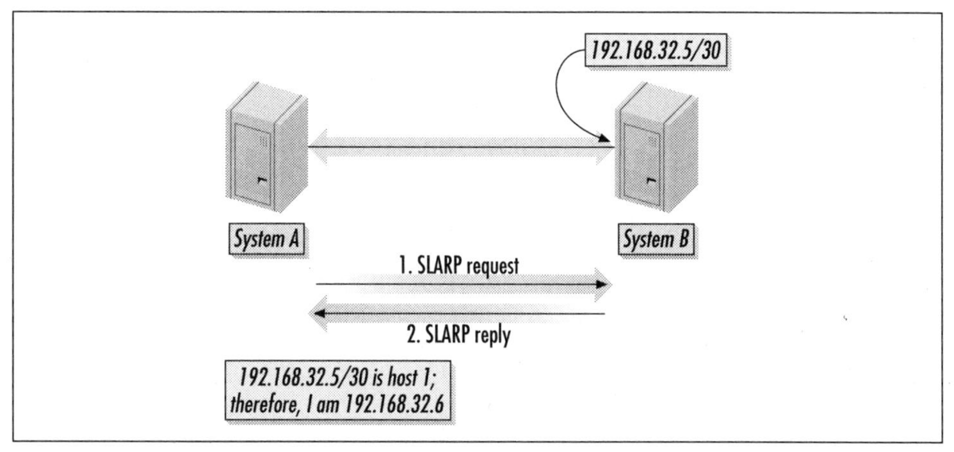

Figure 8-8. cHDLC serial-line addressing model

1	1	1	2	4	4	4	1	1	1
Flag	Address	Control	Type Code	SLARP opcode 0x2	Sender Sequence Number	Last Received Sequence Number		FCS	Flag
				Line Check			Reserved, Set to 0xFF		

Figure 8-9. cHDLC keepalive frame

Polling, sequence numbers, and the keepalive

Most link layer protocols rely on a request/response polling mechanism. HDLC incorporates both the request and response into the same frame by including two sequence numbers after the type code. Each direction on a link corresponds with a flow of keepalive frames. Sending stations transmit the outbound sequence number first and attach the latest inbound sequence number. Figure 8-10 shows the operation of the sequence-numbering mechanism. Each direction maintains a sequence number, with *mysequence* used for the outbound sequence number and *yoursequence* used for the inbound sequence number.

The cHDLC mechanism allows for the first sequence number to serve as a link layer echo request, with the response coming in subsequent frames on the other side. When a SLARP address request is received, sequence numbers are reset to zero.

Sequence numbers are incremented by one for each keepalive frame that is transmitted. Missed frames or frames corrupted in transit are still transmitted and will still cause the sequence number to rise. Unlike some other sequence-number schemes, the sequence number in cHDLC is a measure of the total number of keepalives transmitted, not just those that are processed successfully. Therefore, the difference

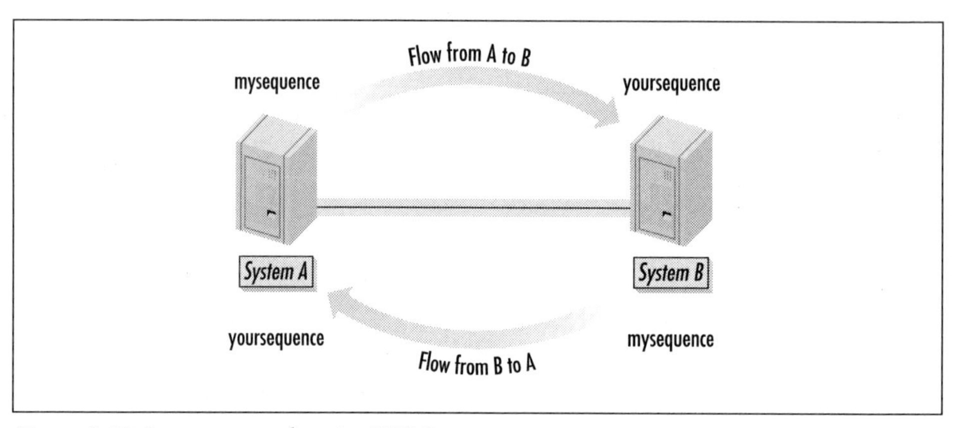

Figure 8-10. Sequence numbers in cHDLC

between the transmitted sequence number and the received sequence number is a measure of the link quality. In normal operations, the transmitted and received sequence numbers should be identical or differ by only one, with the latter case reflecting one keepalive frame "in flight" between the two endpoints. Severe noise or congestion could result in a skew of the window or two; Cisco's engineers decided that three missed keepalives probably means that the link is dead and should be reported to higher protocol layers as dead.

Recovering from missed keepalive packets is simple. The only number that matters to keepalive monitoring is the difference between the transmitted and received sequence number. When a keepalive is missed, the difference rises to two. When the next keepalive after the missed transmission is received, the receiver updates the received sequence number. However, the received sequence number is two larger than the previously received one, so the difference is brought back down below the limit.

Configuring Cisco HDLC

Cisco HDLC is a snap to configure. Only a few options can be adjusted, and not all options are found on all vendors' equipment. Here is a list of options:

Cisco HDLC encapsulation

Cisco IOS defaults to the Cisco HDLC encapsulation on serial links unless otherwise specified. Several vendors default to PPP, but allow HDLC as an option. Cisco's HDLC encapsulation may or may not be compatible with the HDLC encapsulation offered by other vendors. Refer to the vendor's documentation for details.

IP addresses and network mask
: If the deployment is compatible with the SLARP addressing model and both ends support the use of automatic address assignment with SLARP, only one end must be configured. However, not all vendors support automatic address assignment with SLARP, and not all deployments can be supported with the SLARP addressing model.

Keepalive timer
: SLARP keepalive packets are sent at regular intervals. IOS defaults to 10 seconds, as do many other implementations.

Link mode
: ABM is the most likely link configuration because it allows unsolicited transmissions from both endpoints of the link. Most equipment supports only ABM. Most vendors do not provide an option for the HDLC link mode.

9

PPP

Everything secret degenerates; nothing is safe that does not bear discussion and publicity.
—Lord Acton

HDLC, as specified by ISO, falls short of the Platonic ideal of a generic link layer protocol. It cannot multiplex higher-layer network protocols; cannot detect faults, slowly degrading links, or looped-back links; and is not easily extensible or dynamically configurable. Even after Cisco's extensions to HDLC, many of these problems remained unsolved.

To address these shortcomings, the IETF developed the Point-to-Point Protocol (PPP). PPP framing is based on HDLC, but it was designed from the outset to be a multiprotocol link layer. A system of configurable option negotiation allows both ends of a link to choose a mutually compatible set of configuration parameters, even if the configuration differs between the two ends of the link. PPP's negotiation mechanism allows for easy extensibility, a feature that has been exploited by several minor revisions to the protocol after its most recent major standardization in 1994. Well-designed protocols can provide a framework for generations, and PPP has adapted nearly effortlessly to the sea of changes in the Internet since 1994.

A great deal of PPP's success stems from the solid foundation of HDLC on which it was built, but its innovations add tremendous flexibility. PPP can be used on everything from the high-speed SONET links at the core of the Internet to the lowly dial-up links connecting subscribers over common telephone lines. This chapter describes PPP's major architectural features, with special regard to how they apply to T1 links. As a result, this chapter does not pretend to be a definitive work on PPP. Features that are common and practically necessary on dial-up links (asynchronous HDLC, the octet-stuffed interface, authentication, and the different types of compression) are not described at all. See Appendix F for further reading.

Introduction to PPP

PPP is familiar to most Internet users as the standard method of establishing a dial-up connection to an Internet service provider. PPP is far more than an enabler of dial-up connections, though, which is why its presence is required on all IPv4 routers with serial interfaces.* PPP is specified in RFC 1661, and critical framing information is found in RFC 1662. Appendix F lists complete reference information.

Taken together, these two RFCs provide the basic elements needed to move packets from one end of a serial link to another. Three basic components come together to make it all happen:

Encapsulation and framing
: To multiplex several protocols across a network link simultaneously, PPP defines a link layer header that includes a protocol type field. PPP frames are derived from HDLC frames and provide a byte-oriented logical interface to bit-oriented serial hardware. After a link is established, data frames of various protocols coexist with link-management traffic.

Link management and monitoring
: This is accomplished through the use of PPP's *Link Control Protocol* (LCP). PPP may be used with circuit-switched connections such as dial-up or ISDN as well as with dedicated leased lines. Circuit-switched connections often incur charges based on usage, so PPP includes a way to bring up and tear down the physical link. After the link is established, users may wish to monitor the quality of the link to ensure that it is able to cope with the demands placed on it, so PPP includes a link-monitoring mechanism.

Network-protocol management
: Many different network protocols may peacefully coexist on a PPP link. Each network protocol must define an associated *Network Control Protocol* (NCP) to manage and assign network layer addresses and to terminate idle network connections. Most PPP links carry IP, though not necessarily exclusively, so the *Internet Protocol Control Protocol* (IPCP) is an important component of PPP links.

PPP provides a generic link layer that can be used for several different network protocols. Figure 9-1 shows how PPP and its control protocols fit into the protocol stack.

High-speed leased lines present a different environment from low-speed, asynchronous links. Asynchronous links usually require the use of header-compression techniques for bandwidth efficiency, as well as an *asynchronous control*

* RFC 1812, since updated by RFC 2644, documents the requirements for IPv4 routers. That is, of course, not to imply that all routers follow the requirements religiously...

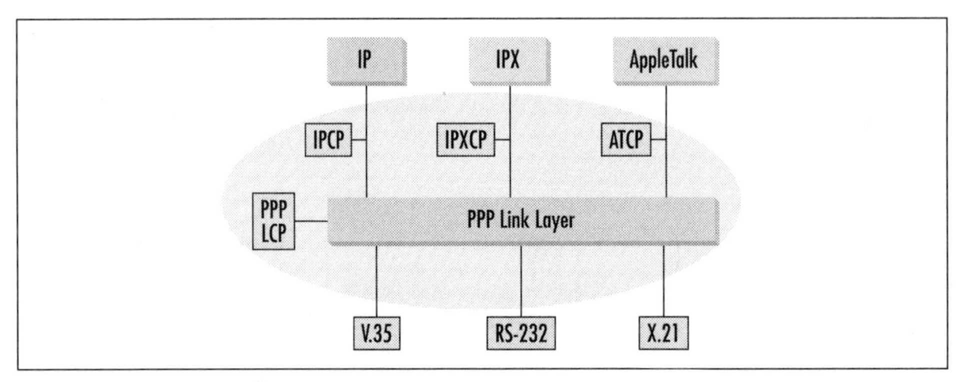

Figure 9-1. PPP protocol suite

character map (ACCM) to avoid interpreting data as commands for the communicating equipment. Synchronous leased lines do not require special treatment, so the ACCM is not used. On leased lines, however, speed is usually high, so the router CPU load of compressing PPP headers is too great on high-speed links. When asynchronous links fail, the modems disconnect. Synchronous serial links may fail in obscure ways, as discussed in prior chapters, which means that link monitoring is an important feature for detecting problems. Both high-speed and low-speed asynchronous links typically use magic numbers to detect looped-back conditions. Arguably, use of magic numbers is more important on high-speed links because leased lines are much easier to accidentally loop back.

PPP Logical Link States and State Machines

The PPP link layer is a nested set of state machines. The state machines do not reflect protocol layers in the strict sense of passing encapsulated packets between layers, but some of the "layers" depend on lower layers for correct functionality. At the bottom is the HDLC layer running directly on the physical link. Above the HDLC layer is the LCP layer, which negotiates transmission options for the HDLC layer. If authentication is required, it occurs before the NCPs complete negotiation of network layer options. The relationship between the state machines is shown in Figure 9-2. Multiple state machines may be involved at the network-protocol layer; Figure 9-2 shows state machines for IP, IPX, and SNA.

Each layer in Figure 9-2 must receive an administrative Open from the higher layer, instruct the lower layer to initialize, and then negotiate parameters for its own layer. At that point, it responds to the Open command with an Up event of its own. To transmit IP data, for example, the link must be initialized for IP. The steps shown in Figure 9-2 are summarized in Table 9-1.

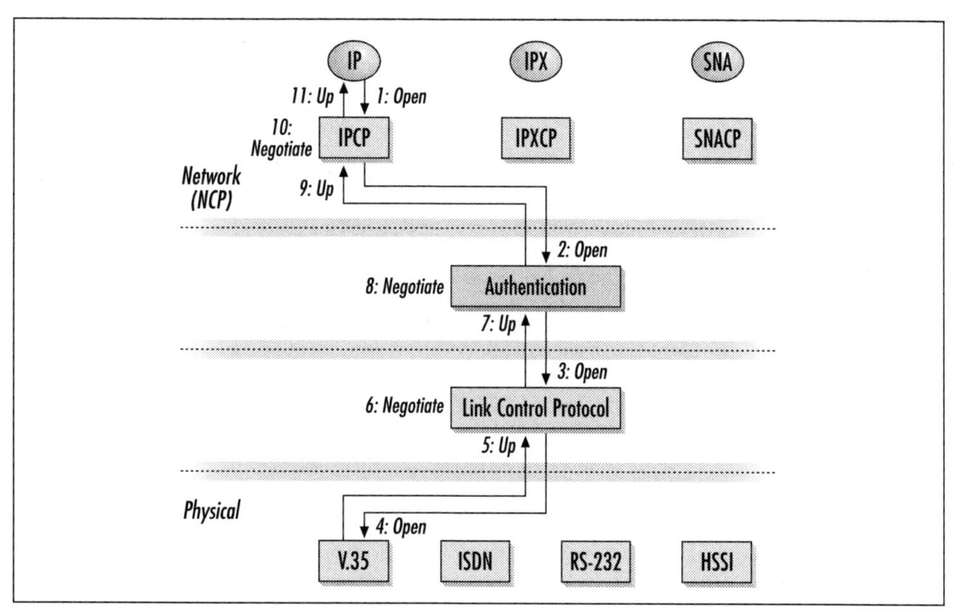

Figure 9-2. PPP layering

Table 9-1. Steps to initialize link in Figure 9-2

Step	Physical	LCP	Authentication	NCP (IP)
0	Down	Down	Down	Down, but data is ready for transmission
1	Down	Down	Down	IP stack sends Open event to IPCP to initialize link for IP
2	Down	Down	IPCP issues Open request to the authentication layer; authentication layer requests that the link be initialized to perform the authentication process	Waiting on IPCP to finish opening operation
3	Down	LCP receives Open request from authentication layer, but requires physical connectivity; LCP issues Open request to physical layer to establish connectivity	Waiting on LCP to make link available	Waiting on IPCP to make link available

Table 9-1. Steps to initialize link in Figure 9-2 (continued)

Step	Physical	LCP	Authentication	NCP (IP)
4	Physical layer negotiation begins	Waiting on physical layer	Waiting on LCP	Waiting on IPCP
5	Physical layer negotiation completes, so physical layer sends an Up event to LCP	Up event received from physical layer, so LCP negotiation begins	Waiting on LCP	Waiting on IPCP
6	Up	LCP negotiates options for the link	Waiting on LCP	Waiting on IPCP
7	Up	When compatible option sets have been selected, LCP sends an Up to the authentication layer	Up event received from LCP; authentication begins	Waiting on IPCP
8	Up	Up	Authentication is attempted by one or both sides	Waiting on IPCP
9	Up	Up	When authentication is successful, an Up event is sent to IPCP	Up event received from authentication layer
10	Up	Up	Up	IP addresses negotiated; Up event sent to IP stack
11	Up	Up	Up	Up event received, so data transmission begins

The series of steps needed to set up the link for transmission of IP packets is shown in Figure 9-2 by the numbered sequence of steps in the diagram. When the IP layer has packets to transmit, but is not initialized, an administrative Open command is sent from the IP layer to the IP Control Protocol. IPCP is responsible for negotiating IP parameters. However, data cannot be transmitted until the lower layers have been configured to transmit data. Therefore, IPCP sends an Open command of its own to the authentication state machine. Once again, lower layers are not available, so the authentication layer sends its own Open command to the Link Control Protocol. LCP then proceeds to initialize the physical link layer by sending an Open command down to the appropriate physical layer. Depending on the physical layer, the Open command may be simple or quite complex. Asynchronous modems must establish circuit-switched paths through the telephone network and establish a carrier signal before returning an Up event. Dedicated leased lines normally have an available physical layer at all times, so they return an Up event much more quickly.

When the Up event has been received from the physical layer, LCP negotiates parameters for link encapsulation. Negotiation of encapsulation parameters is an involved process, as detailed in the next section. When negotiation completes after LCP has established a set of parameters used for encapsulation on the link, LCP will send an Up event to the authentication layer. Authentication may or may not be employed on any particular link. If it is not, the authentication layer simply does nothing. Placing the authentication layer before the network-protocol initialization prevents network attacks from outsiders who are not authorized to use the network. If it does not receive network layer parameters, the host is not on the network and cannot transmit data.

When authentication completes, an Up event is sent to the NCP, which is IPCP in this case. IPCP negotiates addressing and sends an Up event to the IP layer. The IP layer can then transmit the packets that were queued in the beginning. Once LCP is initialized, adding network protocols is much less time-consuming. If IPX packets were to be transmitted, only IPXCP would need to negotiate because the other layers would have previously been completed. Figure 9-3 shows the procedure that would be used once the IP initialization was complete. When the Open event goes down the layers, it reaches the authentication layer. However, authentication was already performed when the link initialized for IP, so the authentication layer can respond immediately with an Up notification. The only negotiation that must take place is the IPXCP negotiation to configure the link for use with IPX. The steps shown in Figure 9-3 are summarized in Table 9-2.

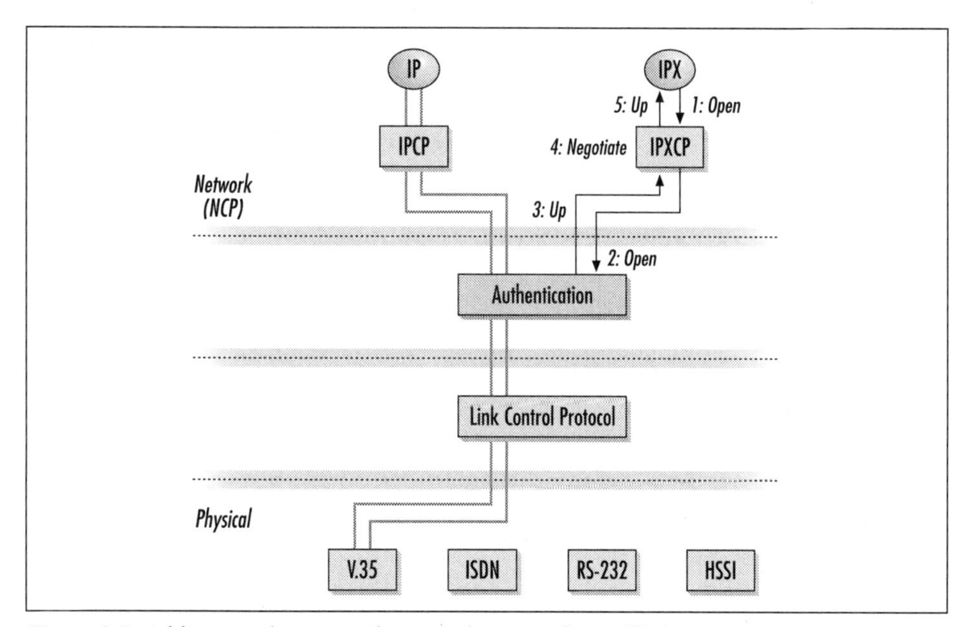

Figure 9-3. Adding another network protocol to a configured link

Table 9-2. Steps to initialize link in Figure 9-3

Step	Physical	LCP	Authentication	NCP (IPX)
0	Up	Up	Up	Down, but data is ready for transmission
1	Up	Up	Up	IPX stack sends Open event to IPXCP to initialize link for IPX
2	Up	Up	IPXCP issues Open request to the authentication layer	Waiting on IPXCP to finish opening operation
3	Up	Up	Authentication has completed, so authentication layer sends Up event back to IPXCP	Waiting on IPXCP to make link available
4	Up	Up	Up	Link is authenticated and IPXCP can negotiate IPX options
5	Up	Up	Up	IPXCP sends Up event to IPX stack; IPX transmission begins

Figure 9-4 shows another view of the link-initialization process, through the PPP link state diagram.

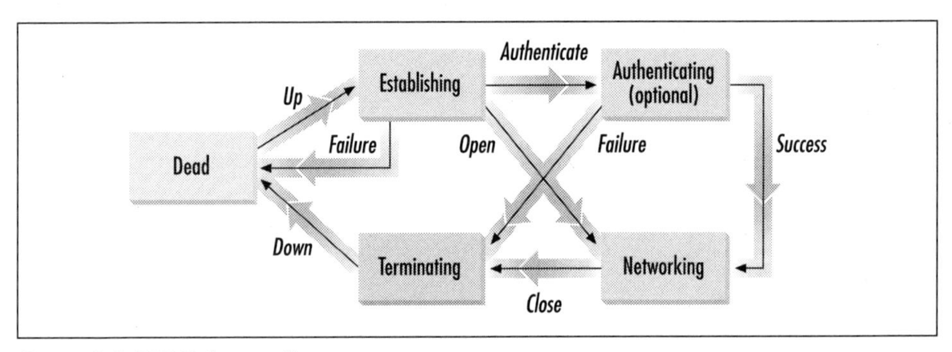

Figure 9-4. PPP link state diagram

Five link states are defined:

1. *Link dead*

   All links begin in the dead phase, which corresponds to a physical layer not ready for data transmission. PPP moves out of the dead phase when a physical signal indicates that the underlying medium is ready to accept data for transmission. On a T1 circuit, this corresponds to the carrier detect signal on the V.35 circuit to the router.

2. *Link establishment*

   After the physical layer is ready to transmit, an LCP exchange negotiates link layer parameters. LCP handles only network layer independent parameters.

Establishment ends only when both sides have signaled agreement on parameters by sending identical LCP configure acknowledgments. If PPP receives a configure request during any other state, it returns to the establishment phase.

3. *Authentication (optional)*

   PPP may enter an authenticating state if authentication on the link is desired. Either side may request authentication during the establishment phase. If authentication fails, the network layer is not initialized and PPP proceeds immediately to the termination phase without passing Go or collecting $200. PPP prevents higher-layer protocol attacks by holding back from initializing the network layer until authentication is successful. Generally, PPP does not use authentication on leased lines because the endpoints are static.

4. *Network layer protocol phase*

   Each desired network protocol is configured through its NCP. Each protocol must be configured independently, although negotiations may occur concurrently. NCPs can be opened and closed independently. Poor implementations may send frames corresponding to a network protocol before its associated NCP has configured the link. Frames received for an unconfigured network protocol are dropped. In addition to IP, NCPs have been designed for protocols such as IPX, AppleTalk, OSI, and others with which many network engineers are not familiar.

5. *Termination*

   PPP links may be terminated for a variety of reasons. The carrier signal for the underlying physical link may be lost due to physical problems with the circuit. As previously noted, authentication failures cause link termination. In the case of an authentication failure, most PPP implementations will disconnect the physical link as well. PPP may detect link quality problems and close the link. Link termination can also occur for administrative reasons. Network users are allowed to terminate the link at their discretion.

   Two common misconceptions about termination exist. One is that each NCP must close before link termination. On the contrary, a single LCP termination request is sufficient to close the link. The second is that closing the only active network protocol leads to link termination. It is possible for a link to be configured and ready without having any active network protocols.

PPP Encapsulation and Framing

Nearly every byte-oriented protocol has borrowed something from HDLC. In the case of PPP, it is the framing format, which is shown in Figure 9-5. PPP frames open and close with the familiar HDLC flag. Following the opening flag, PPP uses a 1-byte address field. The address field is normally all ones, which means it is a broadcast address. Serial lines deliver frames in the order in which they are received, so frames are unnumbered by default. As in HDLC, a single byte set to 0x03 is used to indicate unnumbered information.

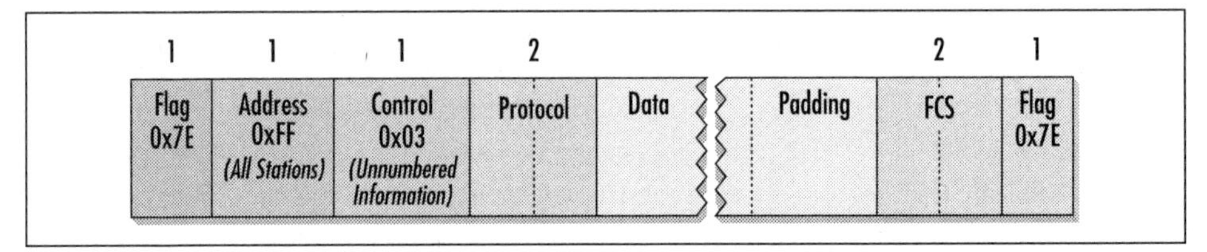

Figure 9-5. PPP framing and encapsulation

The protocol field follows the control byte. Two bytes are used to identify the encapsulated protocol. PPP protocol numbers for higher-layer protocols are assigned so as to ensure that protocol numbers begin with zero and end with one. Additionally, the PPP protocol number for an NCP is obtained by adding 32,768 to the corresponding network protocol. Protocols in the range of 0xc000 to 0xffff are reserved for link layer operations. Table 9-3 shows common PPP protocol numbers.

Table 9-3. PPP protocol numbers

Protocol number (hexadecimal)	Protocol
0021	IP
8021	IPCP (NCP for IP)
002d	Van Jacobson TCP/IP header compression (RFC 1144)
002f	Van Jacobson IP header compression
c021	Link Control Protocol
c025	Link Quality Report

The protocol numbers 802d and 802f are not used because Van Jacobson header compression is negotiated within IPCP. When header compression is in use, 0x002d and 0x002f are used on the link layer frames, but this does not require a dedicated LCP protocol number to negotiate options because that is done within IPCP. A *frame check sequence* (FCS) closes the PPP frame to provide protection against corrupted data.

Transparency

Link layer protocols must provide a transparent path to user data. Bit-synchronous links use the same stuffing procedure as HDLC, described in the previous chapter. RFC 1662 also has stuffing procedures for asynchronous interfaces and serial interfaces that transmit and receive bytes.

Link Control Protocol (LCP)

Data links come in many flavors with widely varying properties. PPP was intended to operate over all types of physical links, so it includes the Link Control Protocol. LCP allows two endpoints of a connection to negotiate encapsulation options when the data link is initialized and can monitor the link quality to ensure that quality is not slowly degrading.

Configuration option negotiation is an interesting departure from other types of negotiations in the networking world. In PPP, the transmitted requests are requests to the remote end on the inbound data. PPP allows the local end to specify what it will receive, not to tell the remote end all the things the local end can send.

LCP uses three classes of packets:

- Link Configuration packets negotiate options.
- Link Termination packets tear down the link.
- Link Maintenance packets ensure continued operation of the link.

The LCP State Engine

LCP is described as a state engine by the standard. A link may be in one of several states, and certain events cause transitions between the states. Figure 9-6 shows a simplified diagram of the LCP state engine. Each state in the diagram is numbered. For quick reference, these states are also described in Table 9-4:

0. *Initial*

   All links begin in the Initial state. Physical links are down, and no administrative actions have been taken to bring them up.

1. *Starting*

   In this state, the physical link is not yet available, but has been administratively activated. This state might correspond to an asynchronous modem waiting for its carrier signal.

2. *Closed*

   This is a complementary state to Starting. Physical connectivity is available, but no administrative action has configured it to pass traffic. This state is common on T1 links because the physical layer becomes active quite soon after the physical plugs are connected.

3. *Stopped*

   This state is similar to Closed in that the physical link is up but the logical link is not configured. It is usually entered when timeouts have expired or link termination has been requested. Reconfiguration of the link will not be initiated, but if the peer attempts configuration, LCP will attempt to renegotiate.

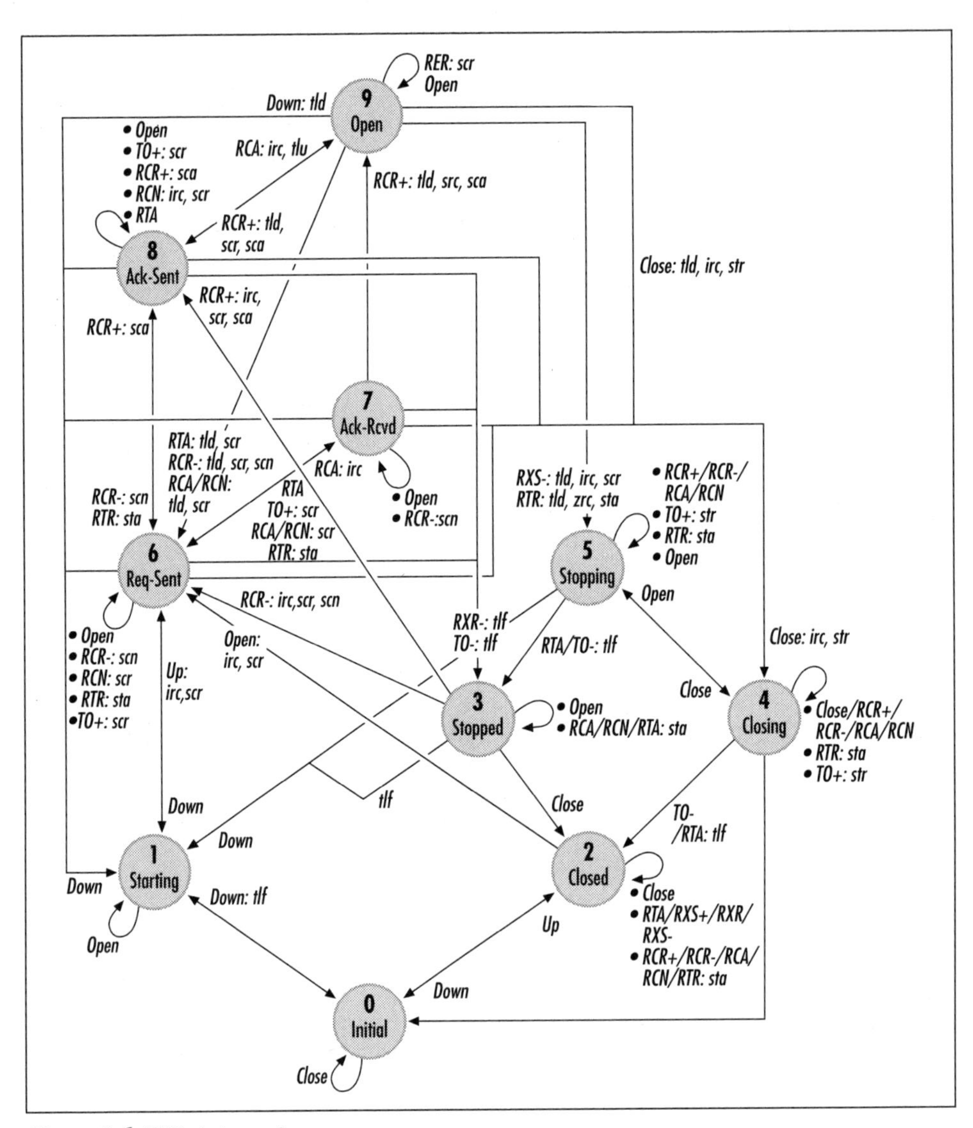

Figure 9-6. PPP state engine

4. *Closing*

   When the link has been administratively closed, a Terminate-Request is sent to the peer and LCP enters the Closing state to wait for acknowledgment. Attempts to reconfigure the link are discarded.

5. *Stopping*

   This state is similar to the Closing state, except that the link is administratively open. When the link is torn down, LCP is in either the Stopped state or the Starting state. Both of these states allow reconfiguration.

6. *Request-Sent (Req-Sent)*

   This state is the common entry point for the configuration exchange procedures in LCP. In this state, a Configure-Request has been sent to the peer, but no acknowledgments have been made.

7. *Ack-Received (Ack-Rcvd)*

   In this state, the outbound link has been configured, but the inbound link configuration is not yet complete. The peer has agreed with and acknowledged the proposal sent by the local end.

8. *Ack-Sent*

   In this state, the inbound link has been successfully configured. The peer has sent a proposal that has been acknowledged. Outbound link configuration is still proceeding.

9. *Open*

   This state is the Holy Grail of LCP states. Once this state has been reached, further negotiations proceed.

Table 9-4. States of the LCP state engine

Number	Name	Description
0	Initial	No physical carrier signal is available and the logical link is administratively closed.
1	Starting	The physical carrier is unavailable, but the link is administratively active. Reconfiguration of the link is possible.
2	Closed	The physical carrier is available, but the link is administratively closed.
3	Stopped	Both the physical and logical links are available, but there is not necessarily any intent to close them. This state can be reached when the link is being terminated or when configuration has failed. Reconfiguration is possible.
4	Closing	The link has been administratively closed; LCP has sent a termination request and is waiting on the acknowledgment before moving to Closed state. No reconfiguration is possible.
5	Stopping	When the peer has administratively closed a link, a Terminate-Request will be received.
6	Req-Sent	A Configure-Request has been transmitted, but no configuration proposals have been acknowledged, either by the peer or locally.
7	Ack-Rcvd	An acceptable Configure-Request has been transmitted and acknowledged to configure the outbound link, but configuration of the inbound link is not yet complete.
8	Ack-Sent	An acceptable Configure-Request has been transmitted and acknowledged to configure the inbound link, but outbound configuration is not complete.

Table 9-4. States of the LCP state engine (continued)

Number	Name	Description
9	Open	This is the goal of the configuration process. When reached, PPP implementations will notify higher layers that it is up. Conversely, when leaving the Open state, it will notify higher layers that it is now down.

In addition to the states in the diagram, events cause transitions. Table 9-5 summarizes all the events defined by the PPP standard. In response to events, an implementation may also take action. To distinguish between events and actions, events are capitalized in Figure 9-6. If more than one event leads to the same action, they are separated by slashes. For example, any configuration packets received in the Closing state do not cause a state change.

If an event leads to an action, the underlying event is listed, followed by a colon and then the action. Multiple actions are separated by commas. For example, reception of any configuration packets in the Stopped state causes LCP to respond by sending a Terminate-Ack.

Figure 9-6 also makes use of multitailed arrows when the same event in different states leads to the same final state. A Down event in the Open, Ack-Sent, Ack-Received, or Request-Sent state returns LCP to the Starting state. Events have different meanings depending on the layer of the PPP suite at which they occur. An Up event at the LCP layer means that the physical layer has come up, but an Up event at the network layer means that LCP has initialized.

Table 9-5. Abbreviations used in Figure 9-6

Abbreviation	Event/ action	Full meaning	Description
Up	E	Lower layer up	This is generated when a lower layer indicates that it is ready to send packets. For LCP, it means that the physical layer is available; for network protocols, it indicates that the LCP layer has brought up the logical link and generated a This-Layer-Up event.
Down	E	Lower layer down	This is generated when a lower layer indicates that it is unable to send packets. For LCP, it means that the physical layer is down; for network protocols, it indicates that the LCP layer has torn down the logical link and sent a This-Layer-Down message to the network layer.

Table 9-5. Abbreviations used in Figure 9-6 (continued)

Abbreviation	Event/ action	Full meaning	Description
Open	E	Link now administratively available	Up events bubble up from lower layers to indicate that traffic is allowed to pass. In contrast, Open events come down from the top as directives to attempt link configuration. In the course of opening a link, it may be necessary to direct lower layers to negotiate and wait for them to complete.
Close	E	Link now administratively unavailable	This directs that the link should be closed to traffic. It is generally followed by actions that attempt to terminate the link, such as sending Terminate-Requests or tearing down circuit-switched connections.
TO–, TO+	E	Timeout with (–) and without (+) counter expiration	To prevent lost frames from stalling the negotiation or termination process, Configure-Request and Terminate-Request packets are timed. In addition to a timer, a restart counter ensures that multiple requests are made before giving up and moving to another state.
RCR+, RCR–	E	Reception of Configure-Request; + indicates options are good, – indicates options are bad	When options are acceptable, a Configure-Ack is transmitted. Generally, acknowledging an acceptable configuration proposal causes a state change. Unacceptable proposals result in the transmission of a Configure-Nak.
RCA	E	Receive Configure-Ack	Configure-Ack is a positive acknowledgment to an earlier Configure-Request. This event may occur due to crossed connections or implementation errors. To move into the Ack-Received or Open states, it is certain that a Configure-Ack would have been previously received, so receiving another Configure-Ack in these states is a serious problem.
RCN	E	Receive Configure-Nak or Configure-Reject	These events inform the peer of unacceptable options in a previous Configure-Request packet. Both cause identical state changes for the state engine, but they have different effects on the next Configure-Request.
RTR	E	Receive Terminate-Request	A peer signals its intent to tear down the link by sending a Terminate-Request.
RTA	E	Receive Terminate-Ack	Terminate-Acks are used to acknowledge Terminate-Requests. They are also sent in the Stopped or Closed states and assist in moving both ends of the link into compatible states for further operations.

Table 9-5. Abbreviations used in Figure 9-6 (continued)

Abbreviation	Event/ action	Full meaning	Description
RUC	E	Receive Unknown Code	When an unknown code is received, this event is generated and a Code-Reject message is sent in response.
RXJ+	E	Receive Protocol-Reject or Code-Reject (non-catastrophic)	Some Protocol-Reject and Code-Reject messages are acceptable and expected in normal operations. New LCP extensions and new network protocols may not be implemented in all PPP suites, so it is normal to reject new messages or new network protocols.
RXJ–	E	Receive Protocol-Reject or Code-Reject (catastrophic)	Catastrophic reject messages prevent further communication, and the link is terminated.
RXR	E	Receive Echo-Request	Echo-Reply frames are generated in response to Echo-Requests.
tlu	A	This-Layer-Up	This is used by LCP to instruct NCPs to commence negotiation and by NCPs to allow data transmission to occur.
tld	A	This-Layer-Down	This signals that the state engine is leaving the Open state and is no longer available for transmission.
tls	A	This layer started	This indicates that a lower layer must perform negotiation before proceeding.
tlf	A	This layer finished	This indicates that links entering the Initial, Closed, or Stopped states and lower layers are no longer needed. If no other network protocols are open, physical teardown occurs.
irc	A	Initialize restart counter	This initializes the restart counter to the maximum number of configuration or termination attempts.
zrc	A	Zero restart counter	This sets the restart count to zero.
scr	A	Send Configure-Request	This transmits a Configure-Request and decrements the restart counter.
sca	A	Send Configure-Ack	This sends a Configure-Ack to acknowledge that the specified set of options is acceptable.
scn	A	Send Configure-Nak or Configure-Reject	Configure-Reject vetoes selected options, while Configure-Nak proposes new values for options that are acceptable for different values.

Table 9-5. Abbreviations used in Figure 9-6 (continued)

Abbreviation	Event/ action	Full meaning	Description
str	A	Send Terminate-Request	This sends a Terminate-Request to close the link. Each time a Terminate-Request is sent, the restart counter is decremented.
sta	A	Send Terminate-Ack	This sends a Terminate-Ack to agree with link termination. Terminate-Acks are also used to synchronize state engines in other states.
scj	A	Send Code-Reject	This sends a Code-Reject to acknowledge an unknown code type.
ser	A	Send Echo-Reply	This sends an Echo-Reply to acknowledge the receipt of an Echo-Request.

Relation of Link Management to the LCP State Engine

To breathe life into the complex state diagram, this section shows how some common operations cause transitions between the different states.

Link initialization

LCP requires administrative action in order to open a link. If no administrative Open is issued, LCP will not perform any configuration negotiations. Instead, it will respond to any polite configuration requests with termination packets. The goal of link initialization is to move from the link-management states into the logical link configuration states. There are two types of link initialization:

Self-initiated
: If the link is currently down, both an administrative Open and a physical-link initialization must occur. No matter what the sequence is, a configuration request is sent and LCP moves to the Request-Sent (6) state. Figure 9-7 overlays a self-initiated initialization onto the diagram of the PPP state engine.

Peer-initiated
: If the logical link was previously stopped, link configuration can proceed from the Stopped (3) state if the peer sends a configuration request. LCP will either acknowledge that request or send a counterproposal, and a configuration request will be sent to propose options to the peer. Figure 9-8 overlays a peer-initiated configuration onto the PPP state diagram.

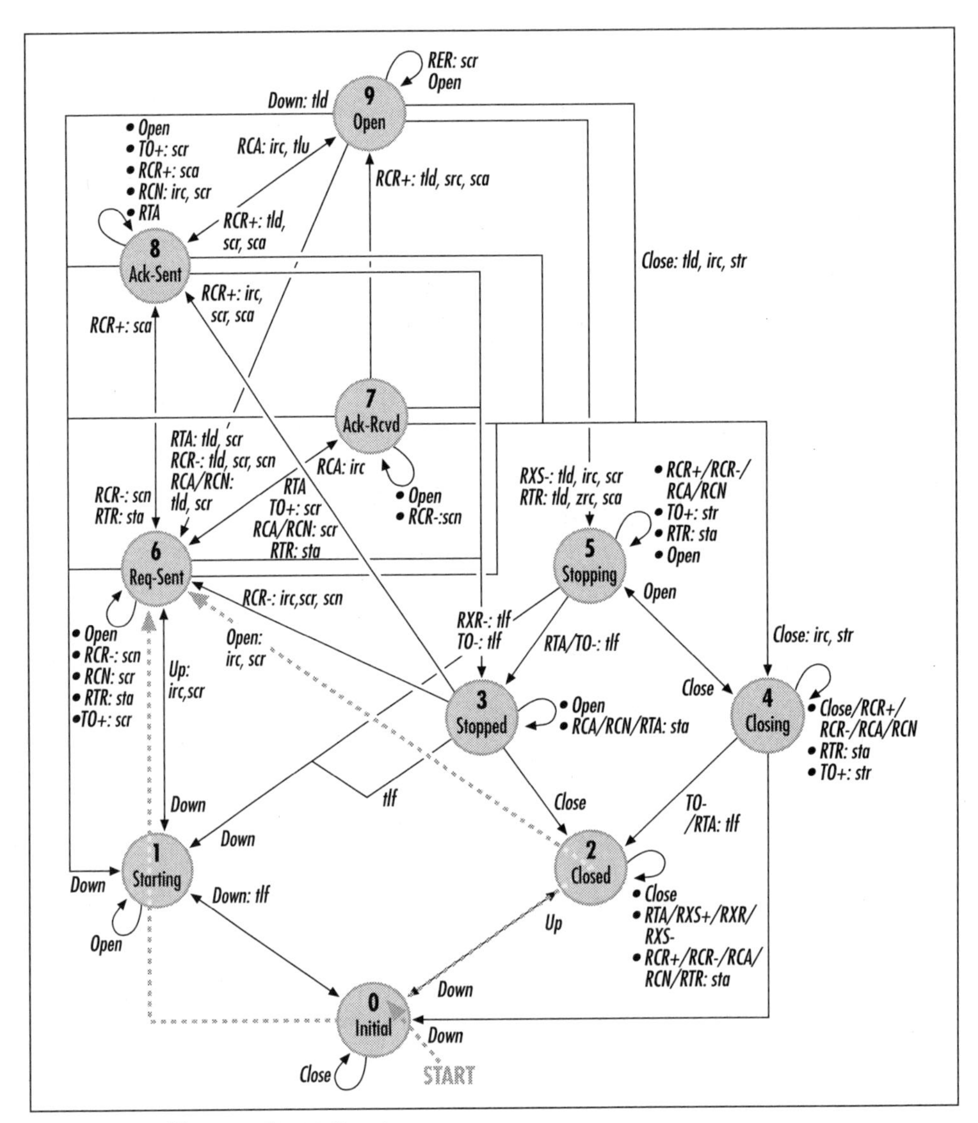

Figure 9-7. Self-initiated initialization

Link configuration

Configuration ends when both sides have agreed on mutually acceptable parameters for both directions. PPP options are like IPSec security associations. Each half of the link can have its own configuration parameters. Configuration requests affect parameters used for the inbound half of a link, and the outbound link direction is negotiated separately. A typical negotiation may go something like Figure 9-9.

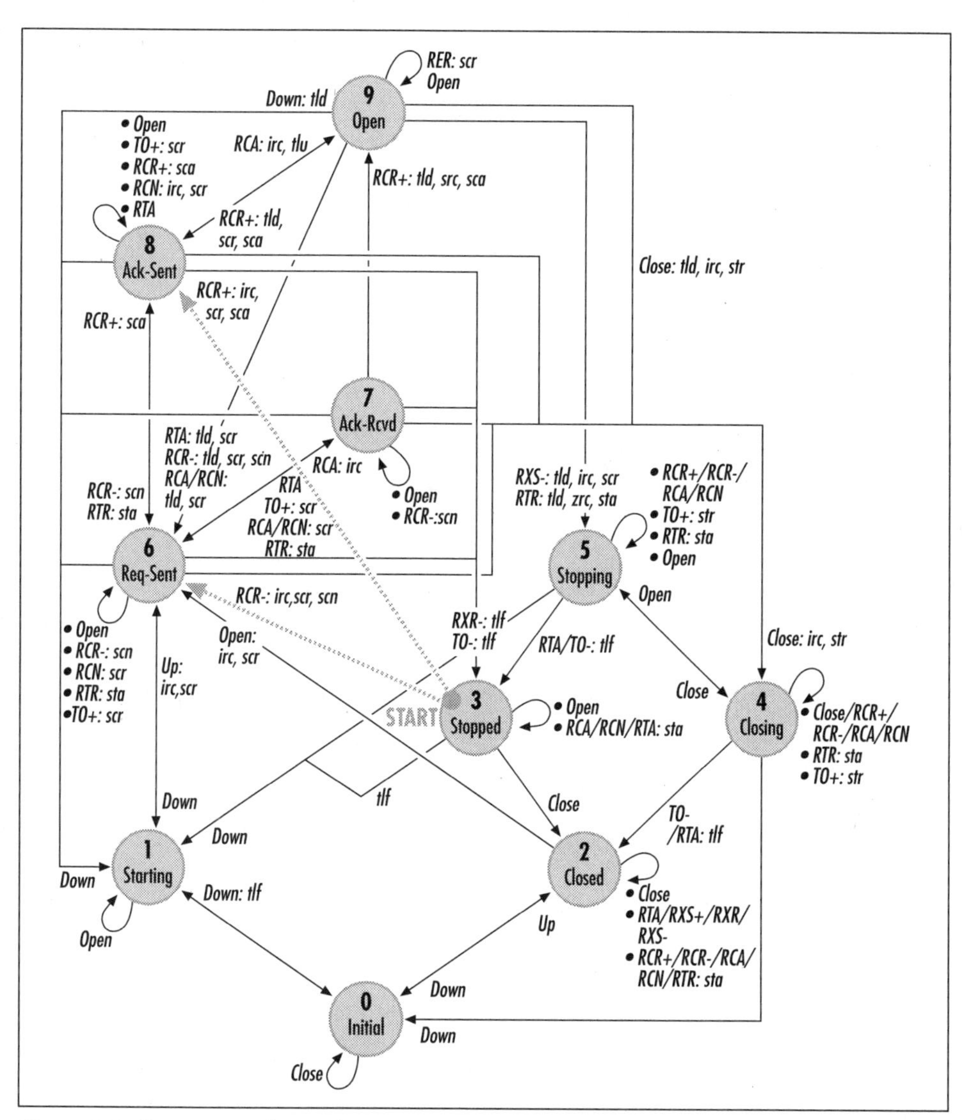

Figure 9-8. Peer-initiated initialization

Initial configuration parameters on the local side are part of the PPP software's initialization files. These options are bundled together in a configuration request and sent to the peer.

Each option in the configuration request is examined for compatibility. Some options may be totally unacceptable, while others need to have values adjusted before they become acceptable. Totally unacceptable options may be due to a policy decision made by the device's owner, such as an ISP requiring authentication on dial-up servers or a software limitation. Show-stoppers are sent back in a configuration reject message.

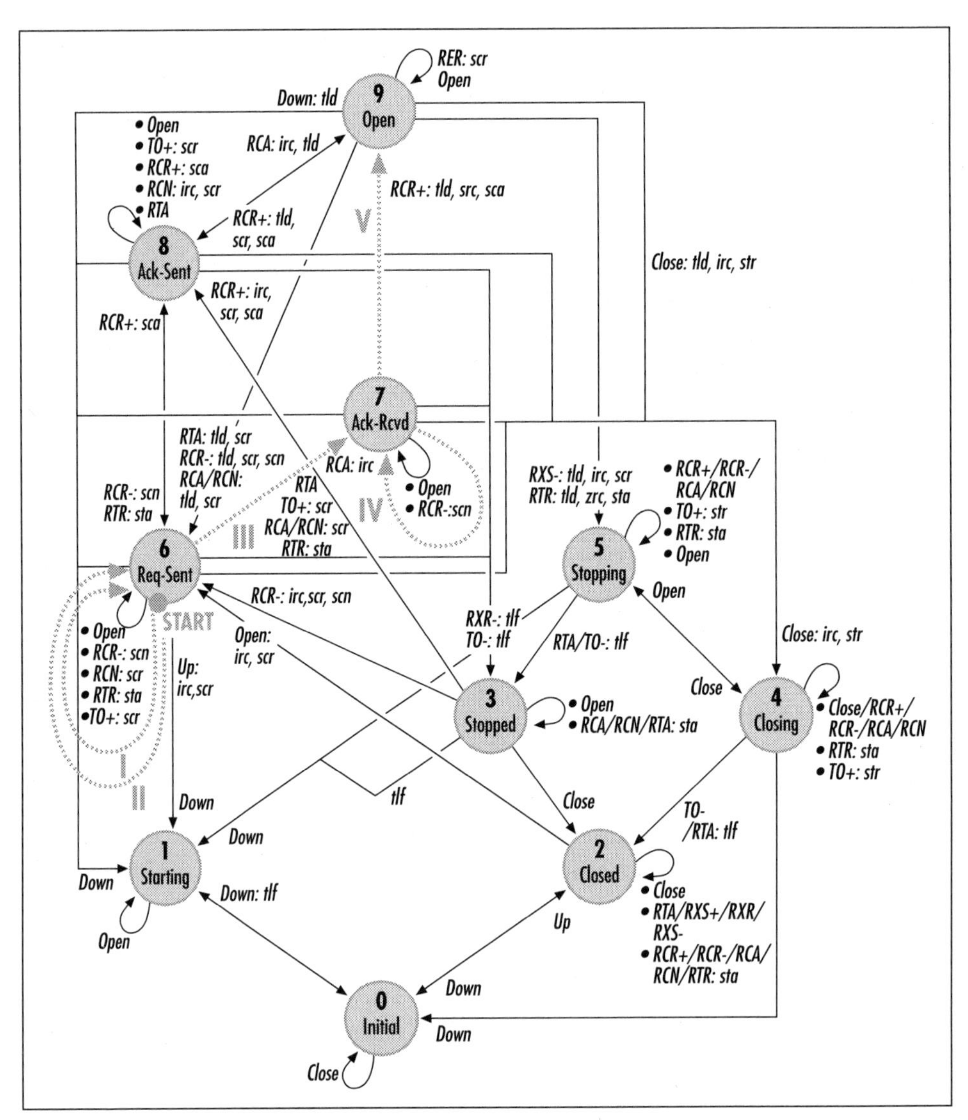

Figure 9-9. Option negotiation

After receiving the Configure-Reject, the local end must revise its parameter set and eliminate the totally unacceptable options in successive configuration requests. A proposal without the unacceptable options will be received more favorably, but it still may not be ideal. No totally unacceptable options will be included because they were previously rejected, but some options may be included with values that should be changed. Unacceptable values will be sent back in a Configure-Nak counter-proposal. Naks include *hints*, which are values that the peer is willing to accept.

Naks present the local side with a choice: the option can either be omitted entirely so that it reverts to its default, or it can be revised to the hint. The former case may result in another Nak with the hint value, or the default may be acceptable. In the latter case, the new configuration request will be acceptable. The peer will reply with a configuration acknowledgment including all options.

Meanwhile, the same procedure has been happening to configure the link from the peer to the local end. Depending on the set of parameters the peer desires, it may take more or fewer cycles, or convergence may never happen. If each side of the link is configured in such a way that no mutually acceptable parameter set exists, the peers will never be able to negotiate away their differences and the link will stay down. In some cases, the incompatibility may be due to parameters that are the defaults of a particular PPP implementation.

In Figure 9-9, the configuration process starts from the Request-Sent (6) state. Step I includes unacceptable options, so the peer sends a veto (Configure-Reject). After revising its options to be acceptable, the peer retransmits the request in step II. That request is also not acceptable because the values chosen for some options were not acceptable. A third request in step III is acceptable, so LCP moves to the Ack-Received (7) state. At this point, the outbound link is configured. Meanwhile, the peer performs a similar process. Unacceptable options received in step IV result in a counterproposal, which means that the next request received is acceptable. LCP moves into the Open (9) state and PPP proceeds to the next layer.

Link termination

There are two types of link termination, which work as follows:

Self-initiated
: An administrative link closure moves LCP from an active configuration state to the Closing (4) state and sends a termination request to the peer. Circuit-switched links are torn down at this point, but leased lines remain constantly physically available under normal circumstances. When the termination acknowledgment is received, LCP moves to the Closed (2) state to wait for another administrative opening of the link. Figure 9-10 shows the steps for a self-initiated termination on the PPP state diagram.

Peer-initiated
: When the peer receives a termination request, LCP acknowledges the request, marks the link layer as down, and moves to the Stopping (5) state. Leased lines remain physically available, so LCP needs to wait for the termination timer to expire. Three events are likely to happen at this point: an acknowledgment of the termination, a timeout, or an administrative closing of the link. Depending on which of the three events occurs, two paths may be taken; these paths are illustrated by the "a" path (termination acknowledgment or

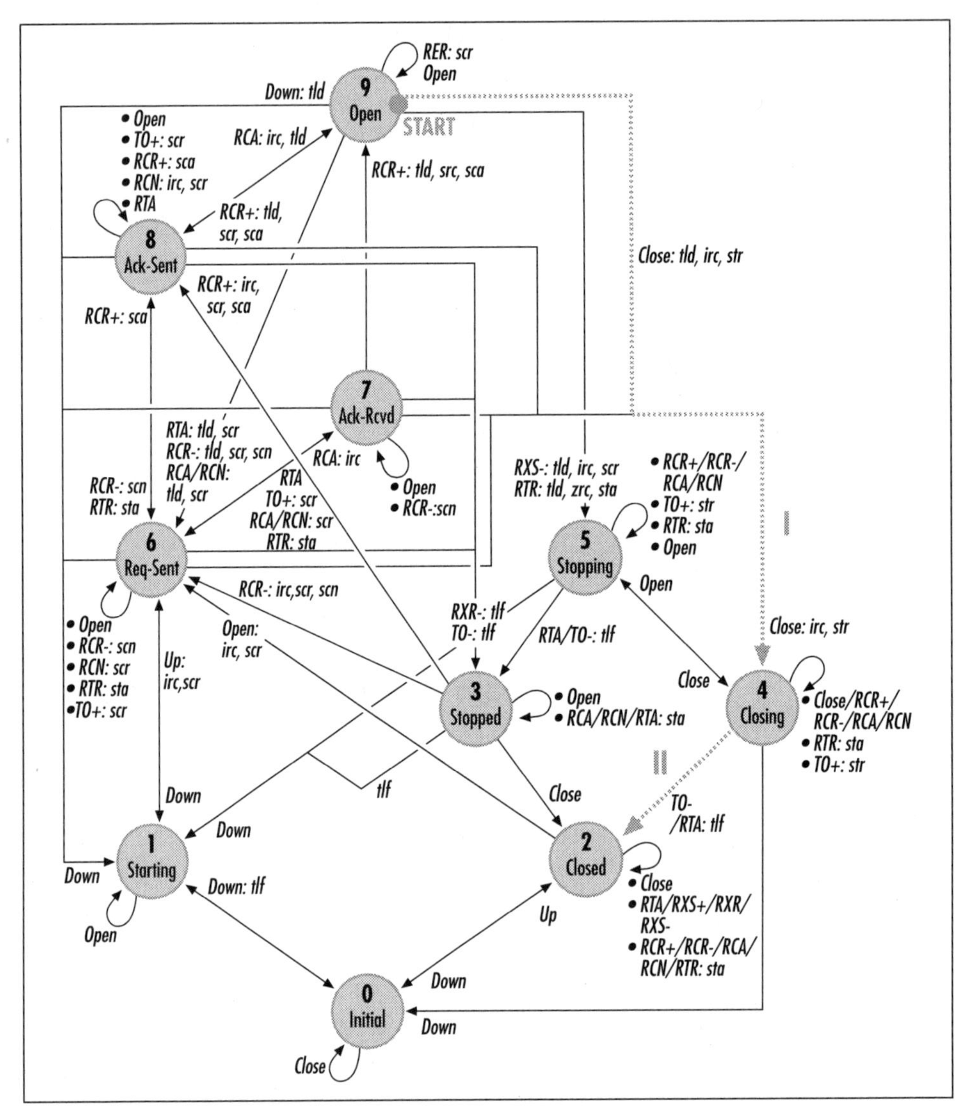

Figure 9-10. Self-initiated termination

timeout) and the "b" path (administrative closure) in Figure 9-11. Following link closure, the underlying physical layer may be torn down, in which case LCP moves to the Initial (0) state. On leased lines, the physical layer typically remains available, so LCP remains in the Closed (2) state.

LCP Frames

LCP is transmitted in standard PPP frames as protocol 0xc021. Although LCP provides for option negotiation, which may affect the stuffing procedures used for

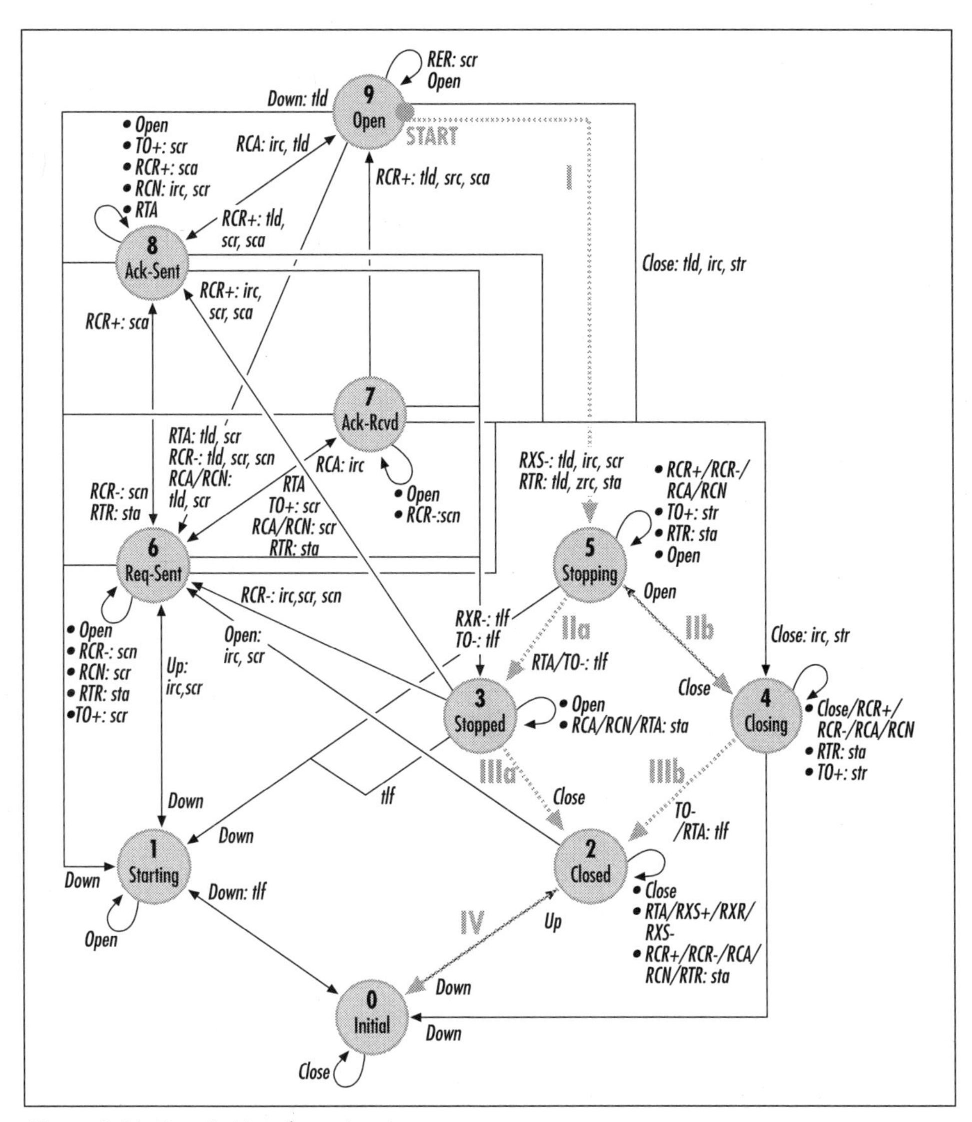

Figure 9-11. Peer-initiated termination

transparency, all LCP frames are sent using the default values for all parameters. This guards against the two ends getting out of synchronization on which parameters to use and LCP becoming unintelligible as a result.

Figure 9-12 shows the format of a generic LCP packet. LCP on the wire uses a header composed of a type code to distinguish between LCP frame types, an identifier, and a length field, which is the length of the LCP data field in bytes. Identifiers are 1-byte numbers used to match up requests and responses. LCP type codes are defined in Table 9-6.

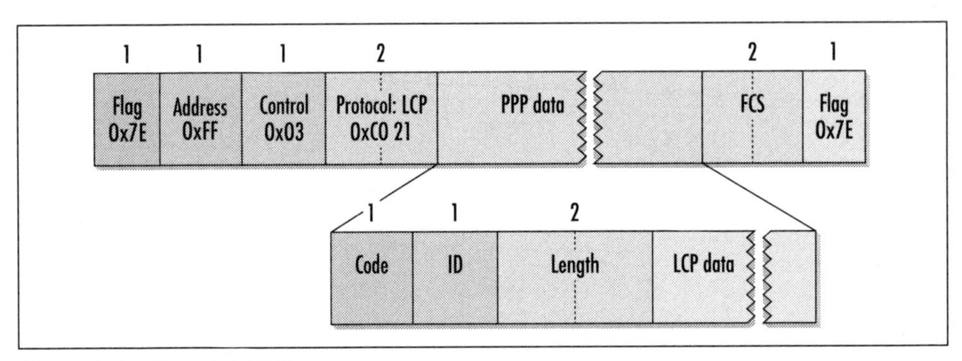

Figure 9-12. Generic LCP frame

Table 9-6. LCP type codes

Code	Type	Description
0	Vendor-Extension	Rarely used vendor extension mechanism specified in RFC 2153.
1	Configure-Request	Proposed set of nondefault configuration options to a peer, with all options enclosed in one frame.
2	Configure-Ack	Acceptance of a set of options; includes a copy of the proposal for verification.
3	Configure-Nak	Counterproposal used when the values assigned to some options are not acceptable. Unacceptable options are modified with suggestions for new values.
4	Configure-Reject	Options that are not recognized or are unacceptable with any value are noted in the data field of the Configure-Reject. Configure-Rejects are also used for Boolean options with unacceptable settings.
5	Terminate-Request	Sent by one side to request that the link be terminated.
6	Terminate-Ack	Used in response to Terminate-Requests to acknowledge that the link should be torn down.
7	Code-Reject	LCP does not use version numbers in packets because unknown codes may always be replied to with a Code-Reject, which is used to indicate that the code in a received LCP packet is unknown.
8	Protocol-Reject	Once a link is opened, any unknown protocols are rejected with Protocol-Reject packets.
9	Echo-Request	Used like ICMP echo request to test link connectivity. Also used in link quality monitoring.
10	Echo-Reply	Sent in response to LCP Echo-Request packets.
11	Discard-Request	The receiver of a Discard-Request throws away the packet. Discard-Requests may be used for debugging or link-connectivity testing.

Table 9-6. LCP type codes (continued)

Code	Type	Description
12	Identification	Identifies sender in an implementation-dependent manner. Text may include software package name and version, hardware platform, or any information the programmer who wrote the PPP implementation would find useful.
13	Time-Remaining	Rarely used message defined in RFC 1570 to notify a peer of the time remaining in a session.
14	Reset-Request	Used in a compression protocol to request reinitialization of compression routines.
15	Reset-Ack	Acknowledges compression reinitialization.

Configuration messages (codes 1, 2, 3, and 4)

Configuration frames are the workhorses of LCP. LCP's main job is to negotiate link options, which is done as the LCP implementations on both sides of the link exchange proposals and counterproposals until they reach mutually acceptable option sets. Each configure packet has an identification number that is supposed to allow matching of requests and responses. The field, generally, can be used for matching requests with responses, though bugs or implementation errors may make this impossible. Do not make any assumptions based on the identification field when looking at LCP traces. Figure 9-13 shows the formats of all four configuration messages.

LCP uses Configure-Request frames to propose changes to the default options on a link. The data portion of the Configure-Request lists a set of nondefault options. Implementations do not need to send default values as part of the option set. Option coding is presented in detail later in this chapter.

Configure-Reject frames are used when unacceptable options are present in the proposed configuration in the Configure-Request. An option may be unacceptable because it is not implemented or because it is unrecognized. It is possible, for example, to create a minimally compliant PPP implementation that simply rejects any nondefault options, though such an implementation would not be very useful.

After receiving a Configure-Reject, a system resends a new Configure-Request with a new identification number, omitting any unacceptable options. These would be options that are acceptable but have values that could be improved upon result in a Configure-Nak. Configure-Naks include only the options that need to be changed. To assist in the negotiation process, Configure-Naks suggest new acceptable values to the peer; these suggestions are known as hints. When a desirable option is not present, it is legal to include the desired (but unrequested) option in an unsolicited Configure-Nak. Not all implementations cope well with unsolicited Configure-Naks.

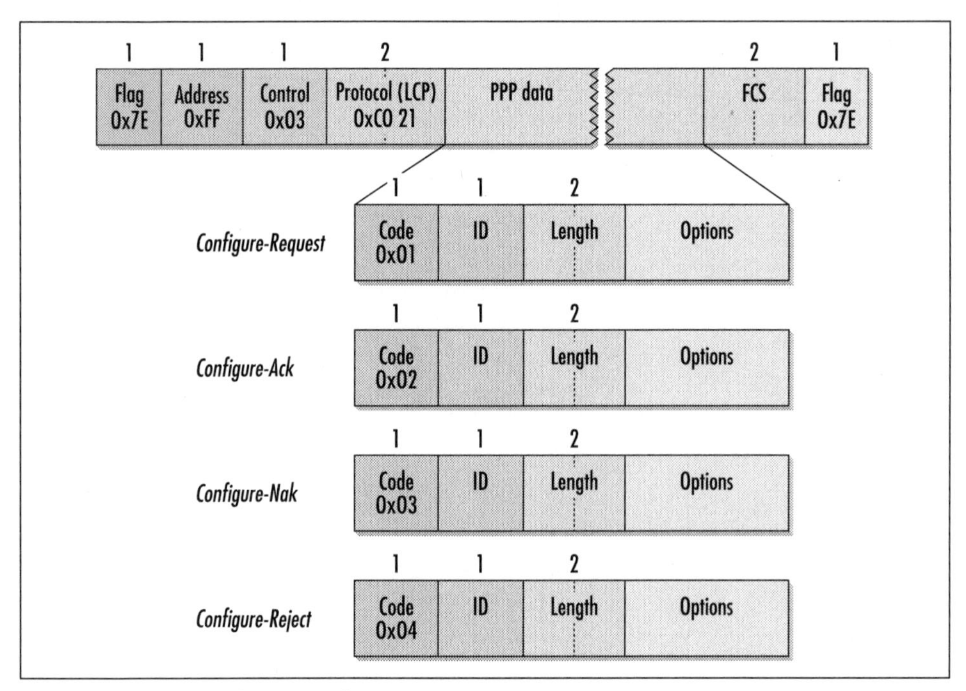

Figure 9-13. LCP configuration frames

When all is well, a Configure-Ack is returned to acknowledge the Configure-Request. It includes a copy of all the options from the Configure-Request for verification.

Termination messages (codes 5 and 6)

Two types of termination packets are used by PPP. Terminate-Request packets are sent to request link closing. After sending a Terminate-Request, LCP waits for a Terminate-Ack. If no Terminate-Ack is received within a timeout period, the link is torn down. This is illustrated in Figure 9-14. Termination requests may include text describing why the link was torn down. Generally, this string is printable ASCII.

Code- and protocol-rejection messages (codes 7 and 8)

LCP uses code- and protocol-rejection messages to indicate to a peer that the received code or protocol is invalid. To assist in identifying the offending message, the received frame is included in the data portion of the rejection frame.

Code-Reject is the mechanism by which LCP avoids using version numbers. Older implementations send Code-Reject messages in response to new protocol options, providing a straightforward interoperability mechanism. In a Code-Reject packet, the data portion contains the packet that was rejected, beginning with its information field. Both the link layer headers and the FCS are omitted, and the entire Code-Reject message must fit within a single frame. If the resulting Code-Reject is larger than the link Maximum Receive Unit (MRU), the frame is truncated to fit.

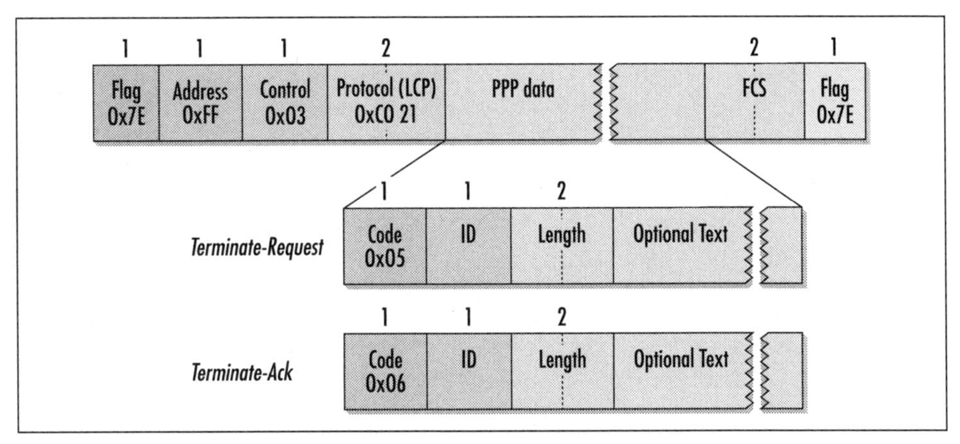

Figure 9-14. Termination LCP frames

LCP must have completed negotiating before sending protocol-rejection messages. Protocol-rejection messages are used to indicate that an entire network layer protocol is not supported by a PPP implementation. If a peer requests to configure an unsupported protocol, a Protocol-Reject message is sent in response. Protocol-Rejects are fairly common and perfectly acceptable in PPP negotiations. They might occur, for example, if a dial-up link attempts to negotiate IP, IPX, and NetBIOS. Dial-in servers at Internet service providers would reply to the IPX and NetBIOS configuration requests with protocol-rejection messages because those protocols are not generally supported by ISPs. Figure 9-15 shows Code-Rejects and Protocol-Rejects.

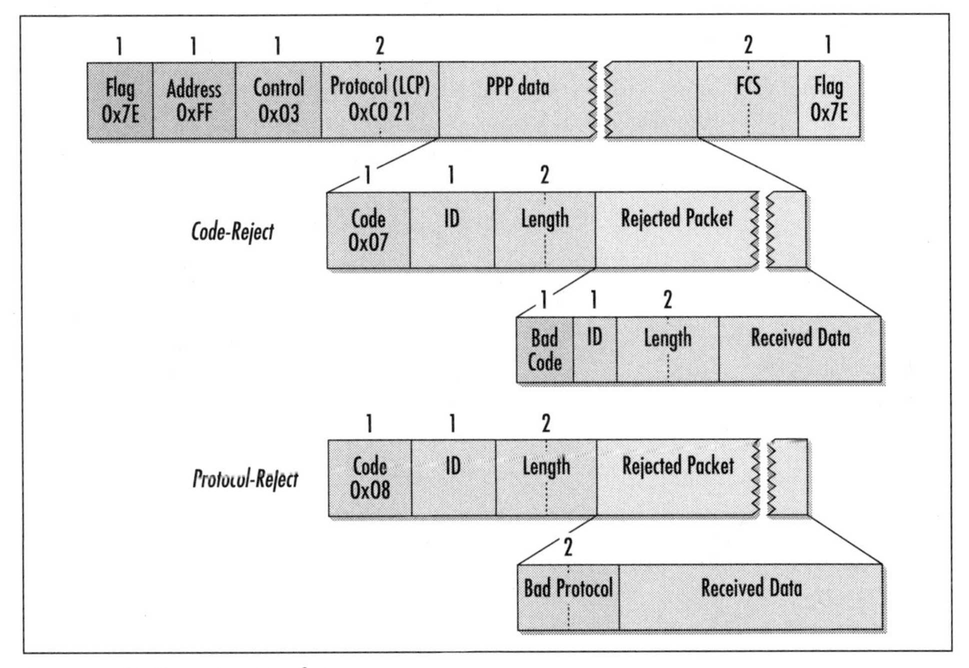

Figure 9-15. Rejection LCP frames

Link integrity messages (codes 9, 10, and 11)

Echo-Request and Echo-Reply are used to check link integrity, much as ICMP echo requests and echo responses are used by network administrators for IP-level connectivity. Discard-Request is relatively uncommon, but may be used during testing. Figure 9-16 shows the format of the link-integrity frames.

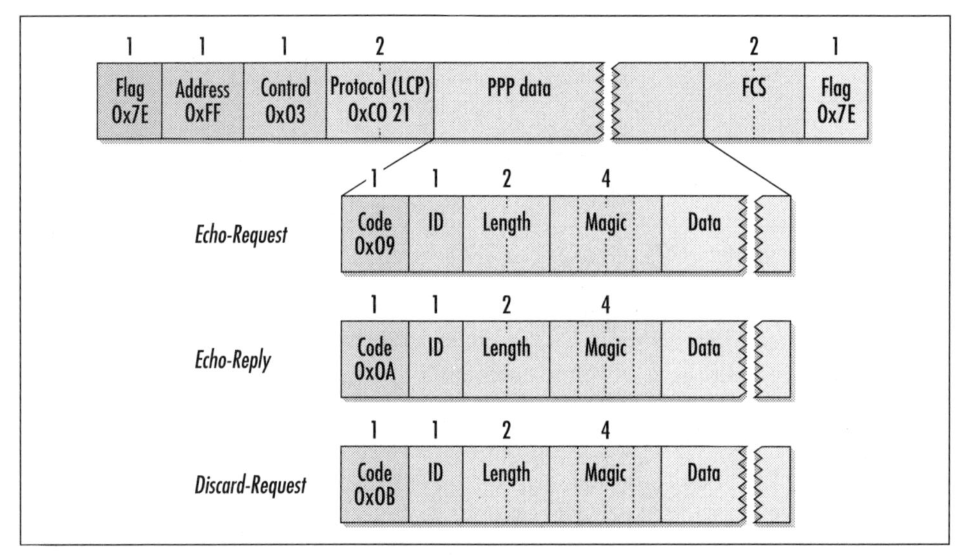

Figure 9-16. Link-integrity frames

Echo messages are used in the same way that SLARP is used for link-integrity verification. Periodic echo requests are sent, and if several transmission intervals elapse without receiving a reply, the link is torn down. Echo timers are nearly always configurable because of the drastic variation in PPP link speeds; a single fixed interval could never hope to satisfy the demands of dial-up users and administrators overseeing SONET networks. Priority is often given to these messages because of their urgency. It is better to drop a frame of user data than to tear down and reestablish the link after a timer was allowed to expire. To provide for link verification, echo requests in Multilink PPP (MP) are transmitted without going through the MP encapsulation routines.

Identification message (code 12)

To aid in debugging, RFC 1570 defined the Identification message. After the customary code, length, and magic number fields, an implementation-defined text string is transmitted. Figure 9-17 shows the format of the LCP Identification message.

When it is implemented, the information field contains information that is helpful in debugging. Most implementations include the sender's software version and hardware platform.

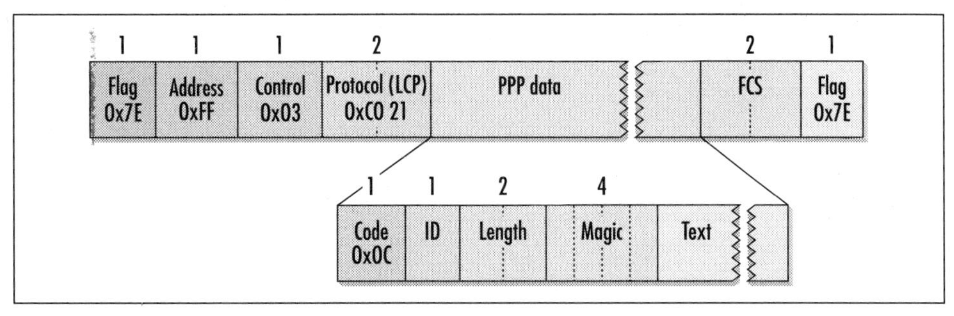

Figure 9-17. LCP Identification message

Other messages

The Time-Remaining (13) message, also defined in RFC 1570, is rarely implemented. Reset-Req (14) and Reset-Ack (15) are used with the PPP Compression Control Protocol. Data compression is useful only when the CPU power on both ends of the link is dramatically greater than the link speed. When T1 access lines are used, the terminating device is usually an access concentrator that must share its CPU power between a large number of incoming connections, so access concentrators typically do not have the spare computational power to run compression.

Code 0 was initially reserved, but was allocated to vendor extension in RFC 2153. Many vendors use unallocated options rather than vendor-extension mechanisms.

LCP Configuration Options

Much of LCP's work is done in negotiating the link-configuration options. Link-configuration options are embedded within the data field of the configuration proposals. Every option has a corresponding type code and is embedded in LCP packets with its code, total option length, and any option data, as in Figure 9-18, which shows several options within a Configure-Request. Some options are enabled or disabled but have no data.

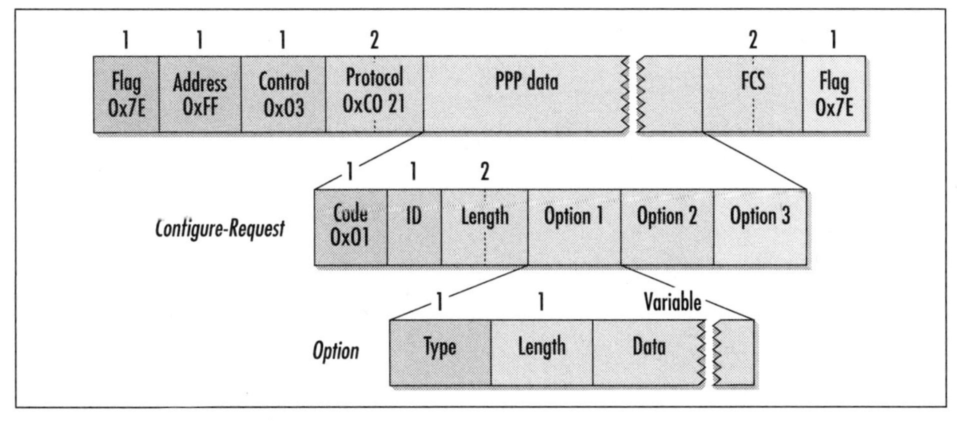

Figure 9-18. LCP option format

Several LCP options are listed in Table 9-7. Common options are treated in more detail in subsequent sections. During LCP negotiation, frames are transmitted with the default options engaged.

Table 9-7. LCP option codes

Code	Type	Description
0	Vendor extensions	In addition to proprietary type codes, RFC 2153 specified vendor-proprietary option codes. In practice, most vendors simply choose unallocated numbers rather than using the RFC 2153 procedure.
1	Maximum Receive Unit (MRU)	An extremely common option used to specify the maximum frame size the peer is willing to receive. Often used to tune performance.
2	Asynchronous control character map (ACCM)	As its name implies, the ACCM is not used on synchronous links.
3	Authentication protocol	Used to select the protocol that will authenticate peers to each other.
4	Quality protocol	Used to negotiate link quality monitoring to ensure adequate link performance.
5	Magic number	Used to guard against bringing up PPP on looped-back physical links.
7	Protocol field compression	Negotiated to shrink the protocol field from 2 bytes to 1 to save transmission bandwidth. Very useful on asynchronous lines.
8	Address and control field compression	Negotiated to shrink the address and control fields in the PPP frame.
11	Numbered Mode	Described in RFC 1663, Numbered Mode allows PPP implementations to maintain a window and ensure reliable delivery.
17	Multilink Maximum Reconstructed Receive Unit	Negotiation of this option sets up Multilink PPP (MP). Appendix B discusses MP in detail.
18	Multilink Short Sequence Number Header Format	Negotiated with MP; see Appendix B.
19	Multilink Endpoint Discriminator	Negotiated with MP; see Appendix B.
22	MP+	Proprietary multilink protocol designed by Ascend and specified in RFC 1934. For details, see Appendix B.

Maximum Receive Unit (code 1)

To indicate the maximum frame-data size an implementation is able to receive, it can transmit the MRU to its peer. The format of the MRU option is shown in Figure 9-19.

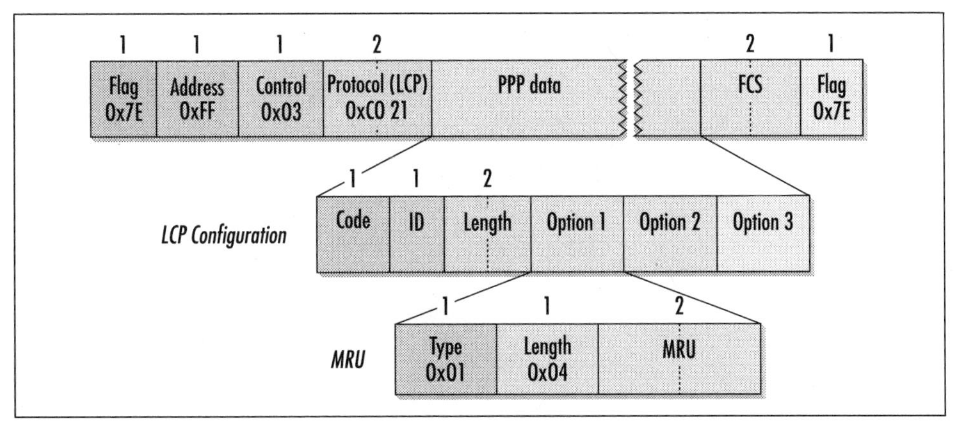

Figure 9-19. MRU LCP option

Headers and framing do not count toward the MRU. All implementations must be able to support an MRU of 1,500 bytes. Systems may request the use of larger MRUs for more efficient transmission of bulk data, but accepting a large MRU does not require its use. An implementation could acknowledge a requested MRU of 4,096 bytes but then only send 1,500 byte frames and still be within the letter of the specification. Smaller MRUs may make interactive traffic more responsive at the loss of bulk data transfer capacity.

Quality protocol (code 4)

PPP provides the capability of monitoring link quality. Quality monitoring is enabled with the Link Quality Monitoring (LQM) option in LCP, shown in Figure 9-20. It includes a quality protocol number and optional data.

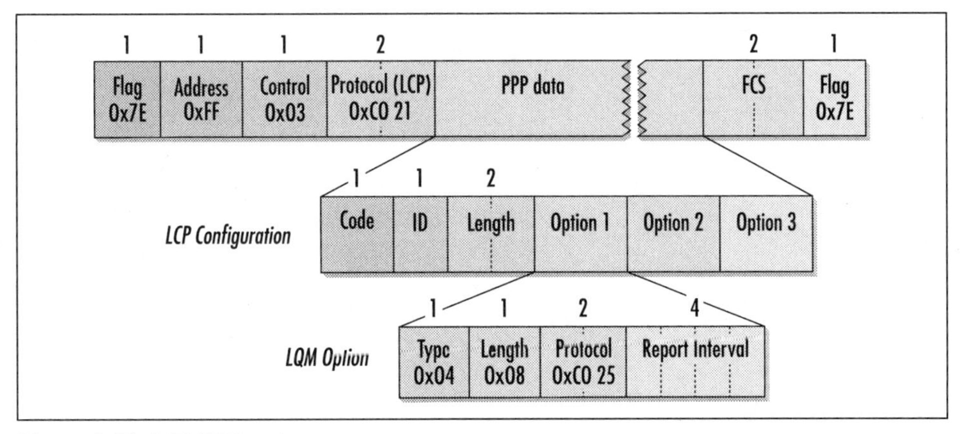

Figure 9-20. LCP LQM option

Only one quality-monitoring protocol is currently defined, and it is not widely implemented. RFC 1989 specifies PPP protocol 0xc025, the PPP LQM protocol.

Figure 9-21 shows the LCP option used to propose the use of this protocol, including the 4-byte report option interval. The specified number is the maximum time between link quality reports. Other quality protocols may be developed in the future, in which case the protocol identifier would be different and the data after the protocol identifier would be defined by the new protocol.

	Address	Control	Protocol	
	0xFF	0x03	0xC0	25
1	Magic Number			
2	Last Out LQRs			
3	Last Out Packets			
4	Last Out Octets			
5	Peer In LQRs			
6	Peer In Packets			
7	Peer In Discards			
8	Peer In Errors			
9	Peer In Octets			
10	Peer Out LQRs			
11	Peer Out Packets			
12	Peer Out Octets			
13	Save In LQRs			
14	Save In Packets			
15	Save In Discards			
16	Save In Errors			
17	Save In Octets			

Figure 9-21. Link quality report

A link quality report is a series of quantities, each of which is 4 bytes long. When quality reports are received, 5 bytes with locally generated statistics are appended. Quality reports allow each side of the link to monitor loss based on the received statistics and take action based on those reports. Any actions that may be taken based on the data are implementation-specific.

Magic number (code 5)

Magic numbers are a part of the initial RFC 1661 specification. The magic number option is illustrated in Figure 9-22. Magic numbers are used to detect looped-back links and data that is mistakenly sent back. Telcos may place loopbacks on T1 lines at any repeater that responds to the loopback commands in the T1 facilities data link. When the telco places these loopbacks, it is often up to customers to detect them and call the telco to take corrective action.

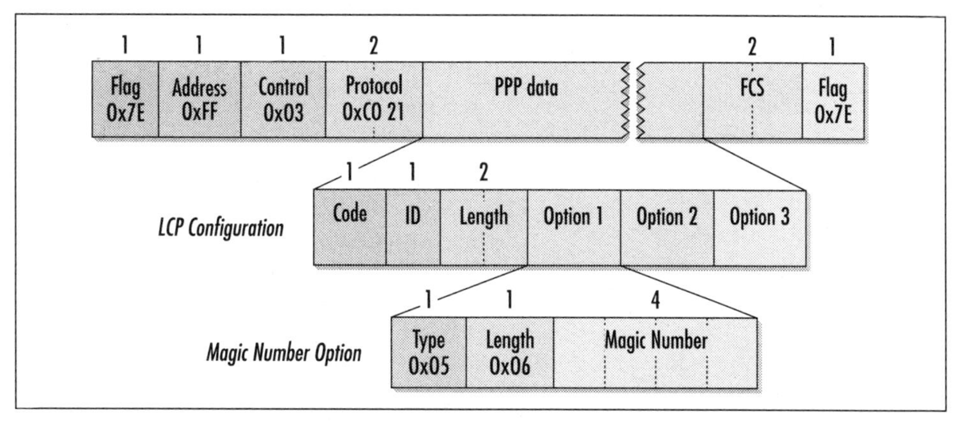

Figure 9-22. LCP magic number option

When negotiation procedures use magic numbers, the goal is a mismatch between the last transmitted number and the last received number. Magic numbers are chosen randomly on both sides of the connection, so any two peers are unlikely to choose the same number. Upon reception of a Configure-Request, the magic number is compared to the last transmitted magic number. If they do not match, it is proof that the line is not looped back.

Matching magic numbers, or magic number *collisions*, indicate that there is a possibility that a link is looped back. To ascertain whether this is the case, a Configure-Nak with a different magic number is transmitted. When the next Configure-Nak is received, the transmitted magic number is compared to the received magic number. Different numbers indicate a straight-through line and configuration proceeds. If a collision is detected, a looped-back line is more likely. Looped-back links are detected by looking for a pattern of send Configure-Request, receive Configure-Request with identical magic number, send Configure-Nak with new number, and receive Configure-Nak with new number. After several cycles, it is almost certain that the line is looped back. Magic numbers may also be used in conjunction with LCP Echo-Requests and Echo-Replies to detect lines that are looped back in mid-session. When systems do not negotiate magic numbers, they use zero instead. Some PPP implementations are unable to use magic numbers.

Numbered Mode (code 11)

RFC 1663 described Numbered Mode as well as the configuration option to enable it, shown in Figure 9-23.

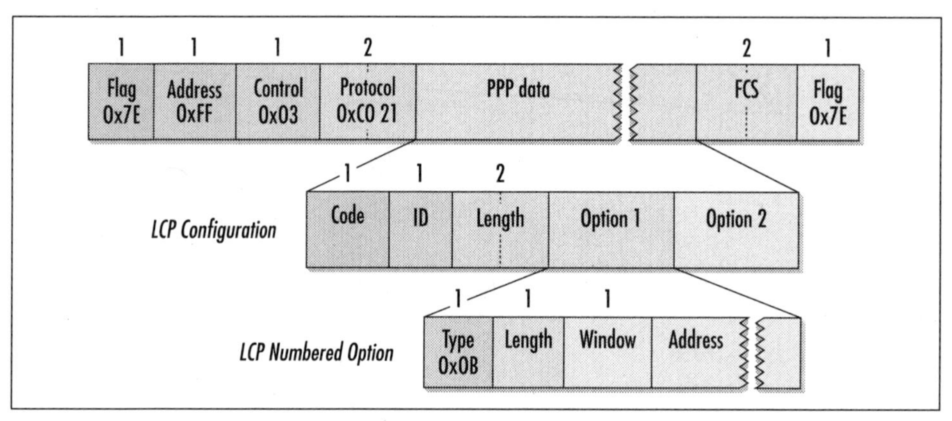

Figure 9-23. LCP Numbered Mode option

RFC 1662 framing specifies that PPP frames be sent using the unnumbered mode control information. Numbered Mode sends frames using Link Access Procedure, Balanced (LAPB). The window is used to select the transmission window size. LAPB is derived from HDLC, and the window is interpreted accordingly. When this option is used, the window size is typically set to either 8, for basic HDLC operation, or 128, for extended HDLC operation. When this option is negotiated, an open PPP link uses the HDLC address specified by the address field to send and receive data links, and the control field is reinterpreted according to LAPB specifications. Sending the appropriate addressing information allows the enabling of ISO-specified multilink procedures.

When Numbered Mode is activated, higher layers are protected from frame loss by a reliable data link layer. As frames are buffered and retransmitted, though, higher layers see a wide variance in delivery time and may adjust transmission rates accordingly. TCP is especially susceptible to variances in transmission time without packet loss. Numbered Mode should be used only when the cost of lost packets drastically exceeds the retransmission cost; PPP with the Compression Control Protocol is one such case. Asynchronous modems incorporate error detection and correction, which makes Numbered Mode redundant overhead. Digital transmission facilities have low enough bit-error rates that it is not usually necessary unless the underlying medium is a noisy radio link.

Multilink MRRU, SSN, and ED (codes 17, 18, and 19)

The MRRU option enables Multilink PPP (MP), specified in RFC 1990. The SSN and ED options modify the behavior of MP, but only the MRRU option can enable multilink operation. MP works by breaking packets into smaller fragments for parallel transmission over several links and reassembling the link layer fragments at the remote end. MP is described in more detail in Appendix B.

PPP Network Control and the IP Control Protocol

After LCP negotiation has completed, the LCP state engine sends an Up event to the NCP state engine. The Up event from LCP triggers network layer negotiations. PPP is commonly used on links primarily devoted to IP, in part because its implementation is required on all IPv4 routers with serial interfaces. RFC 1332 defines the IP Control Protocol (IPCP), which is the Network Control Protocol for IP and has the PPP protocol number 0x8021. Figure 9-24 illustrates IPCP.

Other Network Control Protocols have been specified for a wide variety of network layer protocols. A complete list is found in Appendix F.

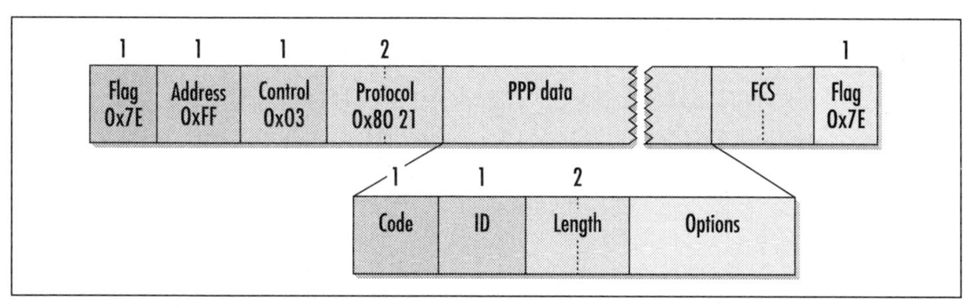

Figure 9-24. IPCP frame format

IPCP shares several common features with LCP, as can be seen in Figure 9-24. Frame formats are identical, but the code list is a subset of LCP's. IPCP uses only codes 1 (Configure-Request) through 7 (Code-Reject). IPCP will trade configuration requests and responses until a common set of options is established. Configuration options are, of course, completely different; they are listed in Table 9-8 and described in the following sections. Terminate-Request and Terminate-Ack are used to tear down the IP layer only. After terminating IPCP, it is possible to reopen IP with another IPCP configuration round. Other network protocols are completely unaffected.

Table 9-8. IPCP configuration options

Code	Type	Description
1	IP addresses	This is an inherited address assignment option from RFC 1172 PPP specifications that has been deprecated and should be used only by older implementations.

Table 9-8. IPCP configuration options (continued)

Code	Type	Description
2	IP Compression Protocol	IPCP may negotiate this option to perform Van Jacobson compression on the link. Negotiating this option enables PPP protocol numbers 0x002d and 0x002f.
3	IP address	This is a new style of address negotiation.
4	Mobile IPv4	This is used to establish mobile IP tunneling by informing the peer of the Mobile IPv4 home agent address.
129	Primary DNS	Microsoft specified this in an informational RFC to allow a peer to specify the primary DNS server.
130	Primary NBNS	Microsoft specified this in an informational RFC to allow systems to request a primary WINS server address from the peer.
131	Secondary DNS	Like option 129, but for the secondary DNS.
132	Secondary NBNS	Like option 130, but for the secondary NBNS.
144	IP Subnet Mask	This option is not a standard and has never been specified in an RFC (even an informational one). Several vendors, however, have implemented it.

RFC 1172-Style Address Assignment (IP Addresses; IPCP Option 1)

RFC 1172, the initial PPP specification, included an address-negotiation mechanism, shown in Figure 9-25, which allowed either peer to specify both its address and the address of its peer.

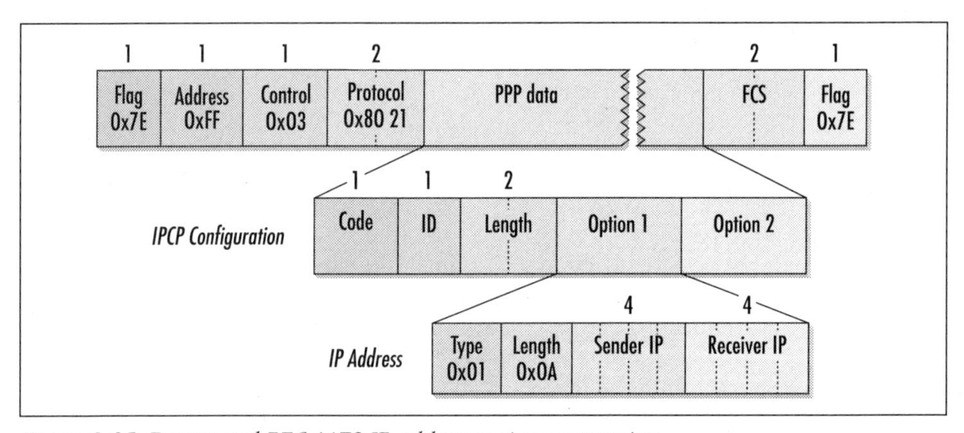

Figure 9-25. Deprecated RFC 1172 IP address assignment option

Use of this option sometimes resulted in poor convergence of IPCP negotiations, so it has been officially deprecated in favor of the RFC 1332–style address negotiation mechanism. This option may be used if a peer sends a Configure-Reject for the new option or sends a Configure-Nak using this option.

RFC 1332-Style Address Negotiation (IP Address; IPCP Option 3)

Modern IPCP implementations use option 3, shown in Figure 9-26, for address assignment. Systems inform peers of the local address with this option. Even though this option is cleaner to use in practice than the earlier IP address code, it is not perfect (see the sidebar, "IP Addressing in PPP"). To receive an address from the upstream service provider's device, this option can be set to zero in the hope that the remote device will send a Configure-Nak with an appropriate value.

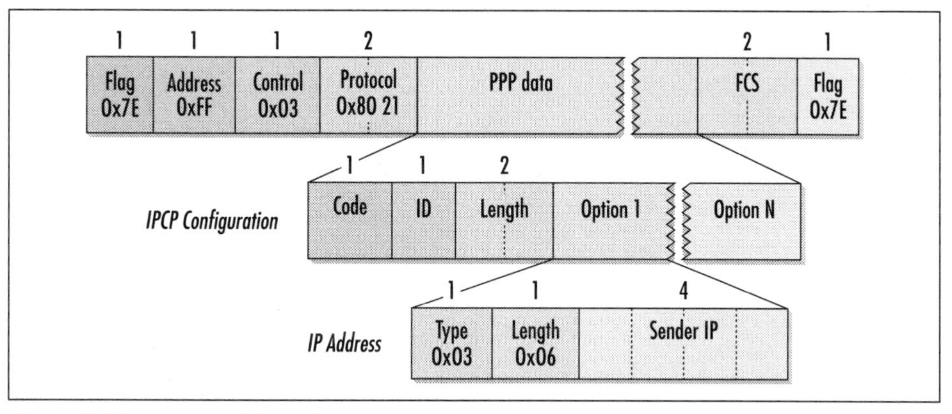

Figure 9-26. Current IPCP address-negotiation options

Microsoft DNS and WINS Server Assignment (Options 129-132)

Options 129 through 132, shown in Figure 9-27, can be used to assign DNS servers and WINS servers, also known as NetBIOS name servers. Two options are defined for primary and secondary DNS servers, with two additional options for primary and secondary WINS servers. These options were published in RFC 1877, which is informational only and not a formal standard.

Hosts use these options in an odd way. As part of the IPCP negotiation process, systems may send cached or preconfigured values for the name servers in hopes that the peer has better values to supply in a Configure-Nak.

Use of these options simplifies the IP stack configuration on Windows dial-up clients, but may pose a problem if an office of dial-up clients moves to a dedicated connection with a leased line. Small routers do not issue requests for these options, nor would it do any good for them to do so. Hosts that formerly used dial-up links must resort to DHCP requests forwarded across the PPP link to learn about name servers once a leased line is installed.

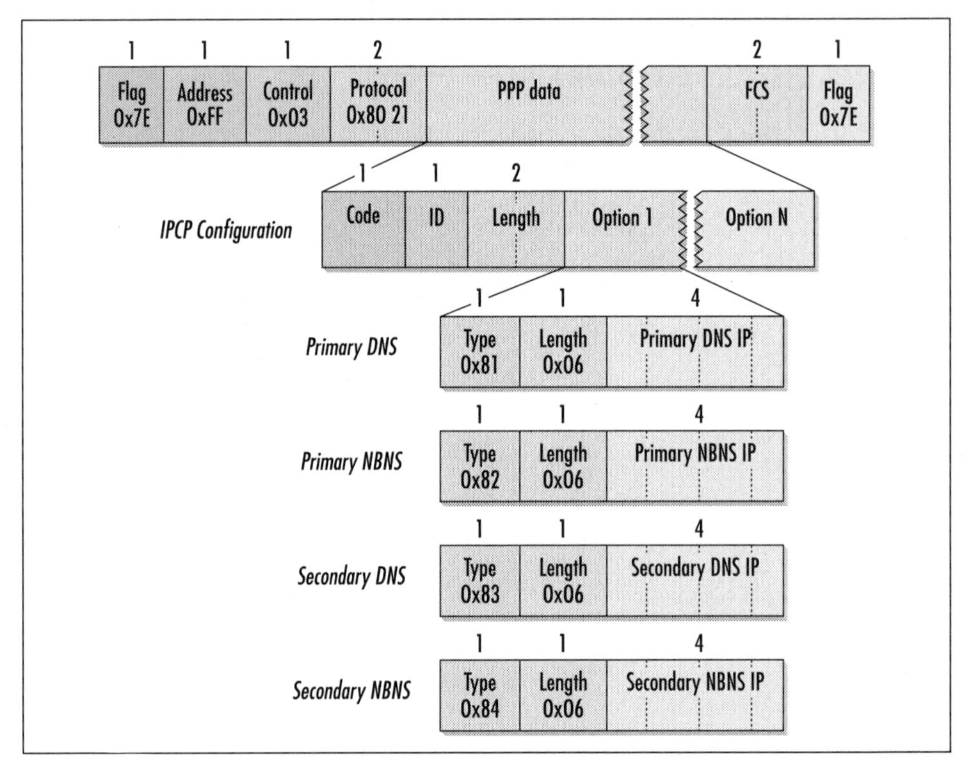

Figure 9-27. Microsoft networking assignment options

Nonstandard Option for Subnet-Mask Assignment (Option 144)

Several proposals have been made to allow IPCP to request allocation of a block of addresses by using configuration option 144. Although it has never been published as an RFC, some vendors have implemented the subnet-mask assignment option.

Configuring PPP

PPP presents a more complex set of configuration parameters to the network administrator than HDLC. Many of these parameters are better suited to use with low-speed asynchronous dial-up lines. Of all the PPP configuration options, you are most likely to need these:

Enable PPP encapsulation

Not all vendors default to PPP encapsulation on serial links, but several do not. PPP encapsulation must often be explicitly configured.

IP Addressing in PPP

Exchanging IP address information is generally a beneficial process. Several types of misconfigurations and crossed lines may be detected long before they are able to cause strange behavior on the link.

IP address messages are exchanged as part of the IPCP negotiation process, which must complete before IPCP can reach the Open state. This behavior may cause a problem when addressing information cannot be negotiated to the satisfaction of both ends.

One of two methods may be used to resolve this problem. One is for a peer to send a Protocol-Reject message to its offending peer. Unfortunately, Protocol-Reject disables the named protocol for the active duration of the link. To reattempt negotiation, the link must be disabled and reenabled. Alternatively, IPCP could enter the Open state and immediately transmit an IPCP Terminate-Request. IPCP must open the link because Terminate-Request packets have no effect until the link has been opened. Many implementations take the open-and-terminate-immediately route because it allows for an attempt at reconfiguration without reinitializing lower layers.

Although there is no distinction between peers at a protocol level, in practice one end of the link is often subordinate; an extreme example is a dial-up PC connecting to a dial-up server at an ISP POP. In these cases, the subordinate IPCP implementation sends a Configure-Request with an address of 0.0.0.0 or, less frequently, an unsolicited Configure-Nak with an address of 0.0.0.0. In both cases, such implementations must be corrected by the server at the other end.

Configure IP networking (addresses, encapsulation, and routing)

Depending on the vendor's implementation, IP encapsulation may be on by default or implied when an IP address is assigned to an interface. ISPs do not usually assign addresses on leased lines automatically. You must also enable routing on the interface. Routing often consists of a simple static default route to the other end of the link, though many WANs make use of dynamic routing protocols.

Magic number negotiation

Magic numbers assist in detecting looped-back links. Leased lines may be looped back at several locations along the path, and administrators do not have access to all of these locations. Loopbacks at the CSU/DSU can be corrected, but any others require phone calls. To make phone calls for looped-back links successful, arm yourself with a PPP log that shows identical magic numbers being transmitted and received.

Link Quality Monitoring (LQM)

This is typically enabled for leased lines. Users do not always have access to leased-line performance data such as the T1 PRM contents. Leased lines may also fail in strange ways or degrade over time. LQM is the best way to catch developing problems before they lead to failed lines.

Several options do not make sense for high-speed synchronous interfaces and probably will not be configured:

Compression options

PPP provides several compression options to save bandwidth on low-speed links. The protocol field may be compressed to a single byte by using Protocol Field Compression (PFC), and Address and Control Field Compression (ACFC) may be used to eliminate transmission of the redundant address and control field information. PPP also provides a framework for compressing the information field.

The trade-off is smaller frames at the cost of increased processing load on the PPP endpoints. High-bandwidth leased lines require prohibitive CPU power at the link endpoints to provide extensive compression.

Maximum Transmission Unit and Maximum Receive Unit

Each end of the link can specify an MRU to indicate the largest frame it is willing to process. Some PPP implementations may also allow specification of an MTU, which is tied into the remote MRU. If the remote end proposes an MRU larger than the MTU, some implementations will Nak the received MRU and suggest the configured MTU.

By default, the MRU is set to 1,500 bytes. On high-latency dial-up links, adjusting the MRU can yield additional performance by tuning the connection for small packets or large packets. Synchronous serial lines typically run at such high speeds that smaller MRUs do not boost performance noticeably for interactive application, so this option will likely be left at its default.

Asynchronous control character map

As the name implies, the ACCM is typically used only on asynchronous links. Synchronous links are designed to be transparent to user data, so they need no escaping.

Authentication (PAP, CHAP)

Leased point-to-point links do not typically use authentication because the ends of the link are known in advance and are fixed. Authentication is needed only when endpoints are not constant or are only intermittently connected.

Demand dialing

Synchronous serial lines have fixed endpoints and are always available, so the notion of dialing does not even apply.

10

Frame Relay

The great artist is the simplifier.
—Henri Frederique Amiel

Development of digital-transmission facilities in the 1960s and 1970s made low error rates a notable attribute of the telecommunications landscape in the 1980s and 1990s. Early packet-switching standards, most notably X.25, compensated for high error rates with extensive buffering and retransmission. Not all protocols that were run on X.25 links had reliability features, so providing reliability in the access protocol was a good thing. As the world moved toward protocols designed with the OSI model in mind, however, reliability and retransmission moved to higher layers of the stack. This resulted in duplicated effort and needless delays, especially on relatively clean digital links.

Frame relay was developed in response to the changing environment and, in many circles, was widely viewed as a direct replacement for X.25 networks. Frame relay specifies only the link layer interconnection of networks. Reliability is minimally addressed through the use of a frame check sequence that allows a frame relay network to discard corrupted frames; sequencing, data-representation errors, and flow control are delegated to higher-level protocols. (For the purposes of this book, most of the "higher-level" flow control related functions are delegated to TCP.)

To view frame relay strictly as a link layer takes too narrow a view; taken as a whole, frame relay provides the standards for building a packet-switched WAN Designing and implementing frame relay networks is far beyond the scope of this book, though, so this chapter looks only at the encapsulation and management features with which frame relay users must be familiar.

Frame Relay Network Overview

Before the development of wide-area packet-switching technologies, connecting geographically distant locations required a choice between several unappealing alternatives. By leasing dedicated capacity from the telephone network, organizations could build "private" networks out of leased lines.* Long-haul leased circuits were (and still are) extremely expensive, and beyond the means of all but the richest institutions. (And even they had to make sure the capacity of traffic between sites justified the expense.) Less well-heeled companies contending with financial reality opted for public data networks built on X.25 or for simple dial-up links, with the possibility of ISDN in the relatively modern telecommunications era. The last option was also often quite expensive, and the capacity was limited. No solution existed for an organization that wished to have fast access without a gigantic budget for network connectivity.

Frame relay filled this void.† A frame relay service gives its subscribers a *virtual* connection between two points because the network that moves the data from one point to the other is shared between all subscribers. Like other data packet–based or cell-based data-transmission methods, frame relay is based on the idea of statistical multiplexing: data is often transmitted in bursts, and it is unlikely that any single party will use the entire capacity for an extended period of time.

Frame relay connections come in two flavors. *Permanent virtual circuits* (PVCs) are "nailed up" and appear to be logically equivalent to a leased-line transmission facility, while *switched virtual circuits* (SVCs) allow subscribers to "dial" remote sites and maintain connections for only as long as necessary. This book focuses on PVCs because they are more widely used and involve far simpler signaling.

PVCs are priced in a manner less sensitive to distance than leased lines, which makes them popular for long-haul connections. The basic setup is shown in Figure 10-1. To access the frame relay network, a leased line is used to connect to a frame relay switch at the frame relay carrier's point of presence (POP). Data is then carried by the carrier's network, often represented (and also referred to) as a cloud, to a remote POP, where a leased access line carries the data to its destination. Customer equipment is referred to as the *frame relay DTE*, while the carrier equipment is the *frame relay DCE*. The frame relay DCE is the entry point to the frame relay network. Some network engineers may refer to the frame relay DCE as the frame relay "server."

* I use the term "private" in quotation marks because the transmission facilities were still owned by the telco and not protected against eavesdropping and tampering.

† Indeed, frame relay was initially billed as a "virtual private" network because it provided functionality that was identical to a private leased-line network. It is interesting to note that connecting geographically disparate offices is now often handled by IPSec-based VPNs over the Internet, which is also a wide-area packet-switching technology.

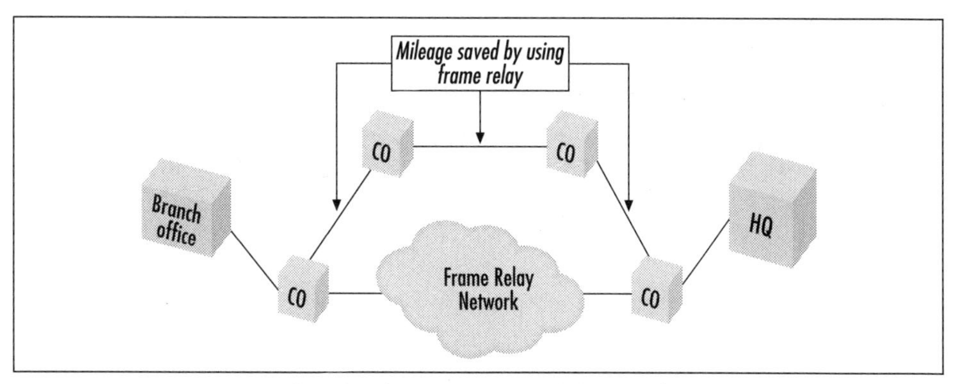

Figure 10-1. Less-expensive long-haul connections with frame relay

Furthermore, frame relay allows multiple virtual connections to use the same physical hardware. A single access link may carry PVCs to several remote locations. To enable the network to distinguish between frames bound for different locations, each outgoing frame is tagged with an address. Frame relay addresses are referred to as *data link connection identifiers* (DLCIs). Figure 10-2 shows this.

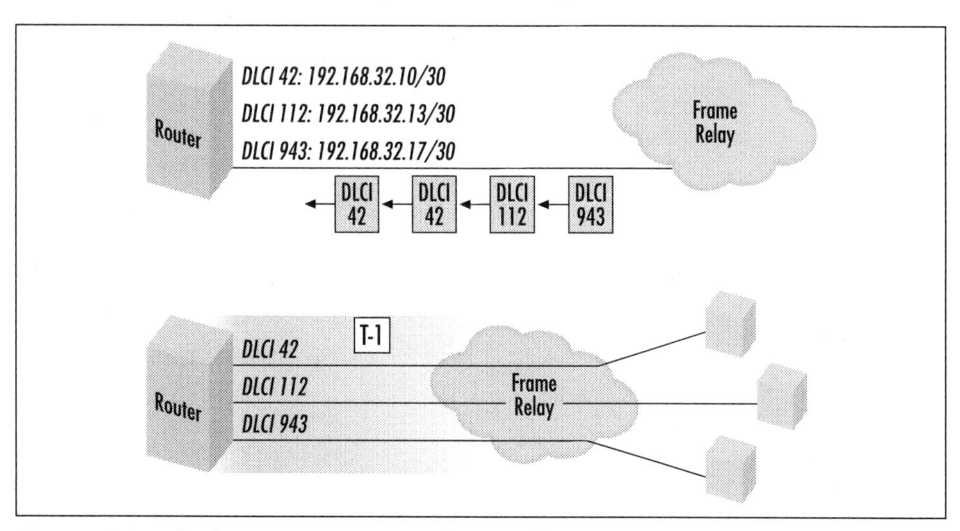

Figure 10-2. Multiple PVC termination and DLCI addressing

Generally speaking, DLCIs have only local significance. Carriers may reuse DLCI numbers throughout the network. DLCIs need not be the same on both ends of a PVC. DLCI values 0 and 1023 are reserved for signaling, as will be discussed later. DLCI values of 1 through 15 are also reserved. Depending on the carrier's implementation, DLCI values typically range from 16 to 991 or from 16 to 1007. If a carrier chooses to use the extended 4-byte header, enough DLCI values exist for each PVC in the frame network to have a unique DLCI. For obvious reasons, these are called *global DLCIs*.

When purchasing frame relay services, there are a variety of options for the *committed information rate* (CIR).* All traffic up to the CIR is given priority treatment, while traffic beyond the CIR is marked as "discard eligible." During quiet times, subscribers may burst above their CIR. When demand for bandwidth exceeds supply, discard eligible traffic will be discarded. If dropping all discard-eligible traffic does not balance demand and supply, traffic within the CIR may be discarded. Low CIRs are often used to sell off as much of the network's capacity as possible. One common tactic is to sell PVCs with a CIR of zero, which allows carriers to wring every last dollar out of the network capacity by selling off access during the quiet periods only.

The Frame Relay Link Layer

Frame relay is a set of link layer functions. Higher-layer protocols chop up user data, and the resulting bite-size pieces are wrapped in frame relay envelopes to be passed to the network for delivery. The link layer protocol associated with frame relay is the Link Access Procedure for Frame Mode Bearer Services—Core Functions, known more simply as the LAPF Core.† LAPF, which is specified in Q.922, is a simpler version of ISDN's LAPD. The basic frame relay frame is shown in Figure 10-3.

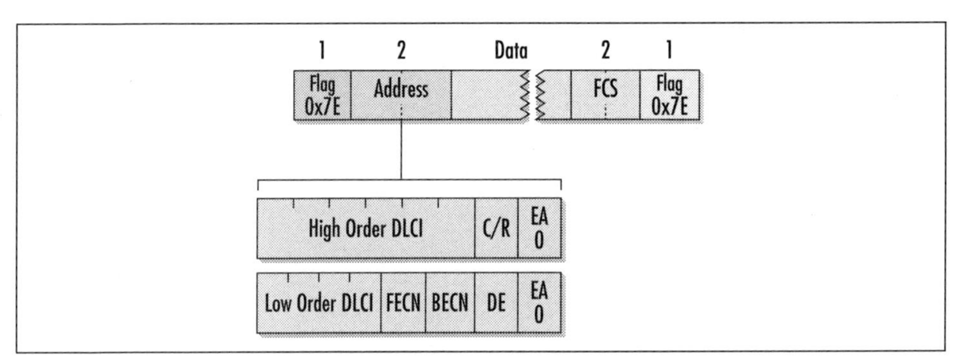

Figure 10-3. The basic frame relay frame

Like the HDLC frame, the frame relay frame opens and closes with a flag. Several header formats may be used, depending on the length of the DLCI. Figure 10-3 shows the most common header format, a 2-byte format that uses 10 bits for the DLCI.‡ At the end of each header byte, the *extended address* (EA) bit appears.

* CIR is not specified as a parameter of the link. Instead, a *burst size* and a *burst duration* are assigned to the port on the frame relay switch, and the relationship between the two gives the CIR. When users exceed the burst size in the burst time interval, the network begins to throttle back traffic.

† ITU documents refer to frame relay as *frame mode bearer services.*

‡ The current Frame Relay Forum UNI implementation agreement, FRF.1.2, allows the use of a 4-byte header, but this is not yet widely supported by vendors.

When it is zero, another header byte follows; a one signifies the end of the header. LAPF does not use the *command/response* (C/R) bit, though higher-layer protocols that require application-specific logic do use this bit.

Frame relay does not include any explicit internode congestion control functions, but it does allow the network to react to congestion in two ways. Traffic in excess of a subscriber's CIR is marked as discard eligible by setting the *discard eligible* (DE) bit to one. Frames with the DE bit set are the first to go in the presence of congestion. Frame relay also notifies the frame relay endpoints of congestion by using the *forward explicit congestion notification* (FECN) and *backward explicit congestion notification* (BECN) bits. FECN indicates that frames moving in the same direction as the received frame may experience congestion, while BECN informs the receiving station that congestion is occurring in the opposite direction of the received frame.

FECN and BECN bits are set by the network in response to congestion as an advisory to the endpoints. When a switch experiences congestion, it may set either the FECN or BECN bit. Figure 10-4 shows the use of the congestion bits for a simple path of three frame relay switches. If switch B experiences congestion, it will set the FECN bit on frames flowing along the indicated path from network X to network Y. Switch B will also set the BECN bit on frames heading toward network X to inform the router connecting network X to the frame relay network of congestion on the path from X to Y.

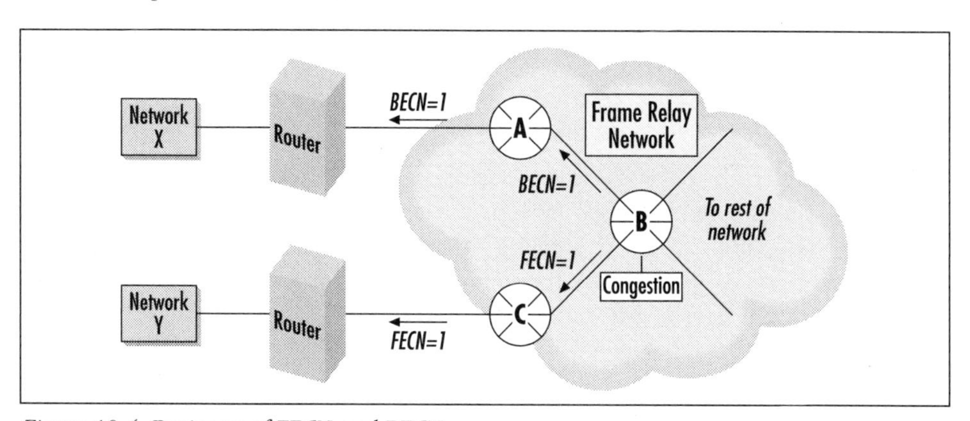

Figure 10-4. Basic use of FECN and BECN

Response to FECN and BECN is vendor-specific. Some vendors may implement flow control, but many do not. Furthermore, the declaration of congestion at a switch is also vendor-specific, and carriers may choose to set FECN and BECN based on a wide variety of criteria. ANSI T1.618 offers suggestions that are typically ignored in practice.

Multiprotocol Encapsulation with RFC 1490

When frame relay was first standardized, the information field did not have any sort of protocol tag, and the only way to transmit multiple protocols across a frame relay network was to use a separate PVC for each one. Service costs, however, are typically dominated by the number of PVCs, so subscribers asked network equipment vendors to develop a multiprotocol encapsulation for frame relay. RFC 1490, which has since been updated by RFC 2427, specifies the most common multiprotocol encapsulation. Even though the current version of the specification is RFC 2427, RFC 1490 was in force for so long that many continue to refer to "RFC 1490 encapsulation"; therefore, this book uses RFC 1490 as well.

When using RFC 1490 encapsulation, the LAPF frame's information field carries a second header that tags the higher-level protocol carried in the LAPF frame. Within the RFC 1490 header, a *Network Layer Protocol Identifier* (NLPID) is used to identify the higher-layer protocol carried inside the LAPF frame. NLPIDs are 1-byte values assigned in ISO/IEC TR 9577. Some examples are shown in Table 10-1.

Table 10-1. Common NLPIDs

Protocol	ISO/IEC TR 9577 NLPID (hexadecimal)
IP	CC
PPP	CF
Q.933	08
Subnetwork Access Protocol (SNAP)	80

RFC 1490 headers begin with 0x03 to indicate that unnumbered information is being sent. After the unnumbered information header, a single padding byte, set to zero, may be sent to assist in aligning other headers on 2-byte boundaries for more efficient processing. To prevent confusion between the padding byte and the NLPID, zero is not used as an NLPID.

Three main types of RFC 1490 encapsulation exist. Two are commonly used with LAN protocols, and the third is used for protocols that do not have an NLPID or Ethernet type code assigned. Because this book is intended to help network engineers connect data networks with T1, the third encapsulation type is omitted. Figure 10-5 illustrates RFC 1490 multiprotocol encapsulation.

Direct NLPID Encapsulation

For routed protocols that are assigned NLPIDs, only two additional bytes are necessary. Because only two bytes are necessary, direct NLPID is sometimes referred to as the *short format*. After the unnumbered information tag, the NLPID identifies

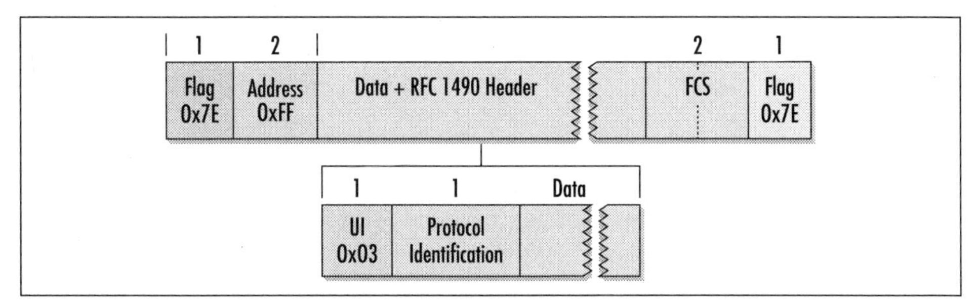

Figure 10-5. RFC 1490 multiprotocol encapsulation

the protocol. The packet follows. TR 9577 defines NLPIDs for OSI and ITU protocols; IP and PPP are the only non-ISO, non-ITU protocols that are given NLPIDs. IPv4 and PPP are assigned 0xCC and 0xCF, respectively. This encapsulation is illustrated for IP in Figure 10-6.

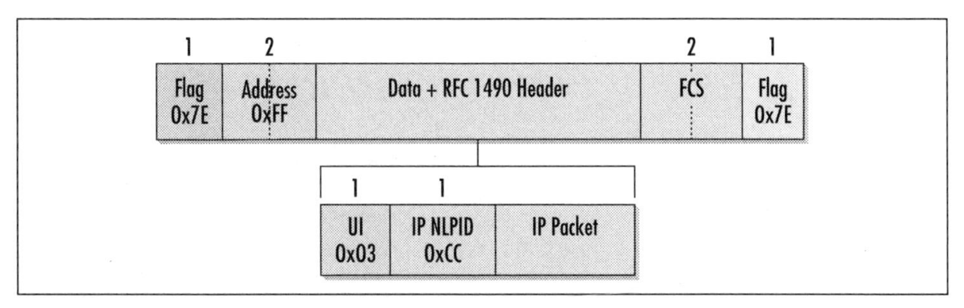

Figure 10-6. RFC 1490 using direct NLPID for IP

RFC 1490 defines direct NLPID encapsulation as the preferred method for transmitting IP across frame relay networks. SNAP encapsulation, described in the next section, could be used instead, but requires more overhead.

SNAP

For protocols without NLPIDs assigned, *Subnetwork Access Protocol* (SNAP) encapsulation must be used. SNAP encapsulation is identified by an NLPID of 0x80. Following the NLPID is a 3-byte *organizationally unique identifier* (OUI) and a 2-byte *protocol identifier* (PID). For protocols with an Ethertype, but without a direct NLPID, an OUI of 0x00-00-00 is used with the assigned Ethertype. SNAP encapsulation is illustrated in Figure 10-7.

IPX is an example of a protocol that needs to use SNAP. It does not have an NLPID assigned, so an OUI of 0x00-00-00 is used with IPX's Ethertype (0x8137).

According to RFC 1490, IP could be encapsulated in a SNAP header over a frame relay network by using the IP Ethertype. No functionality would be added, and the frame relay header would be 5 bytes bigger. RFC 1490 required that stations

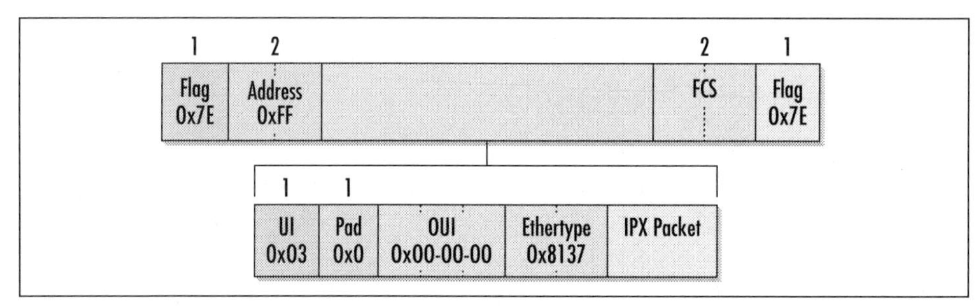

Figure 10-7. RFC 1490 using SNAP

accept and work with both the NLPID format and the SNAP format. RFC 2427, an update to RFC 1490, specifies that the NLPID format is preferred because it is shorter. Although it is no longer required to accept IP frames with the SNAP header, most equipment will continue to do so under the early principle of "be liberal in what you accept."

The Local Management Interface

Network management was conspicuously absent from the first round of frame relay standards. However, the lack of network-management features stunted the growth of frame relay, so the major equipment vendors devised a specification for the *Local Management Interface* (LMI), which eventually became known as the "Gang of Four" LMI.* LMI has three main tasks: link integrity verification, PVC status reporting, and PVC notification.

In many cases, LMI monitoring extends only to the local switch. It does *not* provide end-to-end status monitoring. If the frame relay network fails, LMI will not mark any PVCs as down because it is monitoring only between your router and the edge of the frame relay cloud. With all PVCs reported as active, dial-backup solutions will not kick in. To provide dial-backup, you must run a routing protocol over the PVC.

LMI Annex G provides end-to-end monitoring, but the catch is defining what is meant by the end of a network. frame relay networks often use ATM backbones for transporting frames across wide areas, and the end of a PVC may be either the other end of the virtual circuit or the interface between the frame relay network and the ATM backbone. To be safe, you need to run a routing protocol for each network layer.

* The Gang of Four was composed of Cisco, StrataCom, Northern Telecom, and Digital Equipment Corporation. Arguably, it's now the Gang of Two.

Complementary technology addressed LMI's limitations, and LMI became a great success. Standards organizations enhanced it, but caused a fork in the specification as part of the process. Three main LMI types exist:

Frame Relay Forum LMI
: The Frame Relay Forum initially adopted the original Gang of Four LMI, but this version is largely outdated and is not described in detail.

ANSI LMI or LMI Annex D
: ANSI specified a version of LMI in T1.617 Annex D, which was later revised slightly in T1.617a Annex D.

LMI Annex A (also occasionally ITU LMI or CCITT LMI)
: The ITU made some minor modifications to ANSI's LMI and published it in Q. 933 Annex A. As is common, the Frame Relay Forum has published FRF.1, an implementation agreement relating to the LMI Annex A.

All three flavors of LMI are based on polling mechanisms. Frame relay DTEs may request status updates from the network. Status updates may be a simple keepalive or link integrity verification check to make sure the access line is up, a request for the status of one PVC, or a full status check on all the PVCs configured on a link.

LMI Operations and Timers

Link-integrity checks have sequence numbers and are used much like the SLARP requests in Cisco HDLC. When too many link-integrity checks go unanswered or the window between the transmitted sequence number and the received sequence number gets too large, the access line is declared to be nonfunctional. Link-integrity verification ensures that the link is up before attempting to send data down the PVC. Without link-integrity checks, it is possible to have the *black hole effect*, which occurs when users send data down PVCs and do not receive responses because there is a problem at the frame relay switch.

LMI operations can be modified through five numerical settings. Two timers and three counters can modify the timing of LMI operations. Status-enquiry messages form the basis of LMI operations, and the time between status-enquiry messages is controlled by the Link Integrity Verification Polling Timer. Because that name is a mouthful, whether spelled out or as an acronym, it goes by the simpler name of T391. By default, link-integrity verification occurs every 10 seconds. In addition to controlling the transmission of polling messages, the T391 timer also limits the time of an in-flight status-enquiry request. The timer is started when the status enquiry message is transmitted, and it is stopped only when a status update has been received. If the timer expires, an error is recorded. Significant error counts in short periods of time cause LMI to declare the link dead.

Link integrity verification polling occurs from the user side only. A frame relay DCE watches over the link with its own timer, the Polling Verification Timer, which is assigned the designation T392. Frame relay network equipment must receive status-enquiry messages within the timer period or an error will be recorded.

Most status enquiries are limited to simple keepalive functions. To prevent situations in which the access line and all but one of the PVCs are functional, the Status Polling Counter, also known as the N391 counter, ensures that periodic full status reports are requested by the frame relay DTE. By default N391 is set to 6, which means that the default operation is to request a full status report at every sixth status enquiry. It is important for frame relay devices to check that a PVC is functional before attempting to send data because the default behavior, with T391 set to 10 seconds and N391 set to 6, is to request a full status report every 60 seconds. To accommodate this need, LMI messages may also check the status of a single PVC.

When frame relay devices detect errors through timer expiration, an error counter is incremented. Lone errors do not cause frame relay links to be torn down. Instead, two counters are used to require a significant error density before declaring PVCs down. A Monitored Events Count, N393, sets up sliding error consideration windows. The number of errors within a window of consecutive events must exceed a threshold before LMI marks the link down. Error thresholds may be configured by setting the Error Threshold Counter, N392. By default, the monitored events count is 4 and the error threshold is 3, which implies that three out of four consecutive events must be in error before LMI declares the PVC inactive.

A summary of the timers and counters is given in Table 10-2. Defaults are given from the specifications, but some vendors may have different defaults. Refer to your vendor's documentation for details.

Table 10-2. LMI timers and counters

Name	Timer (T)/ counter (N)	Range	Default	Explanation
Link Integrity Verification Polling Timer	T391	5–30	10	Frame relay DTE status-enquiry interval, in seconds. An error is recorded if no reply is received before the timer expires.
Polling Verification Timer	T392	5–30	15	Frame relay DCE interval for receiving status-enquiry messages, in seconds. Errors are recorded when no messages are received before the expiration of this timer. This timer must be larger than T391, its DTE counterpart.
Status Polling Counter	N391	1–255	6	This counter applies only to the frame relay DTE. For every N391 status enquiry request, a full status report will be requested.

Table 10-2. LMI timers and counters (continued)

Name	Timer (T)/ counter (N)	Range	Default	Explanation
Error Threshold Counter	N392	1–10	3	Number of errors that must be recorded in the monitoring period before LMI declares a link dead.
Monitored Events Counter	N393	1–10	4	Monitoring interval for error counts. Windows of N393 events will be considered for error densities that are high enough to mark the link as dead. N393 must be greater than N392.

LMI Frame Formats

Each of the LMI types has a slightly different frame format. For completeness, this section details them all. In practice, the Gang of Four specification is not likely to be used. In the U.S., the ANSI specification is common, while most of the rest of the world uses the ITU specification.

LMI frames follow the general format of Figure 10-8. After the normal frame relay header, an extended LMI header provides additional information specific to LMI. An address field specifies DLCI 0 for the ANSI and LMI specifications. A protocol discriminator and call reference are included because frame relay was designed so that it could be provisioned over ISDN circuits and LMI signaling could coexist with ISDN signaling. To report data, LMI messages contain a series of *information elements* (IEs). Each LMI operation has a corresponding IE.

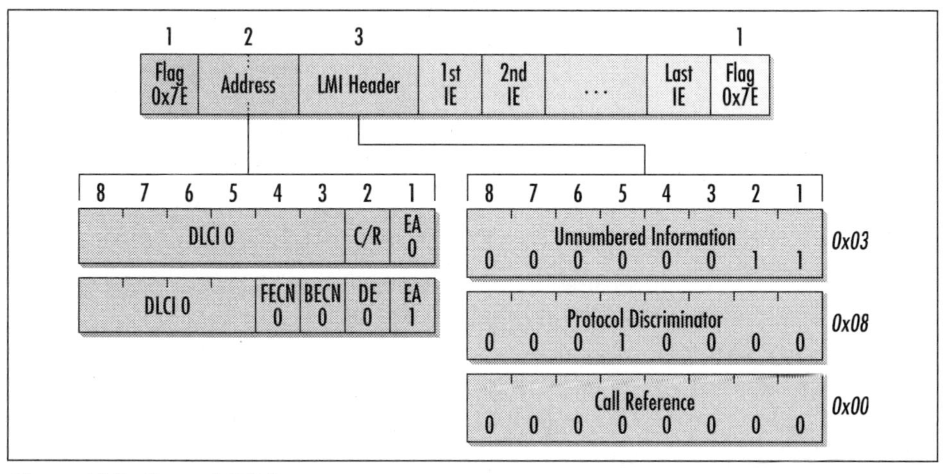

Figure 10-8. General LMI frame

Information elements

Except for some minor differences, information elements are the same between the ANSI and ITU LMI specifications. The main difference between the two is the use of different identification numbers for different report types. This section does not specify the identification numbers, but they are specified in the ANSI-specific and ITU-specific sections.

Three information elements are of interest. The report type IE, shown in Figure 10-9, informs the receiver of what type of LMI message is contained in the frame through the report type identifier in the third byte. It is set to either 0x00 for a full status report, 0x01 for basic link-integrity verification, or 0x02 for an asynchronous check on a single PVC. Like many other information elements, the report type IE contains a length field. All IEs have a 2-byte IE header. The first byte contains a 7-bit identifier, which is a code field to determine the IE type; the eighth bit is presently reserved for future use. After the IE type code, there is a 1-byte length identifier. However, this length identifier counts only the body of the IE, not the header.

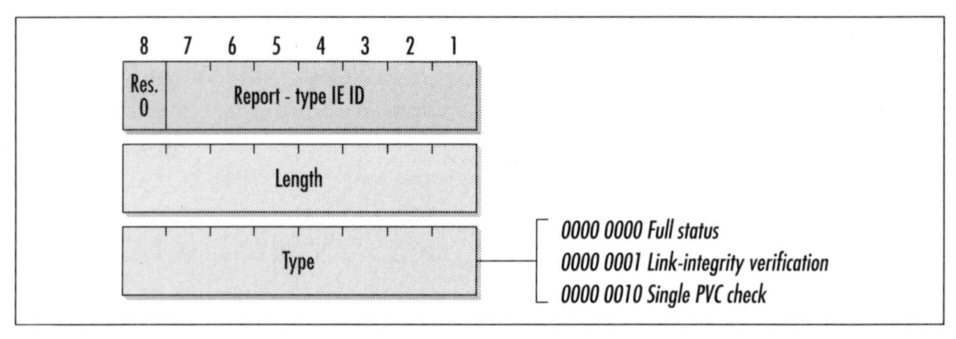

Figure 10-9. Report type information element

Link-integrity verification messages have two sequence numbers, much like the SLARP messages in Cisco HDLC. The link-integrity verification information element is shown in Figure 10-10. After transmitting a link-integrity verification message, it is possible to note that it was received and processed by the remote end as it updates sequence numbers transmitted in the reverse direction.

The PVC status IE, which is used in the full status reports, describes the status of a PVC. It includes a length field to account for the different address lengths present on frame relay networks. Most networks do not use global addressing and have only 10-bit DLCIs, so the IE illustrated in Figure 10-11 has a length of 5 bytes. Using the address extension bit in the same way it is used in the frame relay header allows the report to be extended to DLCIs with much longer addresses.

Shaded bits are spares that are presently set to zero and reserved. In the last byte, two bits, labeled new (N) and active (A), are used to report status. If the PVC is

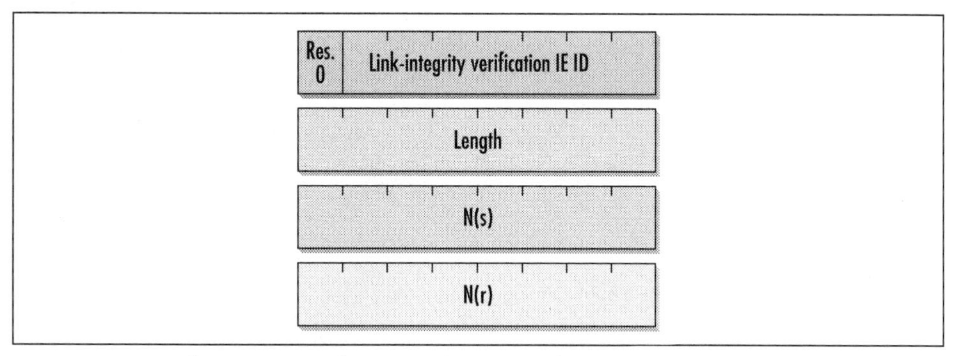

Figure 10-10. Link integrity verification information element

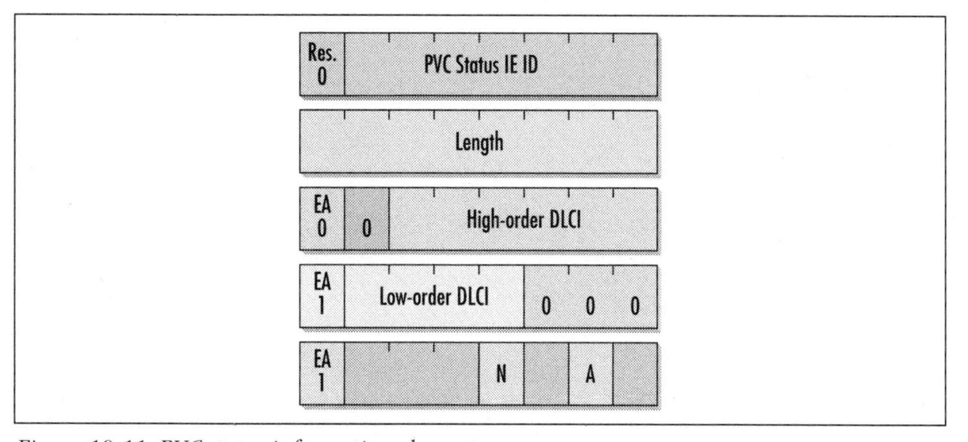

Figure 10-11. PVC status information element

active, the active bit will be set to one. Inactive PVCs are reported with a zero. If the PVC is new since the last report, the new bit will be set to one.

Maximum Number of PVCs on a Frame Relay Line

The maximum number of PVCs that can be configured on an interface is related to the size of the PVC status IE. Full status messages are used by the frame relay DCE to report on the status of all the PVCs configured at a particular frame relay switch port. Status report elements for all the PVCs must fit into the information field of the status reply message.

Two-byte DLCs are associated with 5-byte status elements. If the information field is limited to the 1,500 bytes, a maximum of 300 status report IEs can be fit into the information field. Most providers support a maximum frame size of 4,096 bytes.

LMI Annex D (ANSI T1.617a Annex D)

ANSI's version of the LMI runs on DLCI 0. Flow control bits are always set to zero because LMI messages are transmitted only over the access link, so data frames are meant to carry the load of congestion notification. Figure 10-12 illustrates the basic ANSI LMI frame, along with the three main information elements.

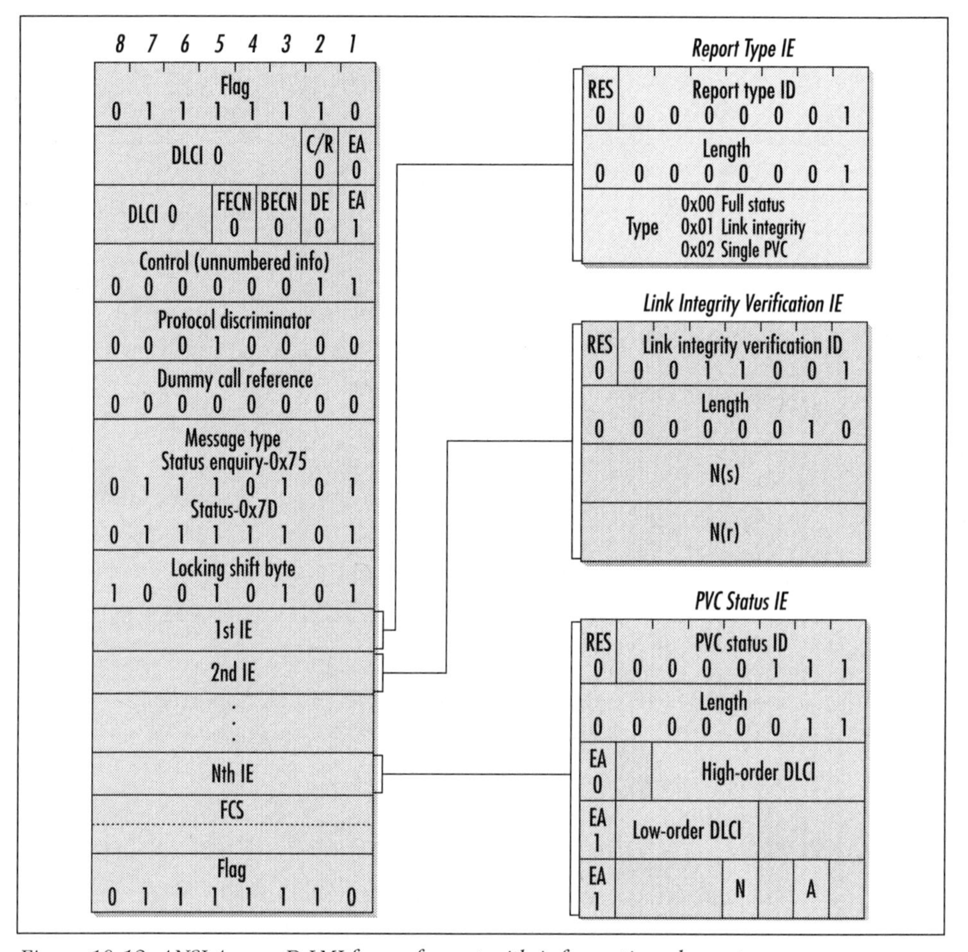

Figure 10-12. ANSI Annex D LMI frame format with information elements

ANSI LMI is transmitted using Codeset 5, which is also known as International Alphabet 5. As part of the use of Codeset 5, ANSI LMI messages include a locking shift byte after the message type, before the body of the information elements. Figure 10-12 also shows the information elements with the identification numbers used in the ANSI LMI specification.

LMI Annex A (ITU-T Q.933 Annex A)

Like the ANSI LMI messages, Q.933 LMI messages are transmitted on DLCI 0. Q. 933 uses Codeset 0, which results in the absence of a locking shift byte. Figure 10-13 shows the equivalent of Figure 10-12 in ITU specifications.

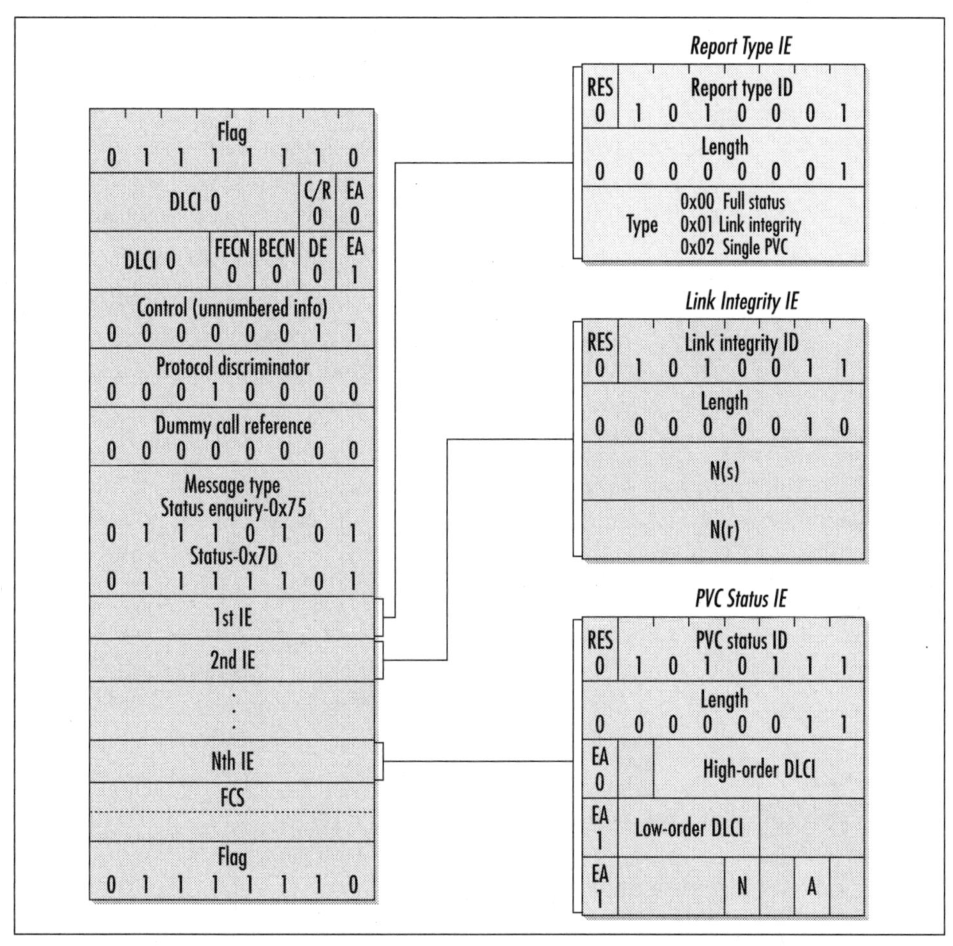

Figure 10-13. ITU Q.933 Annex A LMI frame format with information elements

Differences between the ANSI and ITU LMI specifications are described in Table 10-3.

Table 10-3. Differences between ANSI and ITU LMI specifications

ANSI T1.617a Annex D	ITU Q.933 Annex A
Uses locking shift byte (0x95) between header and information elements	No locking shift byte
Uses 0x01 for report type IE	Uses 0x51 for report type IE

Table 10-3. Differences between ANSI and ITU LMI specifications (continued)

ANSI T1.617a Annex D	ITU Q.933 Annex A
Uses 0x03 for link integrity verification IE	Uses 0x53 for link integrity verification IE
Uses 0x07 for PVC status reporting	Uses 0x57 for PVC status reporting

Configuring Frame Relay

After ordering a frame relay connection, the carrier provides configuration information for the connection. Part of this information is the parameters for the T1 access circuit (encoding, T1 framing type, and so on). The rest of the information is specific to the frame relay link. Here are some things you'll need to know:

LMI type

Frame relay carriers use LMI to monitor the access link and any PVCs associated with it. If there is an LMI type mismatch, the frame relay switch will not activate PVCs associated with the access link and no traffic will pass. When the LMI type is mismatched, the physical line will be up, but the logical link layer will be down.

Some equipment can perform LMI-type autodetection. Some automatic LMI selection procedures require that the LMI DLCI to be known in advance. It is typically DLCI 0, because the Gang of Four LMI is not widely used any more.

Encapsulation

Unless there is an agreement to pass only a certain network protocol, multiprotocol encapsulation must be configured. With rare exceptions, the scheme is RFC 1490. Some vendors may require explicit selection of RFC 1490 encapsulation, while for others, it may be the only mode of operation.

IP addressing

PVCs used for Internet access or private IP networks will have IP addresses. In the former case, the ISP provides IP address information; in the latter, consult the internal address administrators. Because frame relay PVCs are often used as cheaper leased-line replacements, addresses are typically assigned using a 30-bit network mask for a total of two usable addresses.

Local DLCI and address mapping

Each PVC will have a local DLCI, assigned by the frame relay carrier. Depending on the router vendor's implementation and the carrier's network-deployment choices, either you will need to manually inform the access router of PVCs on the link or the carrier will provide that information automatically via LMI.

Likewise, address mapping may be done manually or through automated procedures such as *Inverse ARP* (InARP). InARP, specified in RFC 2390, allows a frame relay endpoint to send a request for IP address information for a specific DLCI.* Not all routers and carriers support InARP.

DLCIs have local significance only, so the remote DLCI is not part of the configuration.

Other router configuration

Frame relay does not make routing decisions—it is purely a layer 2 specification. Once a frame enters a PVC, it heads to the other end. Selecting which PVC to use is left to higher-layer protocols. Depending on the local configuration, extensive routing configuration may be required, or it may be as simple as a single static route to direct traffic toward the remote end of an Internet access PVC.

* InARP is encapsulated with the RFC 1490 SNAP method. However, it defines new request and response codes to avoid interference with other ARP extensions.

11

T1 Troubleshooting

Anything that can go wrong, will.
—Murphy's Law*

As with much else in networking, T1 troubleshooting is often an exercise in patience. T1 is only a physical layer specification with a basic framing protocol used to ensure synchronized timing between the ends. Working with T1s requires patience because of the frequent need to consult with others, whether telco technicians, ISP engineers, or equipment-vendor support personnel. Because T1's popularity exploded quickly, many people are not well equipped to deal with the layers of complexity and multiple organizations needed to bring problems to resolution.

Basic Troubleshooting Tools and Techniques

Before involving others, it helps to pinpoint the problem as accurately as possible. To do this, trace the signal from the origin to its intended destination and make a note of where it fails to propagate. Your main tools are the CSU/DSU's loopback features, though you may wish to involve equipment vendors and service providers.

Loopback Testing

Loopback testing is used to verify that a span is good by "looping" received data back onto the transmit path. Two main types of loopback exist; both are illustrated in Figure 11-1.

* *The New Hacker's Dictionary* (Eric Raymond, Ed., MIT Press, 1996) states that this is actually an *incorrect* formulation of Murphy's Law; rather, it is properly called Finagle's Law.

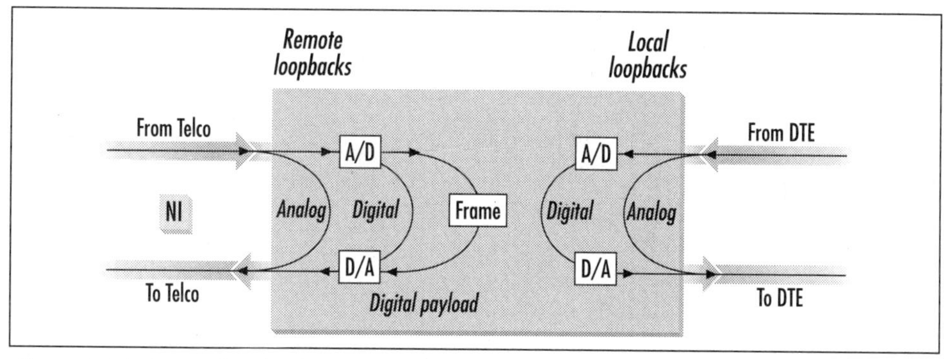

Figure 11-1. Remote loopback and local loopback

Remote loopback sends the data received on the telco network interface back out toward the telco to verify the line between two locations. If data is transmitted across the line, sent back across the line due to a loopback setting, and received intact at the sender, then the main span is good. *Local loopback* verifies the CSU/DSU-to-router connection in exactly the same way.

Some CSU/DSUs offer the option of *digital loopback* or *analog loopback*. Analog loopback simply redirects the incoming signal back out to the origin. Digital loopback, which is more common and more useful, regenerates bits and reconditions the signal before retransmitting it. When testing, make sure that tests are run with digital loopback in place.

Some CSU/DSUs offer loopback at different points of the T1 network interface. A line loopback simply regenerates the received digital signal and returns it to the origin unaltered, including any errors in framing or line code. Data or payload loopbacks extract the digital data and reframe it before retransmission. Refer to your CSU/DSU documentation for details on the types of loopback available.

Loopback can be triggered by equipment without human intervention. Telco employees at network management stations can activate and deactivate loopback capabilities along a line's path to verify its operation on a repeater-by-repeater basis. Circuit equipment may also trigger loopbacks in response to network conditions.

Remote loopback can also be triggered by a smart jack. Many jacks are constructed in such a way that when the plug is removed from the jack, shorting bars connect the input side of the telco NI to the output side electrically. When the plug is removed, the end span becomes a path from the end span repeater to the customer location, then back to the repeater. Using the built-in loopback capability of a jack lets you determine whether the problem lies in the telco's or the customer's wiring.

Running loopback tests

Loopback testing is described in V.54. It allows troubleshooters to determine whether the problem is in the local CSU/DSU-to-router interface, the T1 span itself, or the remote CSU/DSU-to-router interface, as illustrated in Figure 11-2.

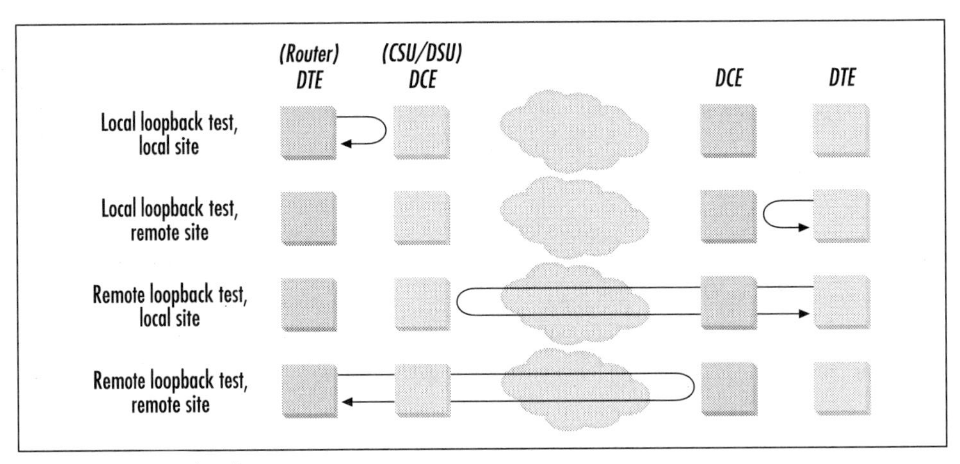

Figure 11-2. Loopback testing

Local loopback tests can isolate the problem in the cable or the timing in the serial circuits between the router and CSU/DSU. Local loopback tests, which are the first two tests in Figure 11-2, need to be run at both the local and remote ends. In a local loopback test, any data sent by the DTE should be looped back from the CSU/DSU to the DTE. When local loopback tests fail, contact the DTE vendor and the CSU/DSU vendor after checking the configuration on the timing of the router-to-CSU/DSU serial link.

Remote loopback tests isolate problems in the carrier's network. In a remote loopback test, the CSU/DSU is configured to take any data received from the network and transmit it back through the network to its source. If all goes correctly, data flows from one router through that router's CSU/DSU and through the carrier's network. When it reaches the remote end, the remote CSU/DSU in loopback mode transmits that data back through the carrier's network, testing the circuit in both directions.

To use loopback testing effectively for problem isolation, follow the three steps in Figure 11-3, which are as follows:

1. Run a local loopback test at the side of the T1 having problems and make sure that any bits the local DTE transmits are received on the DTE's serial port. Depending on the equipment, this may not be a straightforward process. One quick test is to run HDLC or PPP and use the protocol to determine if the CSU/DSU is taking the data received on the serial port and retransmitting it out the serial port. If this test fails, contact the DTE vendor and the CSU/DSU vendor.

2. Remove the local loopback setting on the local CSU/DSU. Call the remote end and have them put the remote CSU/DSU into loopback mode. Alternatively, the remote end may call you and ask you to put your CSU/DSU into loopback mode. The two tests are equivalent because they test the same four wires. The sender should receive anything that is sent out. If this test fails, call the ISP or telco.
3. Run a local loopback test on the remote side of the T1 to verify its DTE-to-CSU/DSU connection.

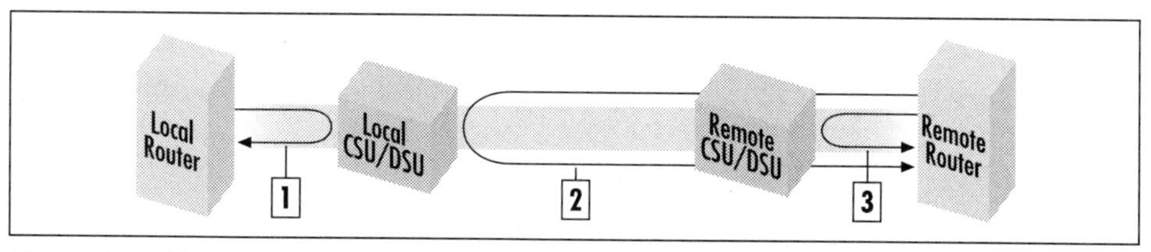

Figure 11-3. Testing strategy

Calling for Help

Sooner or later, you will likely need to call for help on a T1 problem. In this section, you will find some guidelines to make the calls as quick and painless as possible. For the most part, these guidelines are simply common sense. When you report trouble, you will probably be asked for configuration information. If you can gather it before calling for help, you won't need to hang up or put the help desk on hold while you track it down.

Calling the telco

Dealing with telcos can be a trying experience. Like a good number of technical people in a customer-facing role, the people who answer leased-line trouble reports tend to be overworked and totally unappreciated by management. Some basic steps can make the experience more bearable. At the very least, check the obvious places for trouble (is the line plugged in to the CSU/DSU? is loopback deactivated?).

It may help to call the ISP instead of the telco, for two reasons. First, working with the ISP on loopback testing can point to a clear diagnosis for the telco. Second, and perhaps more importantly, if the ISP bundles the line charge with your monthly bill, the ISP is the telco's customer and might have more clout to get problems resolved. Realistically, your ISP will have more pull with the telco than you will, especially if you use one of the larger ISPs.

Before calling your ISP

In most cases, after doing the basic diagnostic checks outlined in the sidebar "Basic Troubleshooting: The Rules of Fixing Things," the first call you should make is to

Basic Troubleshooting: The Rules of Fixing Things

Because of my computer skills, I have often been consulted by friends when they experience computer difficulties. At one point, I was turning on quite a few pieces of equipment (modems do not work without power, monitors do not display screens without power, and so forth), and as a result, I coined the First Rule of Fixing Things: *is it turned on?*

However, that quickly turned out to be inadequate. I was subsequently consulted by a friend who had repeatedly tried to power on whatever piece of equipment with which he was having trouble. He had tried repeatedly to ensure that I could not simply look at him and say "First Rule." I looked over the setup, held up the unattached power cord, and we both simultaneously said "Zeroth Rule?" (We were both taking a thermodynamics course at the time.)

Before pestering any support people, therefore, an excellent basic set of things to check is given by my rules of fixing things:

> Zeroth Rule: is it plugged in?
> First Rule: is it turned on?

your ISP. With luck, you have retained IP addresses for the remote end of the T1 as well as key hosts at the ISP's backbone (web, FTP, and DNS servers). If you are unable to pass traffic through the T1 from your router, check the router-to-CSU/DSU connection with a loopback test and then call the ISP. If the link layer is refusing to negotiate, obtain a trace and decode it as far as you can. When you call the ISP, be prepared to explain your trace to them. Have the following information on hand:

- Account number.
- Circuit ID, if applicable. Some ISPs will give you separate circuit IDs for each access link you order. Having the circuit ID handy will help the ISP's staff locate the configuration information for the circuit in question.
- Decoded link layer trace and, if possible, a theory about the cause of the problem.

Before calling your equipment vendor

Equipment vendors may be able to help with T1 difficulties. Before placing a call to a vendor's support line, though, it pays to collect the information they will undoubtedly ask for. If you don't have at least a large subset of the following points, most vendors will ask you to gather the information and call back. Incidentally, most of this information is what I recommended saving from circuit turn-up in Chapter 7:

- Line framing.
- Line code.

- Line speed (channel speed and number of DS0s).
- Line build out.
- Transmit clocking and DTE clocking settings, along with the make and model of the DTE.
- Link layer information, such as frame relay DLCI or ATM virtual path and virtual circuit IDs.
- IP addressing and routing information from the ISP.
- If the equipment supports it, grab a packet trace of the attempted line negotiation. Obtaining this data is platform-specific. On Cisco IOS, run a *debug* trace for the link layer protocol on the interface giving you trouble. Many other platforms have some Unix ancestry and are equipped with *tcpdump*; the best of those platforms include link layer protocol decoding capabilities not present in stock *tcpdump*.
- A network map showing relevant machines and their IP addresses. Maps do not need to be fancy Visio diagrams to be helpful. Simple hand-drawn sketches faxed to the vendor's support staff can be just as good, but much quicker to draw.

Top T1 Trouble Spots

Cabling is often a blind spot for a data network administrator new to T1 equipment. Specifications require the use of shielded, grounded cabling to limit interference. Purchasing manufactured cables ensures that the cable meets the exacting performance requirements for T1 wiring. Extend the demarc only if you must, and hire an experienced installer.

Of the configuration issues that tend to afflict network administrators, line build out is the easiest to find. If LBO is set incorrectly, it will often appear that the local end cannot send any data at all. When a line is installed, the carrier should supply the LBO value, but installation personnel may not always do so. New T1 installations commonly use smart jacks, so a 0 dB signal is appropriate.

Many of the complex trouble reports tend to involve clocking as a solution. In most cases, the transmit clocking on the network interface is derived from the received clock, but not all CSU/DSU vendors make this the default option. Data port clocking can be more complicated because both the CSU/DSU and the DTE on the serial link are involved. Short serial links usually allow the CSU/DSU to supply the DTE clock on the data port. Long serial cables or routers with external latency may require correction with clock inversion or by having the router supply an external clock with the transmitted data.

Depending on the vendor, the support people may be available for conference calls with your ISP and the telco. To take advantage of conference calls, be sure you have circuit IDs and contact information handy for both the telco and the ISP. An added benefit of involving vendors early in the process is that their support staff is familiar with telco jargon and the methods telco personnel use when working through a trouble report. Equipment vendors often take pride in product quality and are more than willing to assist in diagnosing trouble with their products.

Troubleshooting Outline

The master troubleshooting flow is shown in Figure 11-4. First, make sure that the appropriate cables are plugged in. (I once called in a trouble ticket when the CSU/DSU had not been plugged into the demarc.) Also ensure that the line is administratively enabled.

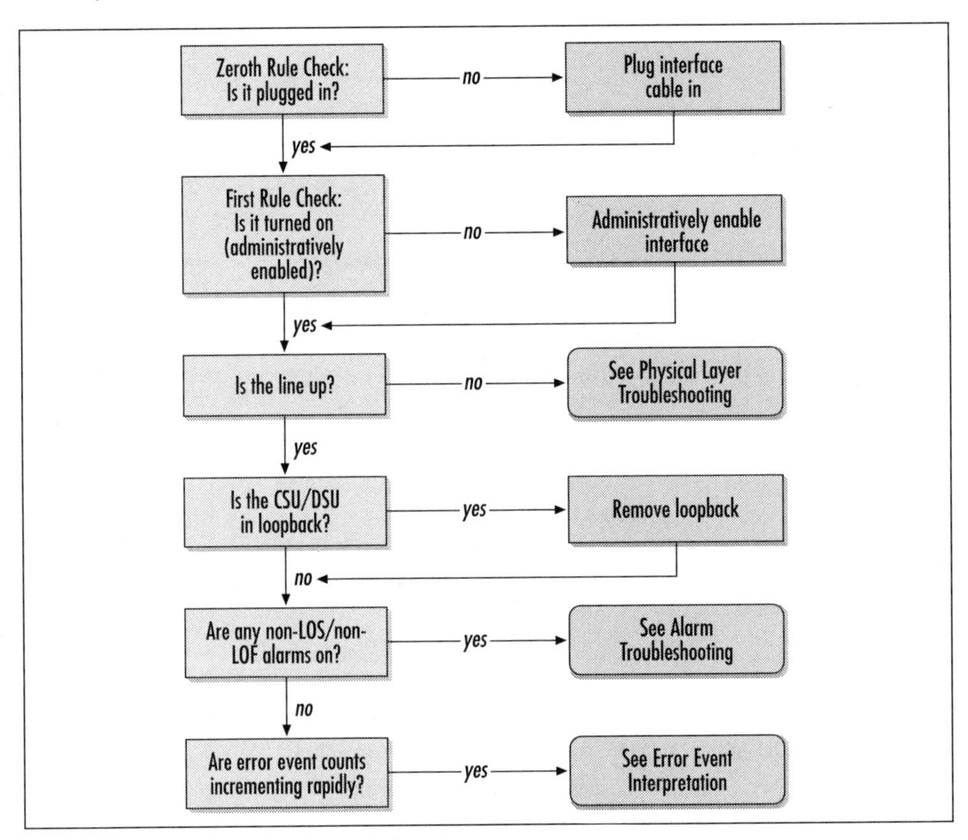

Figure 11-4. Master troubleshooting flowchart

Physical Layer Problems

Physical layer problems prevent the link layer from coming up. Generally speaking, physical layer problems are easy to spot because red lights will glow on the CSU/DSU until you have fixed them. Figure 11-5 shows the physical layer flowchart.

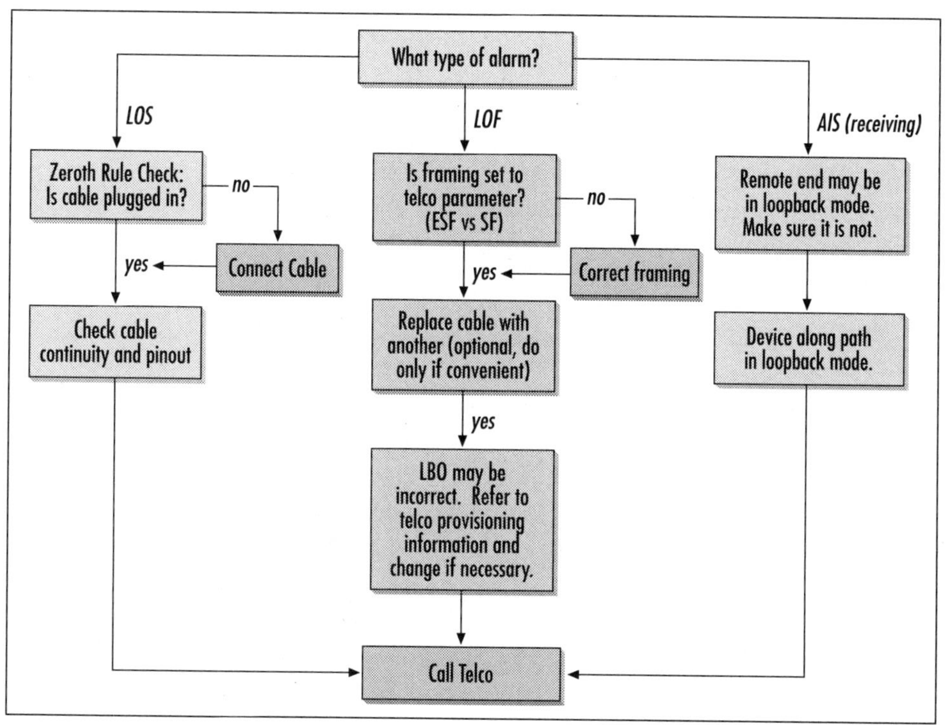

Figure 11-5. Physical layer flowchart

Red Alarm

A red alarm indicates a hard down on the T1 span. Either no incoming signal is being received (LOS—loss of signal) or no framing synchronization is possible (LOF—loss of framing). Do not be concerned if the red alarm is not cleared for several seconds after you plug in the cable. CSU/DSUs have a big job to do in sorting through the stream to find the framing bits and lock on to the synchronization pattern. Table 11-1 associates symptoms with possible causes and suggested actions.

Table 11-1. Physical layer diagnostic table

Symptom/indication	Cause	Corrective action
No incoming signal (LOS)	Cable is not connected to telco network	Plug in the cable.

Table 11-1. Physical layer diagnostic table (continued)

Symptom/indication	Cause	Corrective action
	Cable is bad	Test the cable and replace it if necessary. Be sure to include the extended demarc in any testing. Extended demarc cabling should be appropriate for T1 applications. (Category 5 unshielded twisted pair is not!) Occasionally, the transmit and receive pairs may be reversed.
	T1 circuit is bad	Run a loopback test and call the telco/ISP with the results.
Unable to synchronize framing (LOF)	Wrong frame format is selected	Check and correct the frame format, if necessary. Most new lines use ESF.
	Wrong line code is selected	Check and correct the line code, if necessary. Most new lines use B8ZS.
	Cable is bad, so no framing pattern is detected	T1 uses RJ-48 cabling. Use of straight-through Cat 5 cabling may lead to excessive crosstalk, which may corrupt framing bits.
	Other end is not transmitting correctly	Check LBO and run loopback tests if the remote end is in yellow alarm. Otherwise, call the ISP/telco.
Receiving alarm indication from upstream (AIS) devices	CSU/DSU at one end is in loopback mode	Check to ensure that the CSU/DSUs at the end-user locations are not in loopback mode. Equipment sold to service providers (ISPs) may be able to send loopback commands through the network. Ensure that this option is disabled.
	Device along path is in loopback mode	If the ends of the line are not in loopback or sending loopback commands, an intermediate device is probably in loopback. The T1 service provider can trace the signal.

Differences between RJ-48, RJ-45, and straight-through Cat 5

Both RJ-48 and RJ-45 specify an 8-pin modular jack. Regular Cat 5 cabling will fit into an RJ-48 jack without difficulty and may even work for relatively short cabling runs. On longer runs, though, use of Cat 5 UTP (Ethernet) cabling for T1 circuits may cause problems. If possible, avoid cable runs of more than 50 feet between the NIU and the CSU/DSU.

Ethernet uses two pairs: pins 1 and 2 are used to transmit, while pins 3 and 6 are used to receive. Standard Ethernet wiring twists the component wires in a pair together because twisted pairs have lower electromagnetic field emissions and are less susceptible to interference from external fields. RJ-48, the pinout used for T1, uses two different pairs: pins 1 and 2 are used for reception, and pins 4 and 5 are used to transmit.

Some Cat 5 patch cables are wired incorrectly, with twisted pairs formed by pins 1 and 2, pins 3 and 4, pins 5 and 6, and pins 7 and 8. When an incorrectly wired patch cable like this is used for a long NIU-to-CSU run, pins 4 and 5 will be in separate pairs, leaving the transmit pair vulnerable to interference.

Even if the Cat 5 pairs are twisted together correctly, Cat 5 is, by definition, unshielded wiring. T1 pulses may be much smaller than Ethernet pulses by the time they arrive at the CSU/DSU and can be swamped by any noise much more easily. If you observe problems with Cat 5 cabling, replace it with shielded twisted-pair cabling (with the appropriate pairs twisted together, of course).

Yellow Alarm/RAI

The most significant thing about a yellow alarm is that framing is received correctly, so the remote transmit to local receive path is good. A yellow alarm is sent by the remote end to indicate that it is not receiving a signal. The problem exists somewhere in the local transmit to remote receive path. SF sets the second bit in each channel to zero to indicate RAI, while ESF transmits RAI over the facilities data link. A troubleshooting flowchart is shown in Figure 11-6.

Improper transmissions to the telco network that do not allow the remote end to synchronize framing patterns cause yellow alarms. Two main culprits are LBO and clocking.

A common culprit: LBO

LBO ensures that pulses hit the first repeater with the appropriate amplitude. Pulses that are too big or too small will not be processed correctly and might even be reflected at jacks. For smart jacks, LBO should usually be set to 0 dB. If possible, check with the telco technician who installed the line or the documentation retained from the initial line turn-up. When LBO is set incorrectly, pulses may be too small to register or so powerful that they reflect and interfere with successive pulses. One way to determine whether LBO may be a problem is to vary the LBO setting on the CSU/DSU and see if the yellow alarm state changes. If it does, that is a clear indication that LBO needs to be reconfigured.

Other diagnoses

To make sure the CSU/DSU is functioning correctly, create a loopback plug by taking an RJ-45 plug and linking pin 1 to pin 4 and pin 2 to pin 5. Insert the loopback plug into the telco-side CSU/DSU and adjust the line build out to 0 dB. The CSU/DSU should establish physical connectivity to itself and clear the yellow alarm.

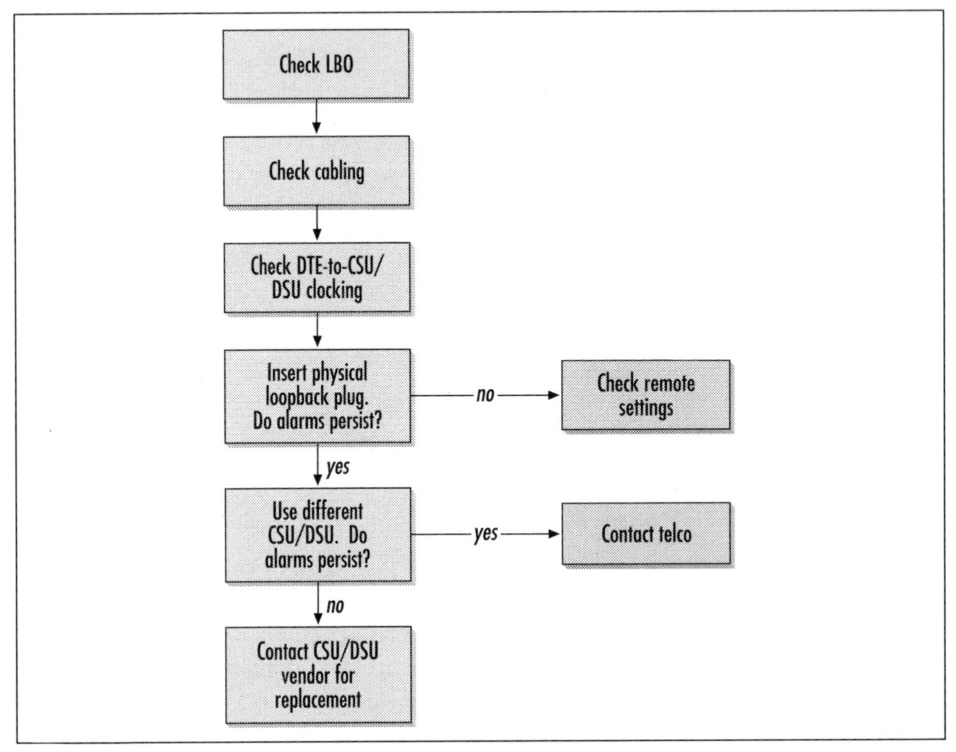

Figure 11-6. Yellow alarm/RAI flowchart

If alarms persist, the DTE-to-CSU/DSU cable may be bad, or the CSU/DSU itself may be bad. If spares are handy, try swapping out the cable and the CSU/DSU in turn to see if the problem is repaired. In a router with integrated CSU/DSU ports, connect the line to a different port and repeat the test. On routers with multiple line cards, select a port on a different line card if possible. If the line comes up with a different CSU/DSU, contact the CSU/DSU vendor for a replacement under any applicable warranties or support agreements.

Troubleshooting table

Yellow alarm causes and corrective actions are summarized in Table 11-2.

Table 11-2. Yellow alarm diagnostic table

Symptom/indication	Cause	Corrective action
Framing received OK, but not transmitted OK	Bad local cable	Test cable with cable tester and replace if necessary, or swap cable with a new one
	Outbound line errors	Eliminate crosstalk due to untwisted transmit pair (see previous section on cabling)
		Line build out is attenuating signal too much; reduce to zero

Table 11-2. Yellow alarm diagnostic table (continued)

Symptom/indication	Cause	Corrective action
	LBO set incorrectly	Verify LBO settings with the telco and adjust accordingly
	CSU/DSU is bad	Connect to a different CSU/DSU or port on the router and contact vendor for replacement
Blinking yellow alarm	Common indicator used to indicate yellow alarm transmission	Correct problem causing transmission of yellow alarm
	False yellow alarm on SF links	Use 23 or fewer time slots, or switch to ESF

Other Error Conditions

Three error counters offer most of the insight into further physical difficulties. Controlled slips, seconds in which the physical layer framing is lost, and bipolar violations can be caused by incorrect configuration or crosstalk. In theory, a functioning line on a well-maintained network should have no errors, and any positive error count can indicate a potentially serious condition. To diagnose the underlying problems indicated by rapidly increasing error counts, use the flowchart in Figure 11-7.

Timing problems (controlled slips)

Clock slips occur due to mismatched timing between the two ends of the T1. Timing can drift for a variety of reasons, but links with multiple telcos are especially susceptible to timing problems because the different telcos may employ different primary reference source clocks.

Bipolar violations (line code violations)

Excessive bipolar violations are the result of one of two problems. An AMI line receiving B8ZS code words will record bipolar violations, and a high bipolar count can indicate that a device that should be set for AMI is instead set for B8ZS. B8ZS devices receiving AMI-encoded data, however, will not generally record line code violations due to excessive zeros.

If the line coding inserts intentional bipolar violations and the local CSU/DSU has an incorrect line code setting, all of the inserted intentional bipolar violations will be flagged. On the other hand, the violations could be caused by a noisy line with crosstalk. After verifying that the line code is set correctly and the cable is good, report the problem to the telco.

Bipolar violations accompanied by clock slips indicate *near-end crosstalk*, often abbreviated by the acronym NEXT. When cable pairs run close together, the mag-

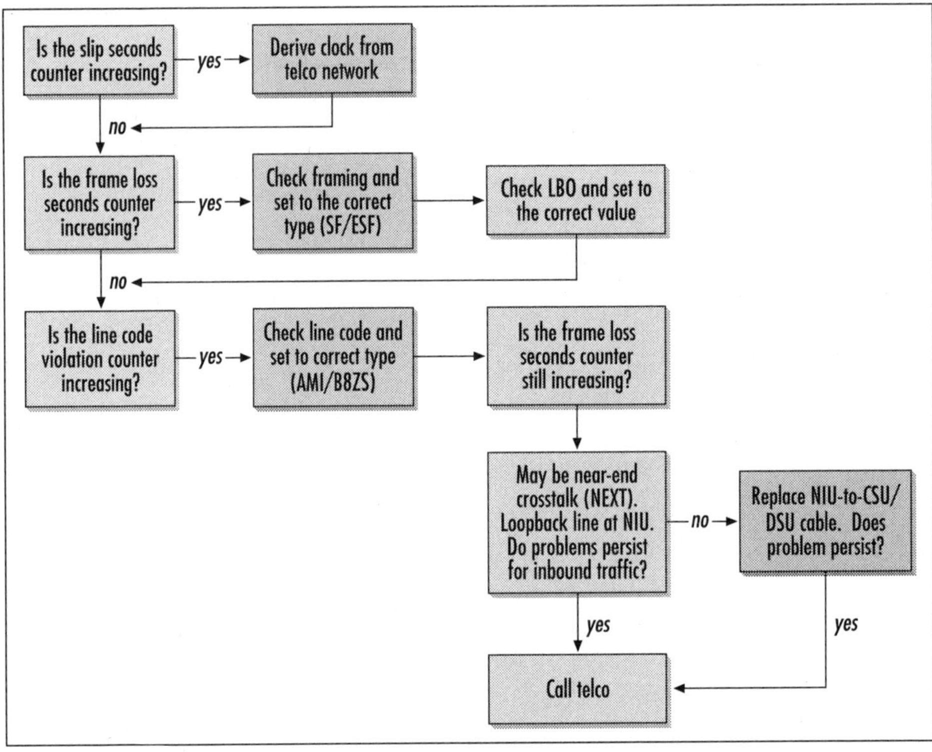

Figure 11-7. Flowchart for troubleshooting error counters

netic fields resulting from a strong signal on one pair may induce a current in a nearby pair. Twisting cable pairs tightly both reduces the emissions and improves immunity to crosstalk. Near-end crosstalk is most often seen at the customer end of a T1 because the transmit power is the strongest there, while the received signal is at its weakest point and susceptible to being drowned out by the crosstalk signal. Standards limit the acceptable crosstalk levels at various points on the T1 line, but staying below the limits imposed by the standards may require the use of high-quality prefabricated cabling. Extended demarcs and the attendant wiring problems are the most common cause of near-end crosstalk. Timing-problem troubleshooting is outlined in Table 11-3.

Table 11-3. Troubleshooting timing problems

Symptom/indication	Cause	Corrective action
Frame slips without BPVs	Timing not synchronized precisely between two ends	Ensure that clocking is configured correctly on the CSU/DSU. Call the telco(s) or CSU/DSU vendor to resolve clocking problems.

Table 11-3. Troubleshooting timing problems (continued)

Symptom/indication	Cause	Corrective action
High numbers of frame slips with BPVs	Near-end crosstalk (transmitted signal appears on receive pair due to magnetic induction currents)	Loop back the line at the NIU to see if crosstalk still affects data. If crosstalk still exists, the problem is on the end span. Call the telco. Test the NIU-to-CSU/DSU cable for near-end crosstalk. The cable may require replacement with shielded cable.

Link Layer Problems

Some apparent link layer problems can be traced to physical layer causes, but others are actually problems in the link layer. The tables in this section will lead you to a physical layer solution where appropriate, because the tables are constructed with symptoms. This section assumes that you have addressed all physical layer problems and that no alarms are present on either side of the link.

General Link Layer Problems

When the link layer fails to come up, a common problem is an incorrect clocking setting. One common scenario is that the line has physical connectivity, but no link layer protocol is able to initialize. This usually occurs because the clocking between the CSU/DSU and the router is not configured correctly. Incoming link layer negotiation requests are received by the CSU/DSU and passed out the serial interface to the router. At the router, the requests are processed correctly and responses are sent. If the transmit clocking on the data port is configured incorrectly, however, the response to the incoming configuration request is not received by the CSU/DSU and thus cannot be sent out the T1 network interface. At the remote end, the line appears to be quiet. No physical problems exist and no alarms are present, but no data comes in, either.

Not all routers can supply a transmit clock with the data. If the CSU/DSU is looking for an external transmit clock from the router, it will not read any incoming data, even if that data is supplied according to the transmit clock from the CSU/DSU. Some routers also have significant internal phase delays and must supply an external transmit clock to avoid frequent corrupted bits when the clock trigger falls on the edge of a voltage transition. To be safe, use external (router-supplied) clocking on the data port if it is supported.

Table 11-4 summarizes basic link layer troubleshooting, independent of any particular link layer.

Table 11-4. General link layer problems table

Symptom/indication	Cause	Corrective action
Link layer negotiations fail, even though physical connectivity is established	Mismatched link layer protocol	Check that both ends are set to the same link layer protocol; some vendors may default to different link layers.
	Remote end not receiving data (verified with packet traces at remote end)	Check that data is transmitted by the operating system on the DTE, then check clocking settings. If no link layer will initialize correctly even when both ends are switched, clocking is almost surely the culprit. Run a remote loopback test and see if the attempted link layer negotiation frames return to the local end. Not all DTEs support sending a transmit clock with the data. It is relatively common for external CSU/DSUs to take clock from the DTE on the data port, but not all DTEs can supply it. If external clocking is not possible, consider using an inverted internal clock signal.
	Specific to a certain link layer	See the following sections.

PPP Problems

Generally, the biggest problem with PPP is that not all WAN switches support all the PPP options. Older software revisions may also require the use of older styles of address negotiation within IPCP. Table 11-5 outlines some possible problems.

Table 11-5. PPP troubleshooting

Symptom/indication	Cause	Corrective action
Debugging information shows lots of Configure Naks (null acknowledgments)	An option in the transmitted PPP configuration request is being rejected	Decode the negotiation to find out which option is being rejected. Not all vendors fully support magic-number negotiation and MTU/MRU negotiation.
	The DTE is receiving the same magic number as it is sending	Magic numbers prevent looped-back lines from coming up. If magic number collisions are occurring, check the line for loopbacks.

Frame Relay Problems

Frame relay depends on the Local Management Interface (LMI) to bring the link up. Unfortunately, three common LMI flavors exist. ANSI specifies one, the ITU specifies another, and the Frame Relay Forum, an industry consortium, specifies a third. Before standards bodies produced any standard, four vendors defined an LMI flavor in a semiformal setting, which became known as the Gang of Four LMI. Most frame relay problems are due to misconfiguration of the LMI type or the local DLCI; frame relay troubleshooting is distilled in Table 11-6.

Table 11-6. Frame relay troubleshooting

Symptom/indication	Cause	Corrective action
Unanswered LMI configuration requests	Mismatched LMI type.	Check the LMI type with the frame relay carrier. Look at the DLCI on which the LMI messages are transmitted—standards-based LMIs are transmitted on DLCI 0, while the original Gang of Four specification was transmitted on DLCI 1023.
	Incorrect DLCI specified at local end.	Check the local DLCI with the carrier. If you can capture and decode an LMI message from the frame switch, it is possible to determine which DLCI the frame cloud is expecting.
	Local DTE does not support the frame relay feature being used.	Frame relay is a complex set of specifications, and equipment vendors may not always support the feature you have tried to configure. Call the vendor and put in a feature request.
LMI does not detect PVC failure	LMI is a local management interface. Even LMI Annex G may not check the full path.	Use a routing protocol to ensure end-to-end checks on PVC integrity and take action to reroute traffic when appropriate.

HDLC Problems

HDLC is by far the simplest of the link layer protocols commonly run on WAN circuits. It does not have any substantial negotiation procedures, which means that any problems observed on HDLC links are likely due to underlying physical problems with the circuit (or a misconfiguration of the link layer encapsulation on the circuit itself). Troubleshooting is described in Table 11-7.

Table 11-7. HDLC troubleshooting

Symptom/indication	Cause	Corrective action
HDLC link is down	Keepalive packets not received by local end	Put CSU/DSU in local loopback mode and ensure that keepalive sequence numbers increment. If they do not, evaluate the clocking configuration.

Table 11-7. HDLC troubleshooting (continued)

Symptom/indication	Cause	Corrective action
		Run local loopback test and capture packets received on the serial interface before calling the DTE vendor.
	Keepalive packets not received by remote end	Transmit clocking is incorrect or cabling is bad.
	Two different vendors' HDLC implementations are not interoperating	Unless specifically stated to the contrary, vendors do not create interoperable HDLC implementations. If one vendor promises interoperability, file a bug with that vendor's support desk.

Access Aggregation with cT1 and ISDN PRI

When have people been half as rotten as what the panderers to the people dangle before crowds?

—Carl Sandburg
The People, Yes

Most of this book has been about T1 as a single data pipe for network access. With the explosion of Internet service providers and dial-up access, though, T1 also became a high-density solution for dial-in POPs. Each T1 has 24 voice channels and can support 23 or 24 dial-in users, depending on the technology chosen. Furthermore, T1 is digital technology that allows higher transmission speeds and lower noise. New 56-kbps modems require that ISPs use digital pipes to their dial-in servers so that only one digital/analog conversion is required.

Two main technologies are used for this purpose. *Channelized T1* (cT1) is a regular T1 that is broken into 24 small 64-kbps pipes. Creating and tearing down phone circuits requires signaling. After bit-robbing, the maximum available throughput is 56 kbps per channel. ISPs may also choose to use ISDN for dial-in POPs. In addition to the familiar 128-kbps Basic Rate Interface (BRI), ISDN has been specified at T1 speeds as the Primary Rate Interface (PRI). ISDN PRI provides 24 64-kbps channels, but only 23 are used for data transmission. The remaining channel is used for signaling messages. ISPs often use ISDN PRIs to provide ISDN services, and large businesses use ISDN PRIs for dial-on-demand backup services. Generally, ISDN PRI is priced so that if more than eight BRIs are required, a single PRI is more cost-effective. Hardware often supports both cT1 and ISDN PRI because the physical interfaces are identical. Different software licenses, however, may be required to use the different feature sets.

Individual DS0s are traditionally used to carry single digitized voice calls. Signaling is an associated detail that depends on the type of link. cT1 signaling is sometimes referred to as *channel associated signaling* (CAS) because the signaling is embedded in the same channel. ISDN, on the other hand, carries signaling information with a separate T1 time slot. The 64-kbps channel from one time slot carries all signaling information for the remaining 23 time slots. Because the signaling channel carries messages for many other time slots, it is sometimes called *common channel signaling* (CCS).

The most important piece of information when using a technology that slices a T1 into its constituent time slots is provided by mapping the component DS0s correctly. Most equipment that provides access to individual time slots requires that the end user specify the signaling method. Lines provisioned for robbed-bit signaling must not be connected to equipment set to use the full 64-kbps line rate because the equipment does not transmit signaling messages correctly. Instead of using the signaling bits for circuit setup, the equipment transmits a data bit. As a result, the telco network receives what appear to be corrupted signaling messages, and calls do not establish. The reverse configuration error is just as bad. Lines provisioned for full-rate data transport on the individual DS0s must not be connected to equipment configured for robbed-bit signaling. When the equipment needs to set up a circuit, it will clobber a data bit on the channel. The telco network will see corrupted data in that time slot, but no circuit-setup message on the actual signaling channel. As in the converse error case, circuits will not establish, and data will be mangled.

Channelized T1

T1 was initially used by telephone companies to digitize voice calls and create a high-capacity trunk line carrying multiple voice calls. Each time slot in the T1 frame has the capacity to support a traditional telephone call. Channelized T1 does precisely that—each of the 24 time slots can be treated as a digital telephone line. Each line has a raw capacity of 64 kbps, but one of the 8 bits must be used for signaling, so the effective maximum throughput of a single cT1 channel is 56 kbps. In contrast, unchannelized T1, the subject of the body of this book, simply treats each time slot as another opportunity to send 8 bits of data to the remote end, and the entire capacity is one big pipe. Figure A-1 illustrates the difference.

For dial-up service providers, such as ISPs or IT departments providing remote access, cT1 is a key enabler of 56k modem connections. Modems send a pristine analog waveform to the CO, where it is digitized and sent on its way. At the other end, a second conversion takes place when the digital signal is reconverted to an analog output and passed to the called party. Any conversion from analog to digital introduces some sampling error and will not perform a perfect conversion. Service providers can use digital lines from the telco to avoid the need of an analog-to-digital conversion on the return path, as shown by Figure A-2.

When Is "56k" Much Less than 56,000?

Compatible digital-access technology at the ISP is only one component of the solution for high-speed dial-up access. Several impediments can obstruct high connection rates.

In the United States, the Federal Communications Commission limits the power output of modems; the power limit restrains throughput to a maximum of 53 kbps. Due to other problems, however, most modems cannot even connect at 53 kbps, so in practice the power-output limitation does not restrict users.

Extensive telco network upgrades are needed to lift most of the barriers. Most old installed copper lines have bridge taps and load coils. Both of these devices limit the bandwidth available on the line and restrict the amount of the audio spectrum the modem can use. As consumer demand for high-speed Internet connectivity has exploded, telcos have begun the long and complex process of removing legacy line impairments. Keeping track of every bridge tap and load coil is an impossibility—it is common to discover an undocumented device when the line quality is too poor to allow for high-speed connections. Local loop upgrades are not popular with telcos. Once in place, they tend to encourage dial-up users to stay online for longer periods of time. Most Internet-access calls are local, though, and do not generate additional revenue for the telco.

Poor inside wiring accounts for many of the remaining problems. Lines may be split to multiple jacks within a building or may be poorly spliced together. Low-quality copper wiring may also limit the bandwidth available from the jack to the demarc, even if the telco has upgraded the local loop.

Modems present several hurdles. One is a design choice made by all modem vendors. The most visible indication of speed is the connect rate of the modem, so modem vendors throw extensive resources into making the initial connect rate as high as possible. Some products will report a high initial connect rate and immediately retrain to a lower, more sustainable speed. Second, modem specifications can be difficult to interpret, so it is not uncommon for two modems to have incompatible firmware. Finally, the miracle of software allows vendors to move some functions off the modem and onto the host CPU, saving the cost of some processing elements on the modem board. Transferring computational tasks to the system CPU, however, brings modem activity into competition with other processing. Other programs running on the computer may interfere with the modem driver's need for CPU time and decrease throughput.

Subscribers dialing in now face an analog-to-digital conversion on the way to the dial-in server, and the return path is purely digital until it hits the subscriber's CO, where the pure digital signal is reconverted to analog. The analog-to-digital

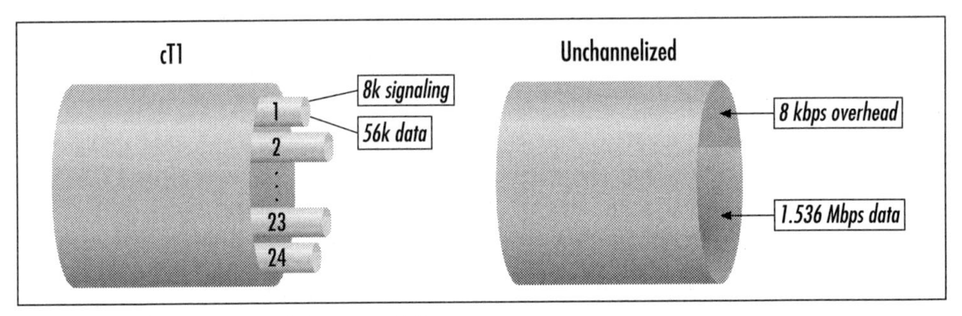

Figure A-1. cT1 versus unchannelized T1

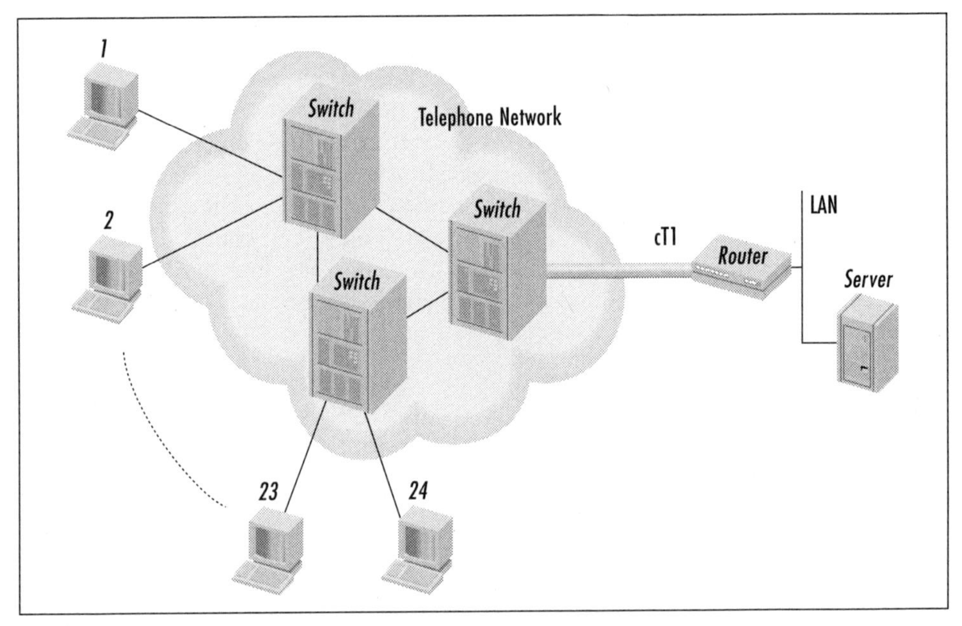

Figure A-2. Dial-up line aggregation with T1

conversion is not as precise. After all, it involves taking a signal with an infinite number of possible levels, analyzing it, and converting it to a series of numbers. Digital-to-analog conversions are far simpler, because they just take a series of numbers and generate a waveform. The sampling uncertainty introduced by analog-to-digital conversions limits effective modem throughput to 33.6 kbps, but the digital downstream signal can run at the full 56 kbps because there is no analog-to-digital sampling error.

As discussed in the earlier sidebar, in practice, the downstream link never reaches 56 kbps. FCC rules limit the transmitted power, which holds throughput to 53 kbps. But that's 53 kbps under ideal conditions, which involve several network upgrades by the telco. These upgrades do not generate revenue, though, so they may exist in some areas but not others. Random noise further reduces throughput. A practical rule of thumb is that "56k" modems usually connect at approximately 40–48 kbps.*

Use of T1 lines for dial-up aggregation reduces costs for ISPs because a rack of 24 modems can be replaced with a small box containing digital modems and the appropriate software. Smaller boxes mean that less rack space is used, and rack space is always at a premium.

Signaling on cT1

cT1 treats each time slot as a separate telephone line. Unlike dedicated lines, telephone lines are circuit-switched and must therefore have signaling to announce incoming calls, release existing calls, and ensure efficient usage of the shared capacity of the telephone network.

No distinct signaling channel exists for cT1; the eighth bit is "robbed" for signaling purposes. Using the eighth bit for signaling was not a problem for voice communications because the effect on the voice was negligible. Data, however, does not have the same qualitative properties—it makes all the difference in the world if a bit is changed. Therefore, when robbed-bit signaling is employed, the eighth bit is always ignored. This means that each time slot has a maximum throughput of 56 kbps.

Use of the robbed bits is quite complex, as it has evolved since the voice-networking days. Older signaling schemes that involved additional wires have also been accommodated into the digital signal format. (Each of the signaling methods presented in this appendix have analog equivalents, but this appendix discusses only their modern digital equivalents.) Initially, signaling schemes are needed to transmit the dial pulses from rotary phones. Modern touch-tone dialing transmits tones down the channel where the switch uses them, and the robbed bits are used only to communicate on-hook (hung up) and off-hook (ready to dial) status. On SF links, each channel has two bits, labeled A and B; ESF has four robbed bits, called A, B, C, and D. Robbed-bit signaling is also called *in-band signaling* or *channel associated signaling* (CAS), because part of the transmission capacity of each channel is given over to signaling.

E&M signaling

E&M signaling is one of the oldest methods available. In its original incarnation, a separate twisted-pair signaling line was used in addition to the multichannel trunk

* ISDN users connecting through a cT1 line are not limited by the analog-to-digital conversion because ISDN is all digital, all the way. ISDN has a capability of 64 kbps, though, so users may be annoyed if they hit a cT1 line that limits them to 56 kbps.

between two switches.* At each end, as shown in Figure A-3, switches had connections for the E and M leads. At one point, *E* stood for *ear* and *M* stood for *mouth*. Switches transmit signaling information on the M lead and receive signaling information on the E lead.

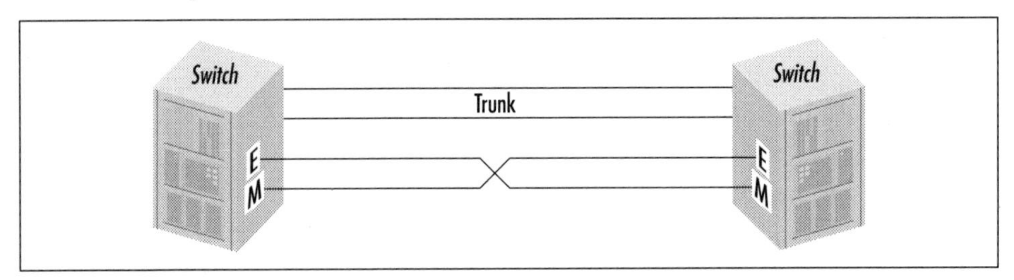

Figure A-3. Basic E&M signaling setup

Digital E&M signaling uses the A and B signaling bits. On ESF links with four signaling bits, the C and D bits have the same value as the B bit. Digital E&M signaling is a fairly simple system with only two states. The signaling bits report whether the equipment is on-hook or off-hook.

Idle state. In this state, nothing is happening. Both ends of the link are on-hook. In this state, shown in Figure A-4, both A and B are set to zero.

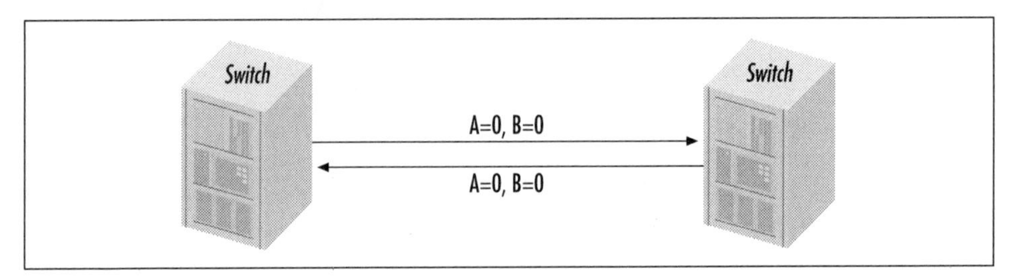

Figure A-4. Digital E&M idle state

Call initiation. When a call comes in, the originating switch must signal to the destination switch that there is an incoming call. The originating switch *seizes* the incoming line by going off-hook. Seizing the incoming line is the originating switch's way of reserving capacity for the circuit to be used for the telephone connection. Off-hook status in E&M signaling is indicated by setting both the A and B bits to one.

Figure A-5 should be viewed as the last leg in a telephone connection. If the caller is attached to the originating switch, Figure A-5 is the complete picture. For callers

* In the context of signaling, *switch* means any device that connects to the telephone network. It can be a CO switch, like the Lucent 5ESS, or a PBX. Devices that terminate telephone calls at the customer's premises, like cT1 equipment, can be considered PBXs for the purpose of the signaling discussions.

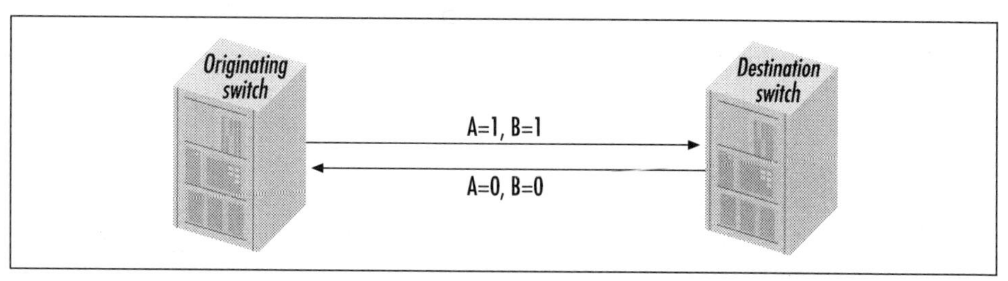

Figure A-5. Digital E&M call initiation

attached to other switches, the telephone network reserves capacity for a circuit from the caller's switch to the originating switch, and the originating switch is responsible for completing the circuit to the destination.

Call completion. Once the originating switch has opened the channel, it transmits call-routing information, such as the telephone number, to the receiving switch. After receiving the telephone number, the destination goes off-hook by setting the A and B signaling bits to one in the reverse direction, as in Figure A-6.

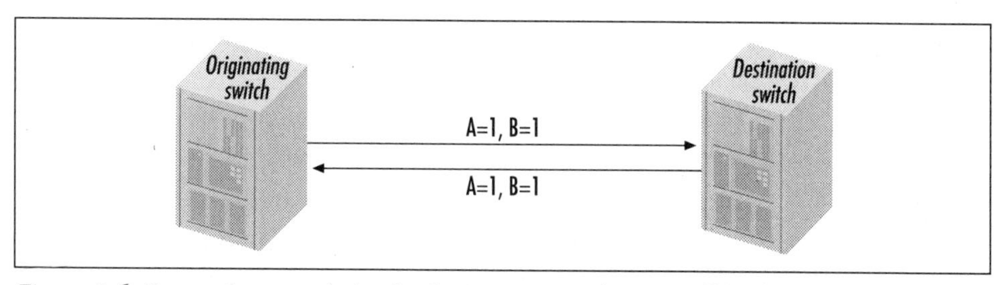

Figure A-6. Connection completion by destination switch going off-hook

For clarity, the entire E&M procedure is shown in Figure A-7. For reasons that will become clear momentarily, this procedure is called *E&M immediate start.* The originating switch goes off-hook for an incoming call, waits for at least 150 milliseconds, and transmits the tones for the destination telephone number. When the destination switch receives that information, it goes off-hook to establish the circuit.

Wink start. If both ends of the switch try to use the same channel at the same time, connections can be fouled up. This problem is known as *glare. Wink start* is used to avoid glare. Wink start turns the two-way handshake in Figure A-8 into a three-way handshake.

After the sending switch goes off-hook, it waits for acknowledgment from the destination switch. That acknowledgment comes when the receiving switch "winks" off-hook for 140–200 ms. When the originating switch receives the wink, it is cleared to transmit the telephone number after a delay of at least 210 ms. Completing the call is identical to the immediate start procedure.

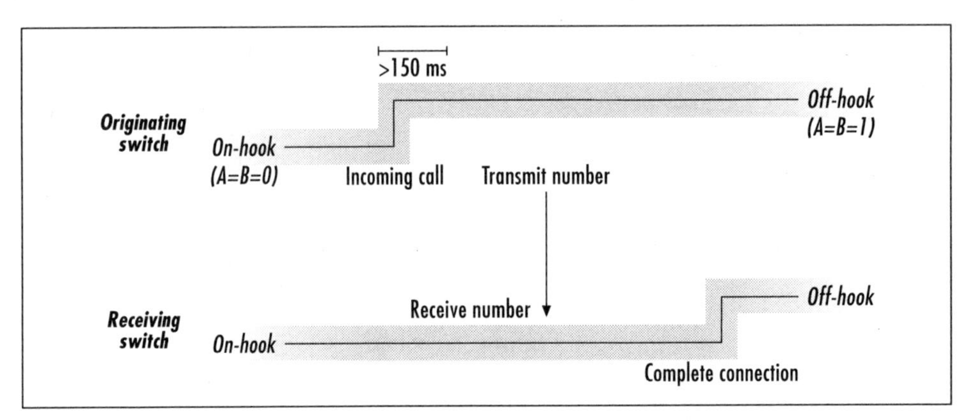

Figure A-7. E&M immediate start

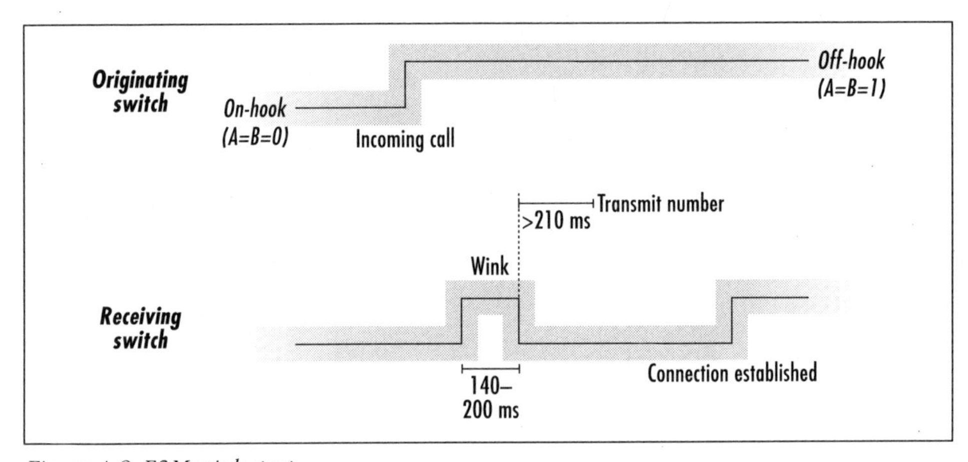

Figure A-8. E&M wink start

Telephone number transmission. Touch-tone dialing works by generating audible tones that switches recognize as numbers. Two common schemes are used: *multi-frequency* (MF) and *dual-tone multifrequency* (DTMF). DTMF is much more common. Devices that perform E&M signaling can be configured to listen for the transmitted telephone number, but they must match the signaling method that the telco uses.

Foreign exchange ground start

A *foreign exchange station* (FXS) is simply a telephone connected to a PBX. Foreign exchange stations are controlled by the local PBX and the remote switch at the CO acting in concert. In situations in which the remote telephone is connected through a PBX, the CO is referred to as the *foreign exchange office* (FXO).

Idle state. In the idle state, the A and B bits that the CO transmits are both set to one; the PBX sets A to zero and B to one on the return path. Figure A-9 illustrates this.

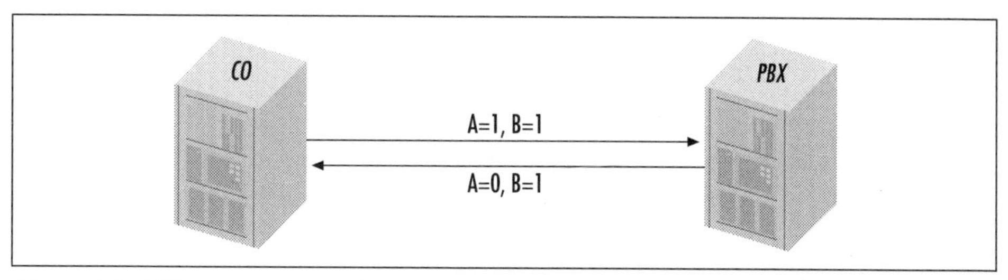

Figure A-9. FXS ground start; idle state

Outgoing call. To initiate a call, the PBX sets the B bit to zero (step 1 in Figure A-10). To indicate the change, the zero is shaded in the figure. The CO switch responds by setting its A bit to zero in step 2. Following the convention of step 1, the zero is shaded. To note that the B bit from the PBX is different from the idle state but unchanged from the previous step, it is shown underlined. Finally, in step 3, the PBX changes both its A and B bits to one.

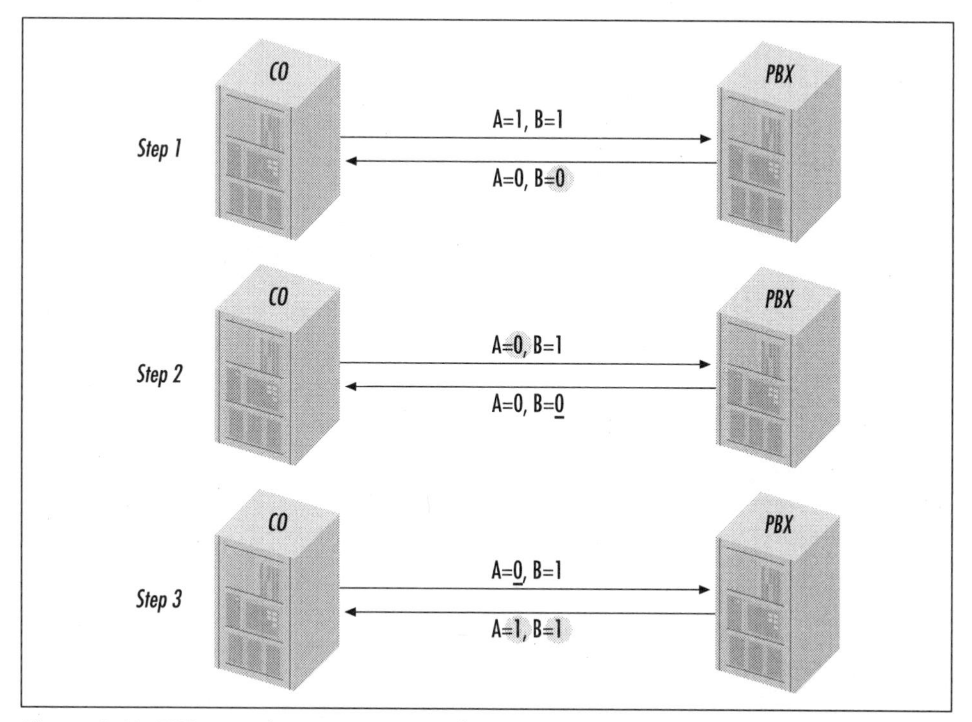

Figure A-10. FXS ground start; outgoing call

Incoming call. When a call comes in from the CO, it signals the PBX by setting both the A and B bits to zero, as in step 1 of Figure A-11. The B bit controls the ring generator and, by setting it to zero, the CO directs the PBX-attached phone to ring. In step 2, the PBX responds by setting its A bit to one. When the switch at the CO stops ringing, it sets the inbound B bit to one and the call is completed.

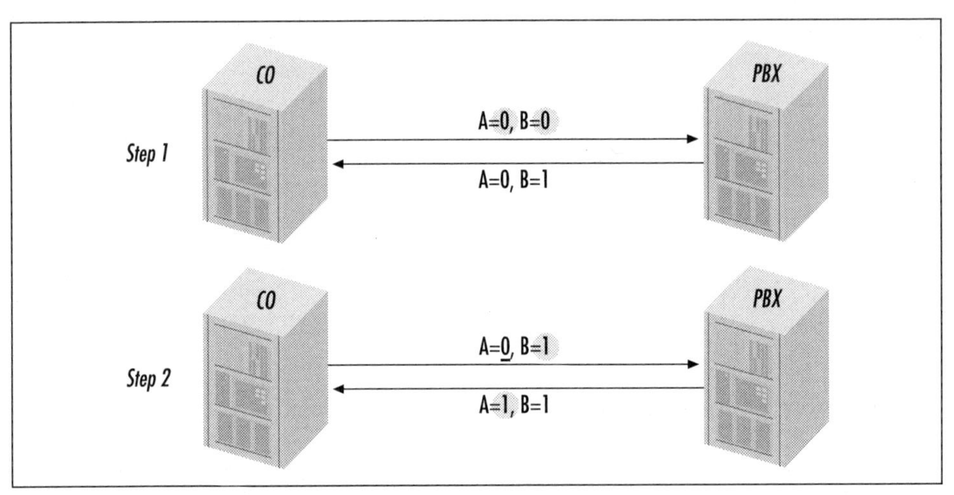

Figure A-11. FXS ground start; incoming call

Foreign exchange loop start

Loop-start signaling may also be used, but the bit settings are slightly different.

Idle state. For loop-start signaling, both the CO switch and the PBX transmit zeros for A and ones for B in the idle state, as shown in Figure A-12.

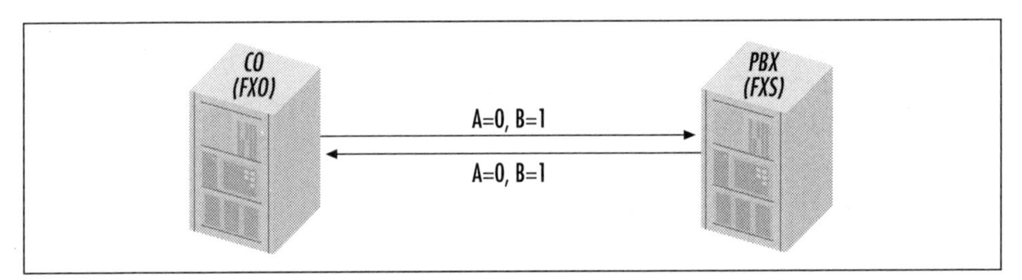

Figure A-12. FXS loop start; idle state

Outgoing call. Outgoing calls with loop-start signaling involve only one step: the PBX must set its A bit to one, as Figure A-13 shows.

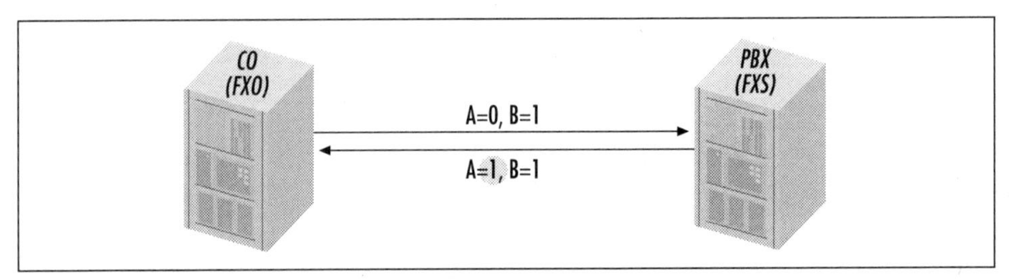

Figure A-13. FXS loop start; outgoing call sequence

Incoming call. As in the ground-start signaling method, the CO begins by setting the incoming B bit to one, which triggers the ring generator at the PBX. When the destination goes off-hook, the PBX sets the outgoing A bit to one to close the loop, and the CO sets the incoming B bit to one to kill the ringer. Figure A-14 illustrates this.

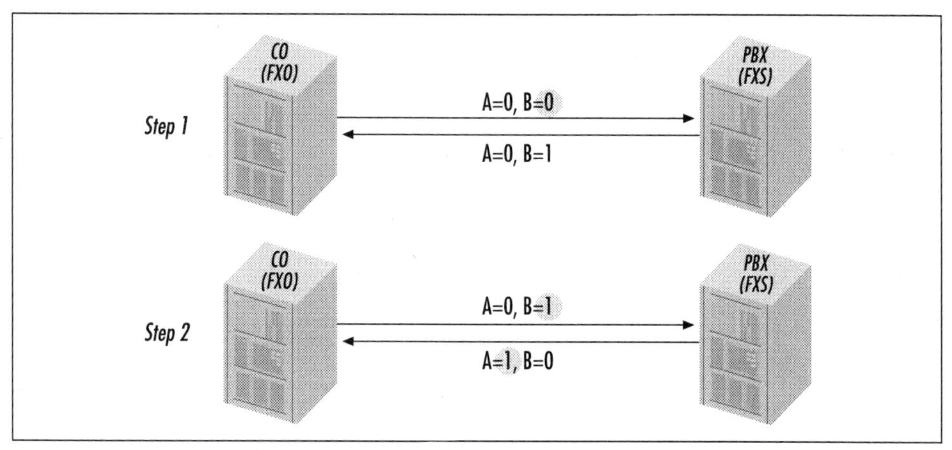

Figure A-14. FXS loop start; incoming call sequence

Other methods

A few locations may use R1 signaling, which is specified in ITU recommendations Q.310 to Q.331. R1 uses robbed bits to signal on-hook and off-hook status, with telephone numbers transmitted by tone sequences.

Some telcos may use *specific access station* (SAS) signaling. SAS signaling is similar to foreign exchange signaling, but the A and B bits are replaced by their complements. Like foreign exchange signaling, SAS signaling can be either loop-start or ground-start.

Ordering, Provisioning, and Record Keeping for cT1

With each channel able to act as a separate line, cT1 has a number of additional parameters that must be specified.

One- or two-way calling

Telcos may allow customers to specify that a line be allowed to accept only incoming calls (one-way).

Telephone numbers, part 1

Each time slot translates to one voice circuit through the telephone network. Each of the 24 time slots can be assigned its own telephone number, or multiple lines can be assigned to *hunt groups*, which allow multiple lines to serve one telephone number. Toll-free customer-service lines are an example of a hunt group applied to voice calls. Depending on the telco, it may be possible to assign some channels to one hunt group, other channels to a second group, and the remainder to a third group. Multiple cT1s can be assigned to a single hunt group, which allows hundreds of modems to be aggregated into a single dial-up number.

Telephone numbers, part 2

T1 does not have a built-in method for transmitting any telephone number–related information as part of the call setup procedure. Two services are of special note:

- Caller-identification services transmit the telephone number of the caller. This may be called Automatic Number Identification (ANI), Calling Line Identification (CLID), or simply caller ID.
- The Dialed Number Information Service (DNIS) transmits the dialed number along with the called number. This can be useful if the dialed number is used to change the system personality based on the telephone number the user dials.

Signaling

As part of the line configuration, you will need to choose a signaling method. Most U.S. lines use E&M wink-start signaling, though E&M immediate start and FXS loop start are also used. Lines that use E&M signaling may also be configured to accept telephone number information with either DTMF or MF signaling.

Different telcos and vendors may refer to the subsets of E&M signaling as simply wink start and immediate start. When used in a cT1 signaling context, loop start and ground start almost always refer to foreign exchange signaling. All high-end vendor equipment supports at least E&M wink start, E&M immediate start, and FXS loop start.

Record Keeping for Later Problems

n the last section of Chapter 7, there is a list of parameters that network administrators should know about each T1 in their networks. Channelized T1 circuits require additional parameters. For completeness, here is a list of parameters that should be maintained on each of the cT1 circuits in your network:

- Telco circuit number and support contact information
- For lines provided in partnership with a service provider other than the telco, a customer number and support contact information
- Line framing (ESF or SF/D4)
- Line code (B8ZS or AMI)

- Line speed (channel speed and number of DS0s)
- Line build out information
- Transmit clocking and DTE clocking settings, along with the make and model of the DTE
- Whether the line is one way (dial-in only) or two way (both dial-in and dial-out)
- Assigned telephone numbers and hunt groups
- Signaling type
- Whether the line uses wink start or immediate start (for E&M signaling only)
- If DNIS (called-number identification) is used, and whether the report is DTMF or MF (for E&M signaling only)

Configuring Channelized T1

The following must be configured when using channelized T1:

Basic T1 parameters
: Basic T1 configuration is identical to unchannelized T1. Framing, line code, clocking, and CSU/DSU configuration are the same. If the link uses ESF, the FDL may also be configured.

Individual DS0s
: Individual DS0s may be configured for different purposes. Some may be dedicated to data, while others are used for voice. Furthermore, different time slots may use different types of signaling. When cT1 is used for dial-up aggregation, all 24 time slots are used for modem support.

Signaling
: Robbed-bit signaling may be implied by configuring 24 time slots, but the type of signaling must be specified. In the U.S., robbed-bit signaling is most commonly done with the E&M wink-start protocol.

Call type and modems
: cT1 can be used to support voice calls or data calls. As it is used in the cT1 arena, *voice* often means "an incoming analog call," which includes dial-in modem access, while *data* means "an incoming digital call," such as ISDN dial-in or Switched56 access. Modems, therefore, usually require configuration of incoming lines for voice. Some products can support assigning each time slot to either voice or data. Products that incorporate digital modems may also require configuration of the digital modems.

Other protocols
: If PPP encapsulation is used for dial-in users, PPP negotiations must be configured. Authentication and IP address assignment are the two big items.

Other network equipment
Dial-in users must be authenticated. Rather than storing authentication information on each dial-in server, it is possible to configure the users to use an external authentication server. Some products may allow user-driven parameters to be assigned by RADIUS servers.

ISDN PRI

The Integrated Services Digital Network, or ISDN, simply extends digital telephone service all the way to the customer. Instead of analog drops from the CO to the customer, it is a complete digital path.

At a high level, ISDN is a composite pipe with subchannels. Subchannels come in two types. *Bearer channels*, or *B channels*, are 64-kbps channels that move user data. User data might be a telephone conversation that is digitized at 64 kbps, videoconferencing, or Internet access. B channels are 64-kbps raw digital pipes that can be directed over the telephone network. To connect bearer channels to endpoints across the telephone network, a second subchannel type, called the *D channel*, provides signaling information. The D channel carries call setup and teardown messages.

ISDN comes in two flavors. Basic rate service is composed of two bearer channels plus a 16-kbps D channel. This package, also known as the ISDN Basic Rate Interface (BRI), is called 2B+D. It provides 128 kbps of user bandwidth in two 64-kbps increments. Each B channel can be used for a digitized telephone call or a raw data pipe. ISDN BRI can be used as two telephone lines, a telephone line plus a 64-kbps data pipe, or a 128-kbps data pipe with no telephone access. Part of ISDN's appeal is its flexibility. It establishes data calls as circuits through the telephone network, so the division between data and telephony can be dynamic. You may start a large download and fire up both B channels for Internet access. However, if you pick up the phone, one of the two will be disconnected so that the B channel can be reallocated for your impending telephone call.

BRI is an excellent tool for remote locations with relatively small connectivity needs. Many institutions need higher line densities. These demanding customers turn to primary rate ISDN service, which is delivered over a T1 line in the U.S. The Primary Rate Interface (PRI) provides 24 raw 64-kbps channels, each of which corresponds to a T1 time slot. Primary rate ISDN can be provided only over B8ZS lines because only B8ZS can guarantee 64-kbps clear channels. Usually, these channels are provided in a 23B+D configuration. Figure A-15 contrasts cT1 and ISDN PRI.

As it turns out, 64 kbps is a huge amount of bandwidth for signaling, especially considering that dial-in users tend to stay connected for long periods of time. ISDN specifications have been developed to allow a D channel to be shared among several PRIs, a configuration called *Network Facilities Associated Signaling* (NFAS).

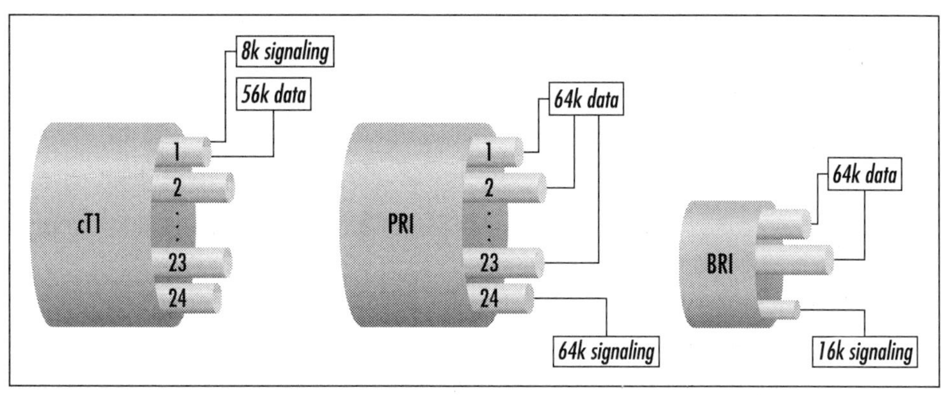

Figure A-15. cT1 and ISDN PRI contrasted

ISDN Signaling

ISDN specifications cover signaling in great detail. While the signaling standards are voluminous and intimidating, the advantage to network administrators is that they are far more uniform than those of cT1.

Q.921 specifies the Link Access Procedure, D Channel (LAPD), which is the data-link layer for ISDN signaling. Q.921 is responsible for framing and basic integrity checking of signaling messages. Within the LAPD frames, messages specified in the Q.930 and Q.931 carry setup information to initiate and end bearer channel connections. No special configuration or additional protocols are needed to perform calling line identification or to determine the called number because all that data is carried in the ISDN call-setup message on the D channel. Using a signaling channel separately from the data-transmission channels is called *common-channel signaling* (CCS). In the ISDN's case, the common signaling channel is the D channel.

Devoting an entire 64-kbps time slot to the D channel provides far more signaling bandwidth than is necessary for 23 bearer channels, especially for long-lived Internet access calls. To allow better use of the available bandwidth, multiple PRIs can share D channels. One PRI line in a group is configured in the traditional 23B+D configuration, while secondary lines are configured as 24B, with signaling on the D channel carrier. If the D channel carrier fails, no connections are possible, because the D channel carries setup and disconnection messages. To prevent a group of lines that share a single D channel from becoming inoperative due to the failure of the D channel carrier, ISDN allows a second line in a group to be designated as the backup D channel carrier. Each line in a group of PRIs must be either the primary D channel carrier, the backup D channel carrier, or a secondary data-only line. Configuration of line type is performed jointly by you and the telco, so make sure that your equipment configuration matches the telco's setup.

To provide all the additional features beyond the simple dialing of plain vanilla telephone service, ISDN signaling is integrated far more deeply with the CO switches. Naturally, the telco is not likely to change the CO switch just for you, so make sure the equipment you purchase will be compatible with the CO switch. In many parts of the country, CO switches support Telcordia's *national ISDN-2* (NI-2) specification for PRIs. If national ISDN features are not available, most equipment provides the ability to interface with the custom ISDN protocols available on the switch model to which you are connecting. In the U.S., the CO switches are commonly the Lucent (AT&T) 5ESS and the Nortel DMS-100, though AT&T 4ESS switches are still in use in many areas.

One oddity of ISDN signaling is that most equipment requests only 56k trunks outside the local exchange. Within a CO, bearer channel connections are 64k, but inter-CO trunks are given only 56 kbps by default. Depending on the telco, you may need to request 64-kbps interoffice trunk channels at the time the line is installed. Some telcos may allow equipment to request 64-kbps channels by dialing the full 10-digit number.

Ordering, Provisioning, and Record Keeping for ISDN

Ordering ISDN is much simpler than ordering cT1 because ISDN has built in so many of the features that users must evaluate and choose between on cT1. Lines must be B8ZS to provide the 64-kbps clear channels, and B8ZS equipment always uses ESF.

Telephone numbers

Like cT1, primary rate ISDN lines are assigned telephone numbers. Most data-oriented deployments use only one number to create a dial-in pool. ISDN specifications define a *Service Profile Identifier* (SPID) that allows switches to treat different bearer channels in different ways. Dial-in services are homogenous, so SPIDs are not used—the first available channel picks up an incoming call. Depending on the telco, you may be allowed to specify the order in which channels are used for incoming calls. It will either be ascending, where the lowest available channel picks up, or descending, where the highest available channel answers.

Signaling

The major unknown for configuring ISDN signaling is the switch type. Indeed, for standalone PRIs, the switch type is the only configuration variable. Service providers using multiple PRIs need to configure shared D channels.

Glare occurs on ISDN circuits when both the CO switch and the customer equipment attempt to use the D channel or one of the B channels. Dial-in services can configure equipment to yield to the CO's wishes.

Record Keeping

In the last section of Chapter 7, there is a list of parameters that network administrators should know about each T1 in their networks. ISDN PRI circuits require additional parameters. For completeness, here is a full list of the parameters that should be maintained on each of the PRI circuits in your network:

- Telco circuit number and support contact information
- For lines provided in partnership with a service provider other than the telco, a customer number and support contact information
- For completeness, note the line code as B8ZS and the framing as ESF
- Line speed (channel speed and number of DS0s)
- Line build out information
- Transmit clocking and DTE clocking settings, along with the make and model of the DTE
- Whether the PRI is one way (dial-in only) or two way (dial-in and dial-out)
- Assigned telephone number
- Channel sequencing (ascending or descending)
- Channel configuration (23B+D or 24B with NFAS)
- NFAS-related configuration (if the line carries a D channel, whether it is the primary or secondary carrier; if not, then it must be associated with the D channel on another line)
- Switch type (NI-2, 4ESS, 5ESS, or DMS-100)
- Interoffice trunk speed selection

Configuring ISDN PRI

The following must be configured when using ISDN PRI:

T1 interfaces
: Because ISDN PRI in the U.S. rides on T1 interfaces, each T1 must be configured before ISDN configuration can be successful. ISDN PRI links must use B8ZS encapsulation for clear channel transmissions. Equipment that supports B8ZS almost always supports ESF framing.

T1 channels
: ISDN PRI need not use all 24 slots of the T1. If some subset is used, it must be configured on the ISDN terminating equipment.

Switch type
: ISDN PRI interfaces are direct digital interfaces to the telephone switch. Customer equipment must match the feature set deployed at the central office.

Most likely, it will be for some variety of *national.** In some spots, though, it could be for an AT&T 4ESS, 5ESS, or Nortel DMS-100.

B channels
: Dial-up servers with ISDN PRI connectivity are assigned to a hunt group, which assigns incoming calls to the next free line. Calls may be in ascending or descending order. Configure your equipment to match the telco's setting.

D channels
: Not all PRIs are 23B+D. D channels may be used to support signaling for multiple PRIs using NFAS. PRIs with D channels in an NFAS configuration may be either the primary D channel or a backup D channel and must be configured appropriately. PRIs without D channels must be associated with a D channel–carrying PRI.

Encapsulation
: Internet service providers use PPP or Multilink PPP encapsulation almost exclusively, but a few institutions may support frame relay, X.25, or V.120 connections.

* Several "national" ISDN (NI) specifications exist. NI-1 is the current BRI specification, and NI-2 is the current PRI specification. Equipment vendors have realized the folly in referring to the exact specification—there is no reason to change the term simply because an additional specification has added functionality. In preparation for the day when equipment might implement several national ISDN specifications, most vendors have changed the designation to simply *national*.

B

Multilink PPP

Art necessarily presupposes knowledge.
—John Stuart Mill
System of Logic

Bandwidth is like disk space or closet space—you can never have too much. Eventually, most observers say, fiber to the home will solve the last-mile bandwidth crunch. In the interim, however, demand for last-mile bandwidth has driven a great deal of research and debate (not to mention marketing) for possible solutions.

Certain types of access channels, most notably ISDN, offer the ability to have different levels of service between two points. Higher capacity comes only with increased cost, but the telco's tariffs may be structured in such a way that it is cheaper to use switched access on demand rather than pay for sufficient dedicated capacity to cover peak usage.

Several approaches exist for taking advantage of multiple links. This appendix focuses on Multilink PPP (MP), the set of PPP extensions that are used to aggregate several distinct lower-layer links into a single logical link. ISPs commonly use MP to offer 128-kbps ISDN access. When MP was first deployed, ISPs with large customer bases often had trouble offering 128-kbps ISDN service because both subscriber calls needed to land on the same access server at the POP. To lift this restriction, a variety of methods were developed to allow individual links in an MP session to land on different devices. This broad collection of techniques is referred to as Multi-Chassis MP (MMP) because it allows links to be spread across multiple access servers.

Multilink PPP

RFC 1990 defines Multilink PPP. The RFC covers several topics, including the encapsulation method for multilink transmission, fragmentation procedures, and the interactions between MP and other PPP components. Specifically, the RFC describes five major components:

- Logical grouping of individual PPP links (a procedure referred to as *bundling*)
- Multilink encapsulation
- LCP extensions for negotiating MP
- Fragmentation of higher-level packets
- Fragment reassembly and detecting lost fragments

MP relies on other PPP components to establish and drop links. When multiple links are to be used, administrators must provide configuration information for all links; MP does not include procedures for automating the configuration of multiple links.

Bundling and the MP Architecture

MP allows several individual member links to be aggregated into a *bundle*. No requirement is imposed on the member links other than the ability to run PPP. Member links may be different circuit-switched paths, such as multiple asynchronous dial-up lines or synchronous leased lines; different channels in a multiplexed circuit such as ISDN or frame relay; or any combination of these. Once member links are established through the normal LCP procedures, they can be dynamically added to an existing bundle. When links are no longer needed, they can be removed from the bundle. Link addition and subtraction may change the throughput of the bundle, but will not affect bundle operations or configuration.

A bundle is simply a logical PPP link layered on top of the member links. Figure B-1 shows the MP layer in the context of PPP. Higher-layer network protocols transmit frames using the MP layer. The MP layer is responsible for fragmenting packets if necessary and distributing those fragments among the available member links.

Bundle-level negotiations are slightly different than those for a member link. Each member link is subject to LCP option negotiations. When the lower-level encapsulation details are complete, each member link presents a uniform PPP interface to the MP layer. With the bundle-level link inheriting the uniform interface, no LCP configuration messages are needed. LCP at the bundle level eliminates all messages other than Code-Reject, Protocol-Reject, Echo-Request, and Echo-Response.*

Protocols may be used at the bundle level or at the member-link level. Bundle-level activity takes place before the MP implementation prepares for transmission across the individual member links. Member-link transmissions bypass the MP layer and use a single member link directly. The contrast between bundle-level activity and member-link activity can be illustrated by the use of LCP echo messages. At

* LCP also allows Discard-Request at the bundle level, though it is not commonly used in practice.

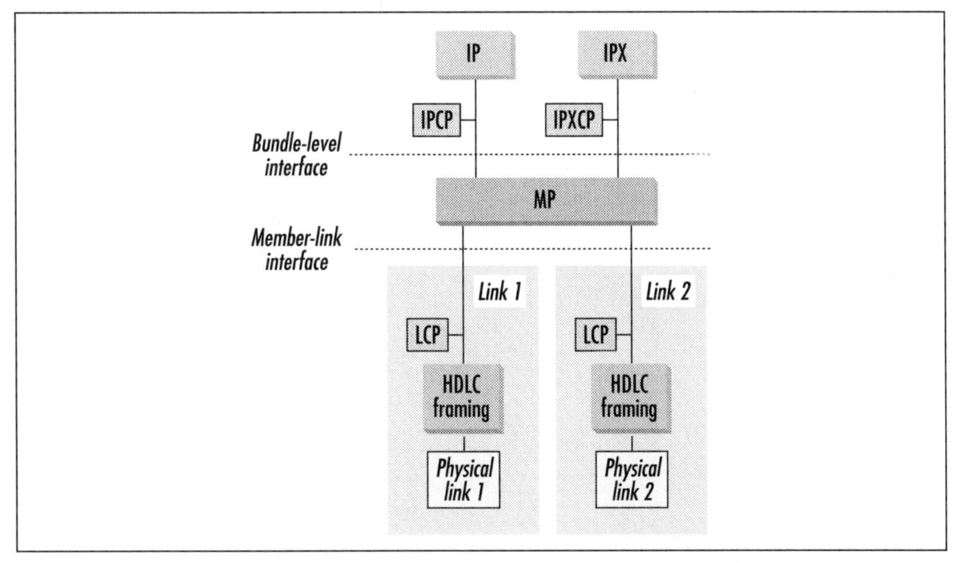

Figure B-1. MP architecture

the bundle level, an LCP Echo-Request verifies that the MP layer can transmit a message out through a member link, the other side can receive it and pass it up to the peer MP layer, and a reply can be generated using the reverse steps. In contrast, an LCP Echo-Request at the member-link level ensures only that a single link is functional. These two approaches are illustrated in Figure B-2.

MP Encapsulation

In a single link configuration, large IP packets are fragmented before being transmitted across the PPP link. MP bundles may safely be given a larger MTU because large packets may be fragmented for transmission across the bundle and immediately reassembled.* Figure B-3 shows the relationship between higher-level packets and MP fragments.

To distinguish MP transmissions at the bundle layer from member-link packets, MP transmits frames with PPP protocol 0x003D, and they have a secondary header with a sequence number. Figure B-4 shows the two variations on the MP frame.

Both frames are transmitted with the normal PPP framing sequence using the assigned MP protocol ID. MP fragments include extension headers of either 2 or 4 bytes. Both headers use a pair of bit flags to indicate the beginning and end of a fragmented higher-level packet. When a higher-level packet is fragmented, the first MP frame sets the B (beginning) bit to one. The last MP frame in a sequence sets

* Even though MP fragmentation is only a one-hop affair, reliable transmission is not often expected in conjunction with MP.

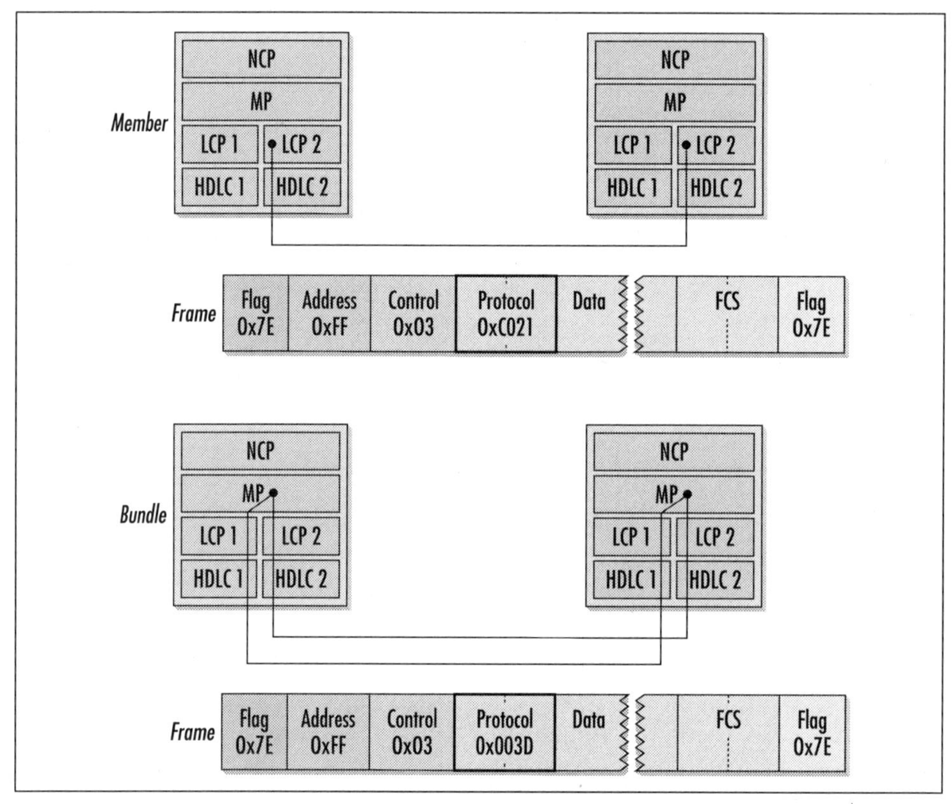

Figure B-2. Comparison of bundle and member link level activity

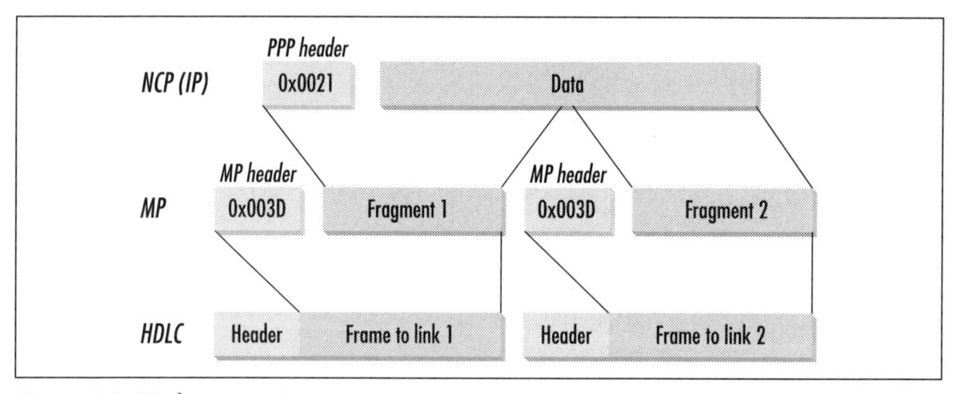

Figure B-3. MP fragmentation

the E (ending) bit to one. If higher-level packets are not fragmented, both the B and E bits are set to one. After the padding zero comes the sequence number, which is either 12 or 24 bits long. Use of the shorter header is negotiated with LCP on the first link in the bundle. Finally, MP frames, like all PPP frames, close with the frame check sequence. MP frames apply FCS error detection only to individual MP fragments. They do not carry out integrity checks on reassembled packets.

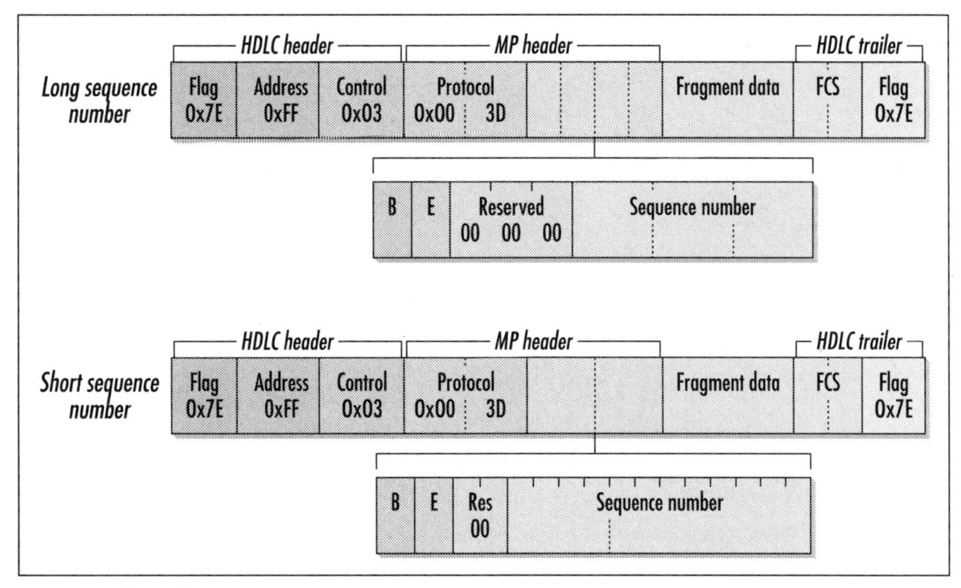

Figure B-4. MP frames

MP Fragmentation and Reassembly

MP imposes no requirements on the member links other than the ability to run PPP. Each MP frame an existing bundle transmits is given a sequence number. When a bundle is established, the sequence number counter is initialized to zero. Each MP frame transmitted on the bundle increases the sequence number by one, even if the frame includes a complete higher-level packet with both the B and E bits set to one. Bundles are higher-level logical constructions independent of member links, so the addition and subtraction of links does not affect the sequence number.

Member links can have different characteristics that may result in out-of-order packet delivery. MP fragments are placed in a reassembly queue and sorted by sequence number. Frames can then be recovered from the fragments based on the values of the B and E bits. Figure B-5 shows a simple example of frame reassembly with no lost fragments.

Fragment loss detection

MP fragment sequence numbers increase for each fragment that is transmitted on the bundle and are unique to each fragment. Sequence number space is shared between all the member links in a bundle.

As fragments are received, they are placed in the reassembly queue. For the purposes of detecting lost fragments, the most important sequence number is the minimum sequence number out of the set of the last-received sequence numbers on each link, a number that RFC 1990 refers to as *M*. M represents the sequence

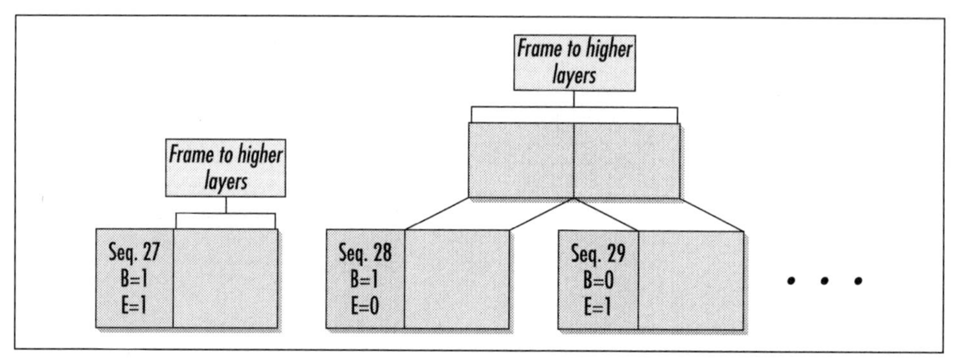

Figure B-5. Reassembly with no loss

number that serves as the cut-off for new fragments. Nothing will ever be received with a sequence number lower than M because sequence numbers always increase.* As an example, consider the fragment queue shown in Figure B-6.

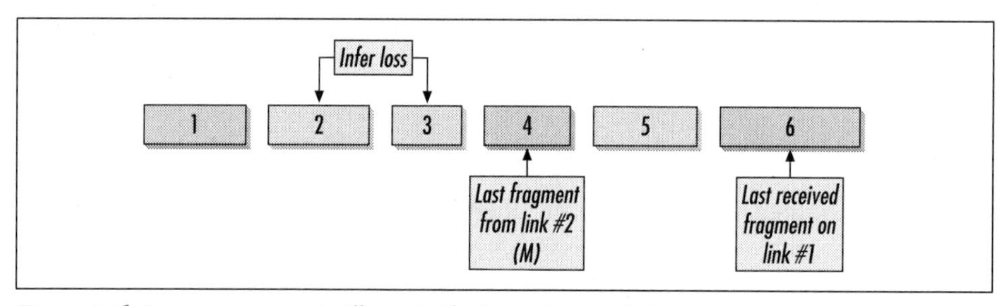

Figure B-6. Fragment queue to illustrate the importance of M

In Figure B-6, packets that have been received and are in the reassembly queue are indicated by shaded boxes, and packets that have not yet been received are white boxes. For simplicity, I'll assume a two-link bundle, though the analysis could easily be extended to any number of links.

At the far right-hand side is the sequence number most recently received on link 1 in the bundle. None of the white boxes will be received on link 1 because fragments are transmitted with increasing sequence numbers. If fragment 1 is the most recently received fragment on the second link in the bundle, there is no reason to suspect fragment loss. Link 2 will continue to transmit fragments in increasing sequence number order and could very well fill in the gaps. On the other hand, if the most recently received fragment is fragment 4, there is no way that fragments 2 and 3 will ever be received because both fragments 2 and 3 have been transmitted on the bundle.

* On long-lived bundles, sequence numbers may wrap around. RFC 1990 states that fragment numbers must not be ambiguous.

When a fragment with the E bit set is received, reassembly may be attempted if all fragments up to that point have been received. Figure B-7 illustrates this condition.

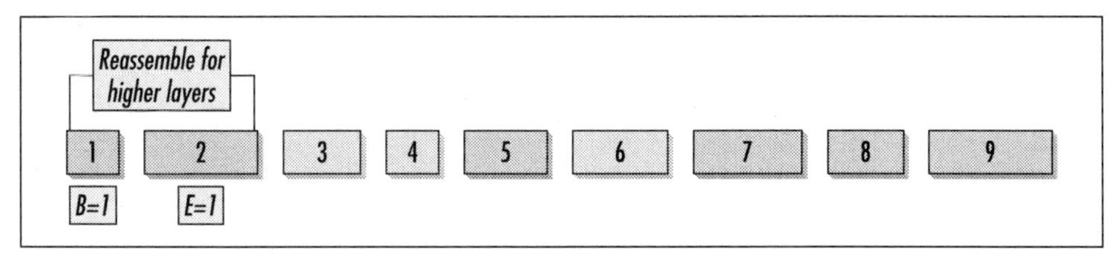

Figure B-7. E bits trigger reassembly

Fragment loss case 1. Fragments are declared missing in transmission if M moves past a fragment with the E bit set that has not been reassembled. Figure B-8 shows fragments from the two links in increasing sequence number order. Any B and E bits are noted on the figure. As in Figures B-6 and B-7, all successfully received frames are shaded.

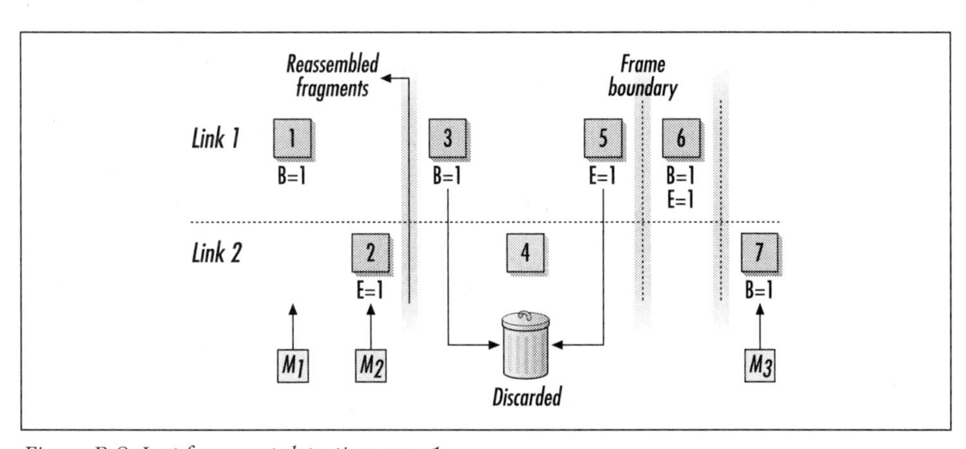

Figure B-8. Lost fragment detection case 1

Table B-1 explains the sequence of events in Figure B-8. Each row of the table corresponds to the transmission of one fragment. Any actions the receiving MP implementation takes are noted in the last column of the table. Figure B-8 uses two links and derives M from the most recent fragment (MRF) number on both member links.

Table B-1. Sequence of actions in detection case 1

Fragment number	Transmission outcome	MRF 1	MRF 2	M	MP receiver actions
1	Successful	1	0	1	Fragment 1 is buffered because it has only the B bit set and is not a complete frame.

Table B-1. Sequence of actions in detection case 1 (continued)

Fragment number	Transmission outcome	MRF 1	MRF 2	M	MP receiver actions
2	Successful	1	2	1	M is initialized to M_1. Fragments 1 and 2 are reassembled into a complete packet and passed to the higher-layer protocol.
3	Successful	3	2	2	Fragment 3 is buffered because it has only the B bit set and does not represent a complete frame. M is updated to M_2.
4	Lost	3	2	2	M is not updated because no sequence numbers change.
5	Successful	5	2	2	No fragment loss can be declared until M moves past 5. Fragment 5 is added to the buffer with fragment 3.
6	Successful	6	2	2	Fragment 6 is an entire packet and is passed to the higher-layer protocol. M is not updated.
7	Successful	6	7	6	M is updated to M_3, and fragment 4 is declared lost. Fragments 3 and 5 are discarded.

One important point should be made about the use of fragment numbers on MP member links. Fragments are always transmitted with increasing sequence numbers, but variations in link performance can lead to out-of-order fragment delivery. MP imposes no restriction on the latency or throughput of member links. Because fragments may be sent over any link, they can be declared lost only when all links have had an opportunity to transmit the frame. In Figure B-8, the receiver is waiting for fragment 4 to arrive. When fragment 5 is received on the first link, it is possible to conclude that the first link will not transmit fragment 4 because that would mean violating the increasing sequence number rule. However, it is not until the second link sends a fragment later that loss can be inferred. Figure B-8 also illustrates the need to buffer previous PPP frames even though subsequent frames may be reassembled. Fragment 6 is a complete frame, but its successful transmission does not mean that the previous frame was lost.

Fragment loss case 2. Fragments with the B bit set are assumed to follow a packet with the E bit set. If M is a fragment with the B bit set, any outstanding fragments with sequence numbers less than M are discarded. Figure B-9 illustrates an application of the second loss detection rule, with the actions taken by a receiver spelled out in Table B-2.

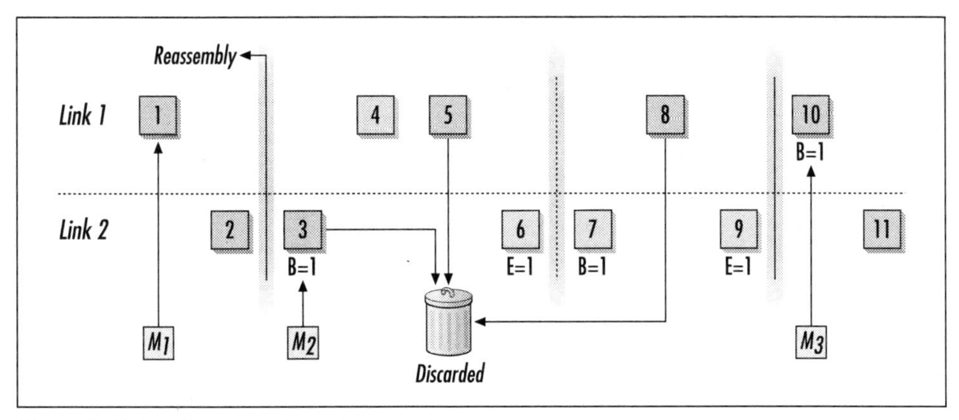

Figure B-9. Lost fragment detection case 2

Table B-2. Sequence of actions in detection case 2

Fragment number	Transmission outcome	MRF 1	MRF 2	M	MP receiver actions
1	Successful	1	0	1	Fragment 1 is buffered because it has only the B bit set and is not a complete frame.
2	Successful	1	2	1	M is initialized to M_1. Fragments 1 and 2 are reassembled into a complete packet and passed to the higher-layer protocol.
3	Successful	1	3	1	Fragment 3 is buffered for reassembly.
4	Lost	1	3	1	
5	Successful	5	3	3	Fragment 5 does not have the E bit set, so it is added to the reassembly queue. M is updated to M_2.
6	Lost	5	3	3	
7	Lost	5	3	3	
8	Successful	8	3	3	Fragment 8 does not have the E bit set, so it is also added to the reassembly queue.
9	Lost	8	3	3	
10	Successful	10	3	3	Because fragment 10 has the B bit set, fragment 9 must have had the E bit set. A frame boundary can be inferred between fragments 9 and 10.

Table B-2. Sequence of actions in detection case 2 (continued)

Fragment number	Transmission outcome	MRF 1	MRF 2	M	MP receiver actions
11	Successful	10	11	10	M advances to 10, which is noted as M_3 in the figure. The second reassembly rule can be applied, so fragments 3, 5, and 8 are discarded.

Practical Considerations. When links are idle, M will not advance, and the fragmentation-loss rules depend on advancing M past the frames with lost fragments. To keep M moving up in the sequence-number space, idle links may transmit empty fragments with B and E both set and no data. Figure B-10 illustrates how this might be useful in an ISDN bandwidth-on-demand scenario.

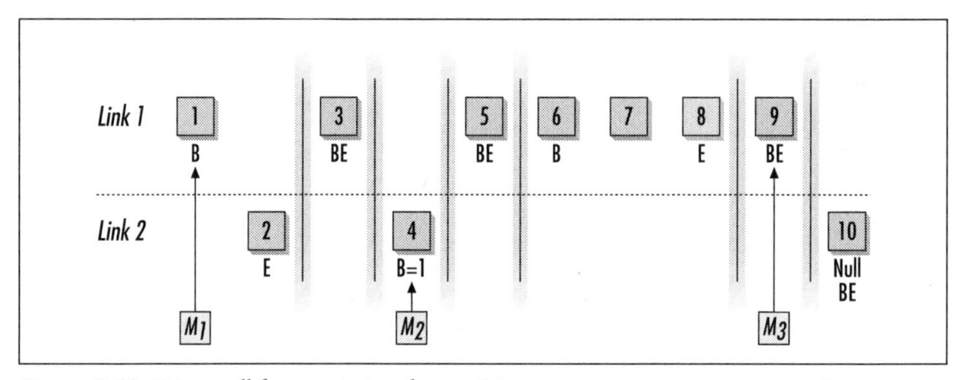

Figure B-10. Using null fragments to advance M

Secondary links are brought up to handle traffic bursts. In Figure B-10, that process has taken place prior to the beginning of the fragment stream shown. Both links are transmitting fragments full-throttle. Fragments 1 and 2 are reassembled, and M is 1 (M_1). Fragments 3, 4, and 5 are all standalone frames that need no reassembly. When fragment 5 is received, M is 4, the minimum of the most recent sequence number on the two links (M_2).

The second link will often go quiet and stop transmitting. Network traffic is bursty. It is common for users to fetch graphics-laden web pages and stop to read them, or for the Internet to limit the incoming stream to less than 64 kbps. In any case, when the second link is no longer needed, everything may be transmitted on the first link. Fragment 8 is lost, but the loss cannot be detected until M moves past 8.

To enable MP implementations to detect this loss, the second link may periodically transmit a null MP frame. Null frames have no data, but they increase the sequence number. Increasing the sequence number also increases M. When fragment 10 is received, M moves to 9 because it is the lesser of the most recent sequence numbers on the two links (M_3). In the case illustrated in Figure B-10, when M moves to 9, the second loss rule applies and fragment 8 can be declared lost. Fragments 6 and 7 are then discarded.

MP as a link layer fragmentation mechanism

PPP does not support fragmentation. If the MTU on a PPP link is too small, IP fragmentation may be used. IP fragmentation, however, is not without its problems. IP fragmentation adds a great deal of overhead because each fragment requires a complete IP header, which can be substantial on slow links. IP fragmentation may also add a noticeable load on the receiver because IP fragments are reassembled only at the destination.* For these reasons, it may be desirable to use link-level fragmentation over the PPP link if the MTU of the link cannot be changed. Although not strictly intended to do so, it is possible to use MP for link layer fragmentation by negotiating MP over the link and setting the MRU to the desired size.

LCP Options for MP

MP, like all other PPP options, is negotiated with LCP. Unlike other protocols that ride on PPP, though, it does not have an associated control protocol. All MP options are negotiated through LCP.

Every bundle in a link must use the same set of LCP options. When additional links are brought online, they will send configuration requests with the values negotiated for any existing links in the bundle.

Maximum Reconstructed Receive Unit (LCP Option 17)

Negotiation of the Maximum Reconstructed Receive Unit (MRRU) option serves two purposes. First, it indicates that the sender implements MP. When it is accepted, the link will be joined to an existing bundle or a new bundle will be created.

In addition to its implied purpose, the option proposes a maximum size to the reconstructed higher-level packets. As such, it looks very similar to the MRUoption. Naturally, it has a different type code. Figure B-11 shows the format of the MRRU option. Like other LCP options, it has a type code and a length, followed by the MRRU.

Both peers must request use of the MRRU to create the logical multilink. The default MRRU value is 1,500, the same as in the single-link PPP specification.

---

* Extremely security-conscious sites may even drop fragments at the border. A number of recent deficiencies in TCP/IP stacks have relied on the ambiguities in the way fragments are handled by different operating systems. To avoid trouble, high-security sites may block fragments entirely and rely on Path MTU discovery (RFC 1191).

Fragmentation Contrasted: IP and MP

To think about MP fragmentation, it is useful to compare it to IP fragmentation. IP fragmentation is a well-known, thoroughly documented procedure. Here are some comparisons:

Basic architecture
: Neither IP nor MP fragmentation employs retransmission or acknowledgment. Loss is detected and fragments are discarded, but it is up to higher-layer protocols to request retransmission.

Fragment break points
: IP fragments must break on an 8-byte boundary, except for the final fragment, which ends wherever it is necessary. MP breaks at any convenient location. The only requirement imposed on MP fragmentation is that PPP present a 1-byte stream interface to higher layers, so MP fragments must break on byte boundaries.

Reordering and fragment overlap
: IP fragments are independent IP datagrams, so they may be routed however the network wishes and are frequently delivered out of order as a result. Refragmentation of fragments may even occur, and it is not uncommon for fragments to have overlapping data segments. MP does not reorder or route fragments. MP fragmentation simply numbers packets, and no overlap between successive fragments is possible.

Discard rules
: IP fragments are discarded when all fragments fail to arrive in a "reasonable" period of time. MP uses the increasing sequence-number rule to infer packet loss. Well-designed MP implementations often use timers in addition to the sequence-number rule as an extra layer of protection, but they are not required.

Unlike single-link PPP, the MRRU must be negotiated because the MRRU option implies the use of MP. Without the MRRU option, there is no multilink. Multilink-capable implementations may attempt to prod a peer into sending an MRRU request by including the MRRU in a Configure-Nak message.

Short Sequence Number header format (LCP Option 18)

On lower-speed links, the shorter MP subheader is used to avoid wasting bandwidth. The option to use the Short Sequence Number (SSN), shown in Figure B-12, is very simple. Including the option indicates a desire to use the SSN.

With a 12-bit sequence number, only 4,096 MP fragments are allowed to be outstanding in the reassembly queues. In order for SSNs to be viable, the product of

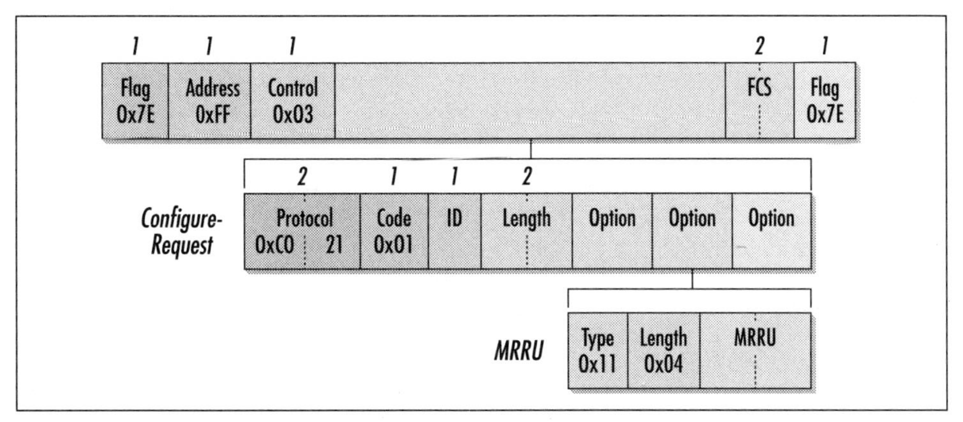

Figure B-11. MP Maximum Reconstructed Receive Unit LCP option

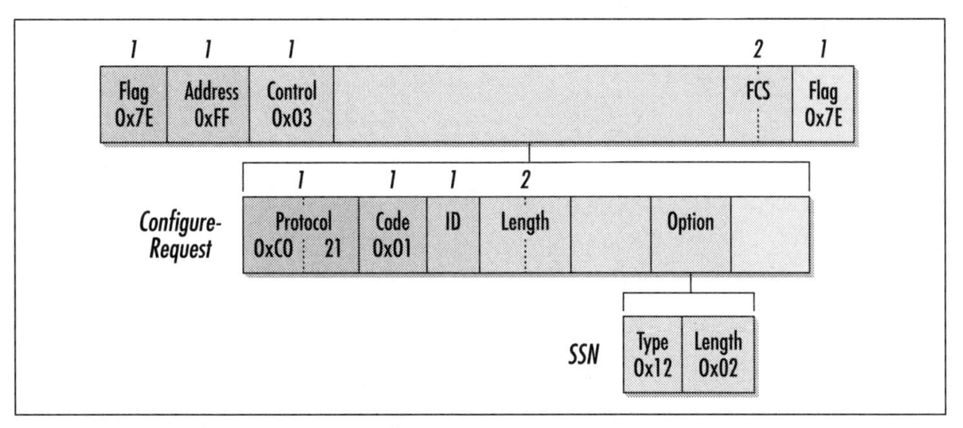

Figure B-12. MP Short Sequence Number LCP option

the time delay over the link (in seconds) and the maximum fragment transmission rate (in fragments per second) must not exceed 4,096. 24-bit sequence numbers allow the product to be almost 17 million and are sufficient for nearly every link.

RFC 1717 was the original multilink specification. It allowed use of the SSN option or the MRRU option to imply multilink operations. Older implementations may still attempt to use the SSN operation to activate multilink procedures. While technically not allowed, some implementations allow it for backward compatibility.

Endpoint Discriminator (LCP Option 19)

The Endpoint Discriminator (ED), shown in Figure B-13, is an unauthenticated field used to help a peer determine the bundle to which a link belongs.

Following the type code and length, the ED consists of a 1-byte class field to determine the type of information given by the ED, followed by the data composing the

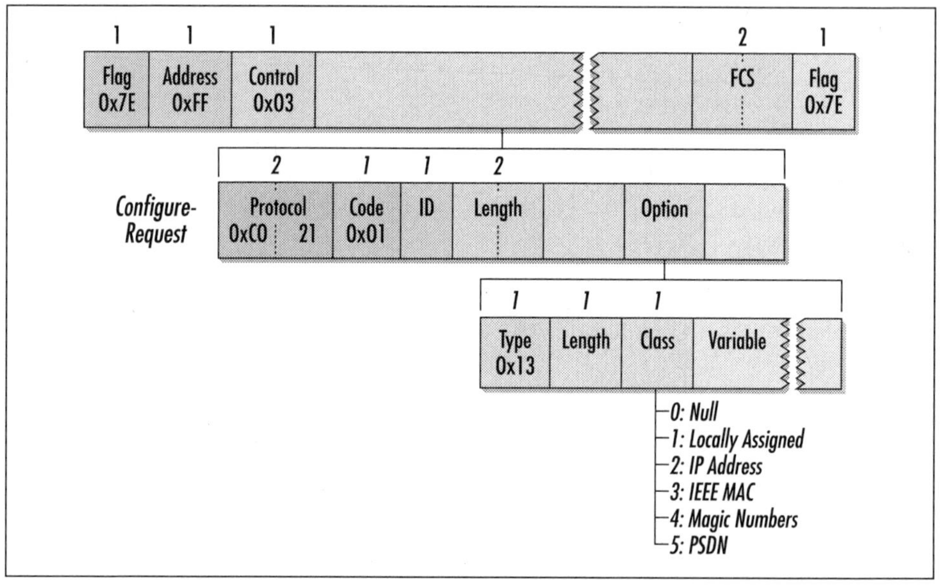

Figure B-13. MP Endpoint Discriminator LCP option

discriminator itself. EDs should be treated as byte strings that must be matched for links to belong to the same bundle. Table B-3 summarizes ED class values.

Table B-3. Endpoint Discriminator class values

Class code	Class name	Length (bytes)	Description
0	Null class	0	Equivalent to, not including, the ED.
1	Locally assigned address	<=20	(Deprecated) This option has been deprecated because there is no way to guarantee unique values, even if device serial numbers are used.
2	IP address	4	An IP address, usually of one of the sender's interfaces.
3	802.1 global MAC	6	The MAC address of one of the sender's Ethernet interfaces, in the 802.3 format.
4	PPP magic number	<=20	(Deprecated) The PPP magic number LCP option is designed to be unique. Up to five can be combined into a single magic number block. Use of this option is deprecated because there is no guarantee that two systems will not come up with the same random bit string.
5	Public switched directory number (telephone number)	<=15	The telephone number, as defined in ITU Recommendation E.164, is used to access the sender.

Bundle Maintenance

Three major operations exist with respect to bundle maintenance. New bundles must be formed for new connections. New links must be added to existing bundles in a manner transparent to the higher-layer network protocols. Existing links must also be dropped from bundles without disturbing higher-layer transmission.

Bundle creation

Figure B-14 illustrates the MP bundle creation procedure. In the course of configuring the link, LCP must offer a value for the MRRU in an LCP Configure-Request message. If a peer wishes to use MP and the MRRU is not included in an incoming configuration request, it may send an unsolicited Configure-Nak suggesting the MRRU option and a value. MP must be negotiated in both directions, which means that both sides must use the MRRU option in configuration requests.

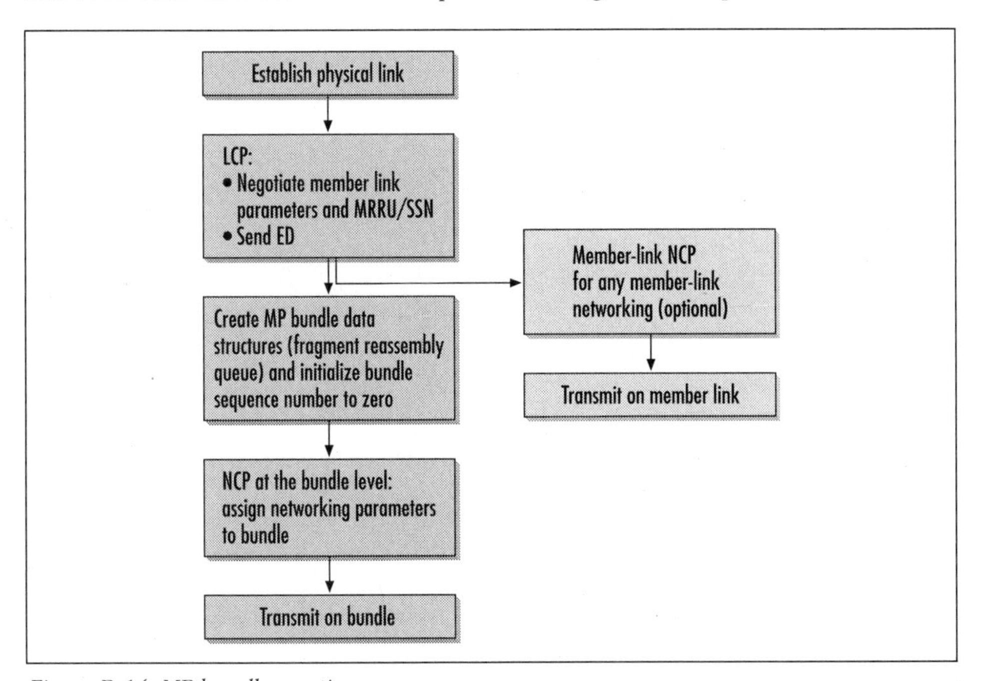

Figure B-14. MP bundle creation

Once the use of MP is agreed upon, the bundle is configured. LCP negotiations are not used on the bundle itself. Instead, each bundle is configured with the following:

- The asynchronous control character map is disabled. The ACCM handles control characters on member links, so the bundle itself is transparent to any data.
- The magic number is disabled. Bundles are established over member links. If anti-loopback protection with the magic number is necessary, the magic number procedures may be used on each member link.

- Link Quality Monitoring is disabled. Several different types of links may be combined into a bundle, so imposing a uniform quality restriction on the entire bundle is counterproductive.
- Address and Control Field Compression and Protocol Field Compression are enabled.

Bundle creation also sets up the various data structures for fragmentation and reassembly and initializes the fragment sequence number to zero.

Adding links to an existing bundle

Figure B-15 illustrates the procedure for adding new links to an existing bundle. MP first verifies that links are in the same logical session by comparing authentication information and the Endpoint Discriminator according to Table B-4. After identifying the existing bundle or establishing a new bundle, the MP parameters (MRRU and SSN) are compared to the bundle. If they are not identical, the link must be reconfigured to match the parameters for the existing bundle.

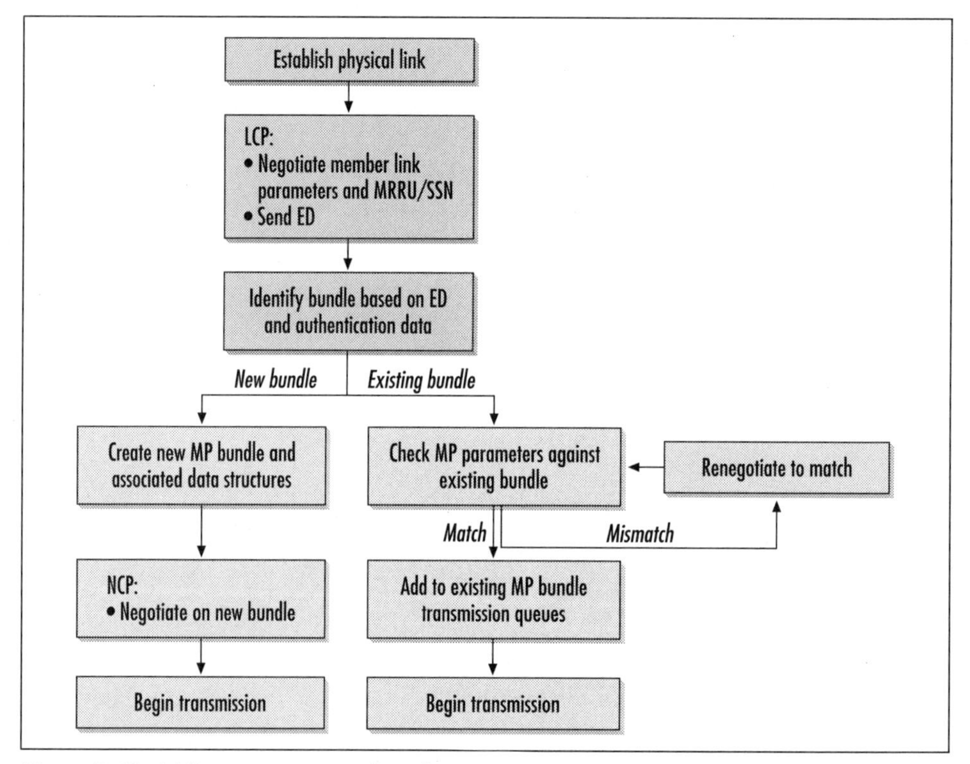

Figure B-15. Adding to an existing bundle

Table B-4. Bundle establishment rules

	Authentication	No authentication
Discriminator	Join existing bundle if both discriminator and authentication match, otherwise establish new bundle	Join existing bundle for discriminator match, otherwise establish new bundle
No discriminator	Join existing bundle for authentication match, otherwise establish new bundle	Join to system default bundle unless overridden by other administrative configuration

Adding links to a bundle is transparent to the MP layer itself. Bandwidth may increase, but the fragment sequence numbers are unaffected by new links.

To speed up negotiation and joining to a bundle, some implementations may use the ED to perform a preliminary bundle lookup. After identifying the existing bundle the new connection is likely to join, it is easy to select compatible LCP options to cut down on the LCP negotiation time. After LCP, the link is authenticated and joined to the existing bundle if the authentication data matches. The only drawback to this method is that it does not support links with the same ED but different authentication credentials; however, this is an unusual situation and is not likely to be observed on dial-up servers providing Internet access.

When links do not supply an ED or authentication credentials, RFC 1990 specifies that they must be joined to a system default bundle unless manual configuration provides an alternative action. Joining nonauthenticated links with no distinguishing information to the same bundle has obvious security consequences, and many implementations will simply refuse nonauthenticated connections. Another common solution is to use manual configuration and additional telco services such as caller ID or ANI to assign new links to bundles based on information supplied by the telco.

Dropping links

Member links are terminated using the standard LCP termination routines. Dropping a link is transparent to the MP session provided that at least one link still remains in the bundle. An MP bundle is a logical construct in software and persists independent of any of the member links, including the first link in a bundle. Some implementations may erroneously terminate all bundle links or otherwise behave strangely if the first link is lost, but RFC 1990 clearly intends for the MP state to exist independently of any lower-level constraints, including the number and identity of any member links. The exception is that if only one remaining link exists, both sides may agree to stop using MP headers and revert to standard PPP encapsulation.

Whenever possible, links should be removed gracefully from the bundle using LCP termination frames. To prevent problems with fragment reassembly, it is common to send a Terminate-Request to the peer and process fragments until a Terminate-Ack is received. (Fragments may not be requeued on other links because fragment sequence numbers must increase, and queued fragments would have earlier sequence numbers than fragments that had already been transmitted.)

Strict RFC constructionists may be disturbed by this interpretation because RFC 1661 and the state engine imply that the administrative closure of a link moves LCP from the Open state to the Closing state. (See Figure 9-6.) In any state other than Open, PPP should discard incoming data frames. However, the strict reading of the RFC would cause any fragments already in the transmit queue on the member link to be discarded. As an efficiency measure, several MP implementations allow the Terminate-Request to be used as a queue flush measure. When the peer receives a Terminate-Request, it marks the link as down, transmits any queued fragments, and sends a Terminate-Ack.

Multi-Chassis MP (MMP)

Classic MP requires that all links in a bundle terminate into the same device. Fragments are held in memory and must be conveniently accessible to a single MP process for reassembly. For a circuit-switched path between two endpoints under the control of the same company, it is easy to have one machine on both ends of the link.

Service providers are a different story.* No matter how large vendors can make dial-up termination equipment, there will always be a service provider who needs something bigger. Service providers may have several large routers terminating subscriber PPP sessions in a POP. All the T1 circuits providing the dial-up lines are aggregated together into one hunt group. At peak times, the dial-up channel a user receives is likely to be almost random; ISDN subscribers can easily have bearer channels terminated on separate routers.

Two major approaches have been developed to allow MP sessions to terminate correctly into a massive POP. One is to use additional protocols to direct subsequent calls to the access server already terminating the existing MP sessions. Alternatively, access servers can communicate with each other to determine when new sessions have landed on the "wrong" access server. The first chassis to accept an incoming session and instantiate the bundle becomes the *bundle head* and is responsible for performing bundle-related operations such as fragment reassembly. Layer-two tunneling techniques are used to forward fragments from the "wrong" access server to the bundle head. The bundle head is the logical peer of the dial-in user, even though the routing of MP fragments to that peer takes a somewhat circuitous path. A high-level sketch of the two approaches is shown in Figure B-16.

* By service provider, I do not mean ISP exclusively. IT departments at large companies are often called on to run massive dial-in setups and function, in effect, as service providers for employees.

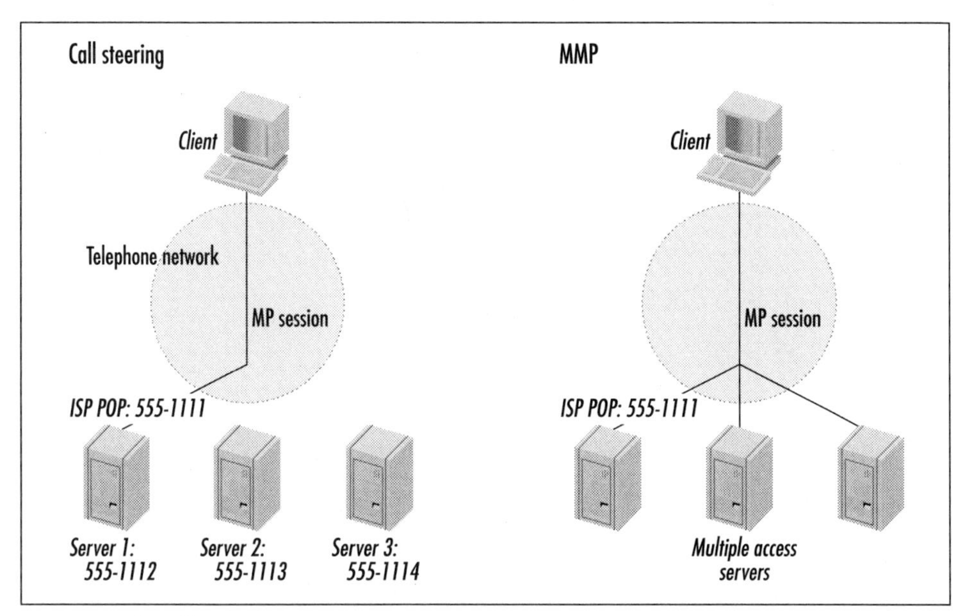

Figure B-16. Call steering versus MMP

In the call-steering approach, the first link in the bundle is terminated by one of the many access servers at the POP. Incoming sessions to the main POP phone number, 555-1111, are routed to the first free line into the POP. If a call is terminated by the first access server, it responds with a message to help additional sessions be routed to the terminating server. One method of doing this is to use the *Bandwidth Allocation Control Protocol* (BACP) to inform the dial-in client of a different number for additional sessions. In Figure B-16, the access server would tell the client to use 555-1112 for any additional sessions.

Multiple chassis can terminate the member link calls when layer-two techniques are used, so the client can use the main POP number for any additional sessions. If additional sessions are split across multiple access servers, the layer-two forwarding of fragments will ensure a seamless MP session. Layer-two techniques are often referred to generically as Multi-Chassis MP (MMP) because sessions may be terminated across several access servers. This is not to imply that MMP is a standardized protocol—in fact, all the major vendors have taken different approaches.

A Generic Approach to MMP

The biggest advantage of the layer-two techniques is that they make better use of the available hardware. When using a call steering technique, the danger is that a call will land on a busy server. When the client tries to establish additional secondary MP links, the busy server will not be able to accept them. However, spare capacity might exist at the POP. This is shown in Figure B-17.

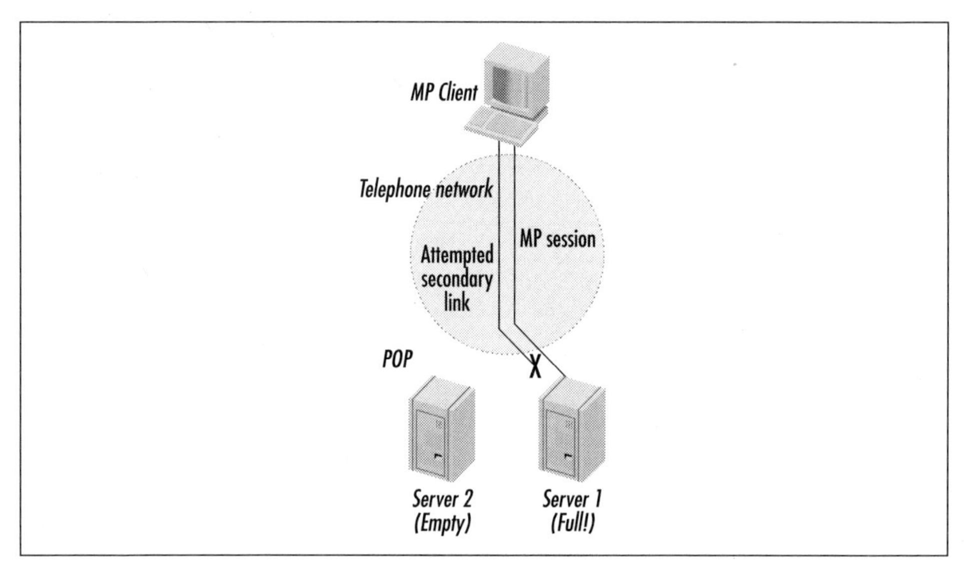

Figure B-17. Problems with call steering when only one chassis is busy

The layer-two techniques separate the MP processing into two distinct parts. Call termination is not special, and any piece of hardware in the POP can perform it. It is only the MP fragmentation and reassembly that must be done centrally at the bundle head; the telephone calls may be received by any of the units in the POP.

In Figure B-18, the first session is terminated on the first access server. MP fragments are passed over an internal bus to the MP termination routines, which run on the first access server's CPU. When the client places a second call in the MP session, it lands on the second access server. After using discovery protocols to determine where the MP termination point is, the second server opens a layer-two tunnel to the MP termination point. (As an implementation detail, the first server may create a virtual call termination interface to the layer-two tunnel.) All the MP termination functions are performed centrally on the first access server, but it does not need to have the capacity to terminate every call from the POP.

When using a layer-two approach, the devices at the end of a link can handle bandwidth control. Both ends of the link have the same information and can use different algorithms to determine if additional capacity is necessary. To prevent link thrashing—where links are repeatedly established by one end and torn down by the peer—it is recommended that only the system that initiated the link establishment be allowed to remove it.

The major networking vendors each have a proprietary MMP technology. Nortel's Multilink Multinode Bundle Discovery Protocol is published in RFC 2701 and will be considered in more detail in the next section. Lucent's Multi-Chassis MP (MCMP), derived from the Ascend MAXStack protocols, is not documented in a detailed form. Cisco's solution is the Stack Group Bidding Protocol (SGBP). Propri-

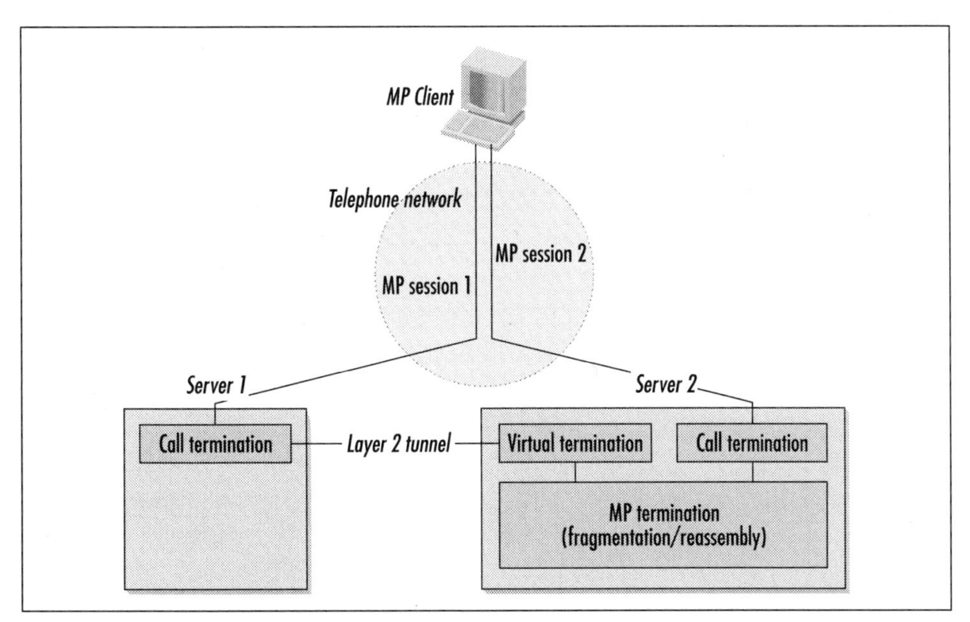

Figure B-18. MMP details

etary approaches can survive in MMP protocols because customers typically try to have homogenous POPs. With only one vendor in the POP, customer demand for an open, interoperable MMP standard is limited.

Nortel Multilink Multinode Bundle Discovery Protocol

Of the common techniques, only Nortel's is specified in the level of detail required for an interoperable implementation. RFC 2701 details the operation of Nortel's Multilink Multinode PPP Bundle Discovery Protocol. All the proprietary layer-two approaches to MMP are similar, but I will describe Nortel's in detail because it is documented publicly and interoperable implementations are, in theory, possible.

Incoming call handling

When a call comes in, LCP establishes the data-link layer and basic MP negotiations. When that is done, there are three possibilities for the call:

1. It is the first call from that user, and the bundle head should be created.
2. It is a secondary link for an established bundle, and the bundle head is on the chassis terminating the call. In this case, the new link can simply be added to the existing bundle using standard MP techniques.
3. It is a secondary link for an established bundle, and the bundle head is on another chassis. In this case, the call-terminating device uses an L2TP to send

the MP fragments to the bundle head. The bundle head performs all fragmentation and reassembly.

Incoming calls are handled according to the procedure depicted in Figure B-19, as described in the following steps:

1. The terminating server checks for an existing bundle head on its own chassis. If a bundle head exists, it is case 2 and the new member link is attached to the existing bundle.
2. As an optimization, a new link for a bundle terminated on a different, but known, chassis can simply be added to the remote bundle using the existing L2TP tunnel. This step is optional.
3. The location of the bundle head is unknown, so the terminating chassis will use the discovery procedure to attempt to find the bundle head. If discovery finds a remote bundle head, the L2TP tunnel is established and the new member link is attached to the remote bundle head. If discovery fails, it indicates that this is the first link and a bundle head is created.

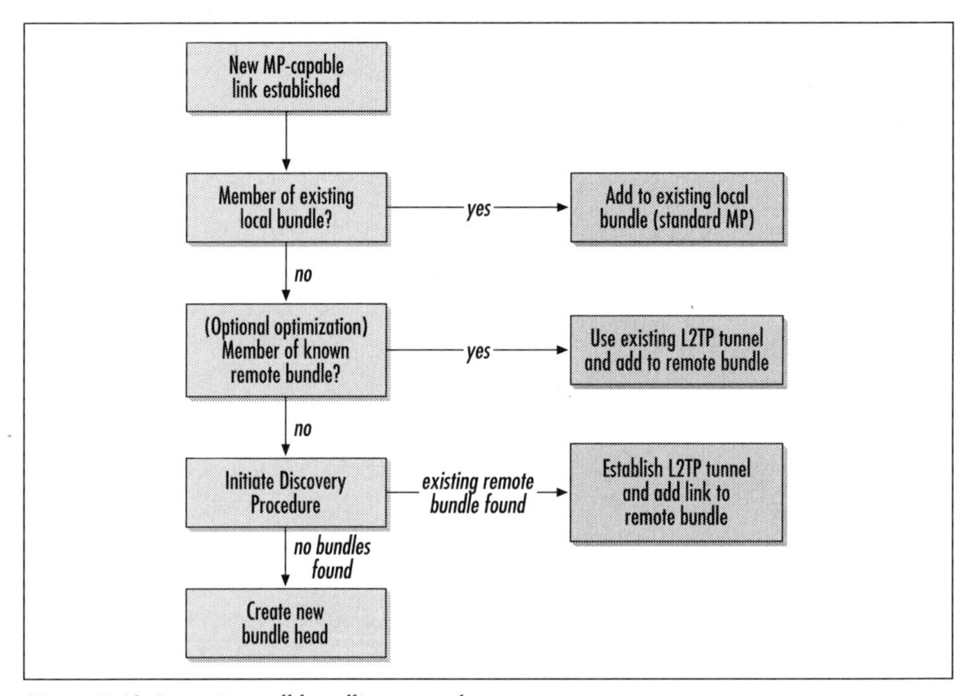

Figure B-19. Incoming call handling procedure

Bundle-head creation must be a fast process. Power users may attempt to initiate two ISDN B channel connections almost simultaneously, and it is preferable if one of the two stations establishes a bundle head that the other may join. However, if two simultaneous bundle heads are created, an election procedure is used to determine which chassis will terminate the bundle head.

Discovery message format

The format of the discovery messages is shown in Figure B-20. The data in a discovery message begins with a 2-byte type code. Queries use a type code of 1, and responses use a type code of 2. A 2-byte length field is set to the length, in bytes, of the endpoint ID field at the end of the message, not including the header. Election procedures make use of the 4-byte random ID, which will be discussed in the next section. To make endpoint ID matching as fast as possible, a 2-byte hash is included in the message. Hashes are a way of speeding up endpoint matching; the calculation method is detailed in the RFC. At the end of the discovery message is a variable-length endpoint ID, which is a concatenation of several MP attributes. Although the fields in the endpoint ID have individual meanings to the MP process, for two incoming calls to be included in the same bundle, the discovery protocol treats them as a string of bits that must match bit-for-bit. In order to ensure that the endpoint ID will result in a bundle match for calls terminated by different access servers, all access servers joined together in a group must have identical EDs.

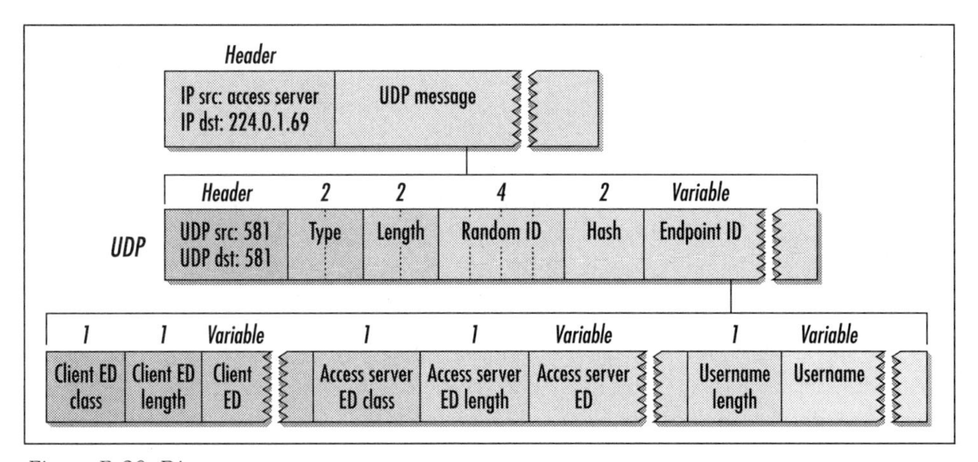

Figure B-20. Discovery message

Discovery protocol operations

The discovery protocol is a request-response protocol, similar to ARP in its operation. When an incoming call goes all the way to step 3 and the discovery process is initiated, queries are sent out to the IANA-reserved multicast address of 224.0.1.69 on the IANA-allocated UDP port of 581. In the rare cases when multicast facilities are not available, the limited all-ones broadcast address, 255.255.255.255, is used instead. By default, messages are transmitted with a time-to-live (TTL) of 1. To guard against message loss, the discovery protocol retransmits messages if no response is received. The number of messages and the query interval are configurable.

If a query arrives that matches a link that already has a bundle head established, the server with the bundle head responds with a unicast response message. Response messages simply have different type codes. This case is shown in Figure B-21.

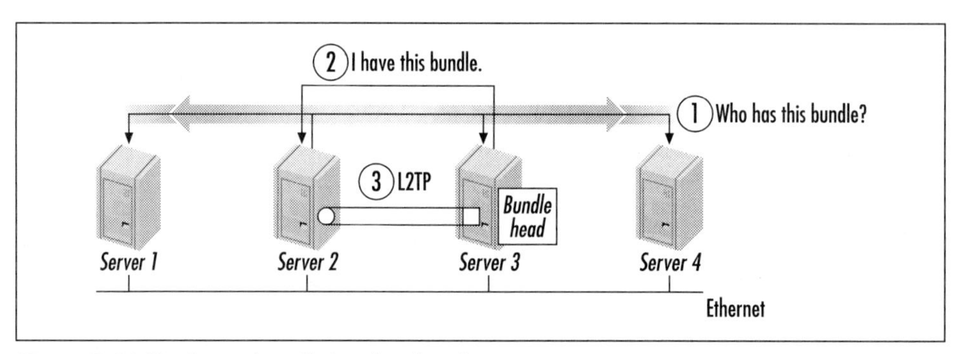

Figure B-21. Finding a bundle head with a discover operation

Quick responses are important. At crowded POPs, busy access servers may need to search through hundreds of bundles to determine whether the query matches any of the bundles they are currently terminating. Comparing endpoint identifiers can be a slow process involving byte-by-byte comparisons. To speed up the process, the hash is generated and sent along with each query message. When a query is received, the endpoint identifier in the query message need only be compared to the bundles with matching hashes. Hashes do not guarantee a match, but they reduce the number of slow byte-by-byte comparisons servers must perform.

If no response is received after transmitting all the queries, it is assumed that no bundle head exists. Once a server begins creating a bundle head, it will multicast a response message to minimize the number of elections by informing the other servers that it will be responsible for the bundle head.

Robustness is built into the discovery protocol through a retransmission counter, which may range from 1 to 15. By default, three transmissions are used before concluding that there is no bundle head. Lower retransmit counter values increase the probability that a lost packet will result in the creation of multiple bundle heads. Higher values increase the amount of time necessary for the discovery procedure to time out and begin creating a bundle head. No retransmission counter is used for lost reply messages. Instead of retransmitting, missed reply messages are simply triggered by retransmitted queries. In addition to the number of queries, the query interval may be configured. The interval is expressed in tenths of a second and may range from 1 to 15.

When bundle heads are established simultaneously, ties are broken by the random ID field. Each new connection to a server is given a unique identification number. IDs can be reused only when there is no longer a danger that they can be confused with an earlier connection. For this reason, it is recommended that the ID field increase with time. If the random ID field is the same, the tie is broken by the lowest IP address. When the tie is broken, the victor sends two response messages and forms the bundle head. Other servers receive the response message and update their bundle lists with the location of the bundle head.

Other Proprietary MMP Solutions

All MMP solutions take the general shape of the Nortel solution described previously. A method of bundle discovery finds the bundle head, and a layer-two encapsulation method sends fragments from the call termination point to the bundle head.

Lucent (Ascend) Multi-Chassis MP (MCMP)

MCMP is not a documented protocol, but it is similar to the Nortel solution. When a new link is established to a MAX access server, several requests are sent out within one second. These messages are sent as a link layer broadcast (with an IP destination of 255.255.255.255) on port 5151. To avoid overwhelming other devices on the link, an Ethernet multicast address from an Ethernet OUI assigned to Ascend is used (01:c0:7b:0:0:1). When another system owns the bundle, it responds to allow the terminating MAX to join the new call to an existing bundle. If no response is received, the bundle is created locally. The Ascend solution also includes an encapsulation method to ferry MP fragments from the terminating node to the bundle head.

Cisco MMP and the Stack Group Bidding Protocol

Cisco's MMP solution is also similar to the Nortel solution. Instead of L2TP, it uses Cisco's Layer Two Forwarding (L2F) protocol to encapsulate MP fragments. Assignment of the bundle-termination function is based on Cisco's Stack Group Bidding Protocol (SGBP). Rather than creating bundle heads where they initially land, SGBP allows systems to bid for the right to create them. Bidding means that bundle heads can be distributed relative to CPU power, which allows higher-powered routers to be used for bundle head functions. MP fragmentation and reassembly require higher CPU power than terminating the telephone call. SGBP allows network administrators to design POPs with simple access servers to terminate incoming calls and high-powered routers to terminate MP sessions. When high-powered routers are used only for MP session termination, Cisco refers to them as *offload MMP servers*. These are illustrated in Figure B-22. Cisco has not published detailed specifications of their MMP solution.

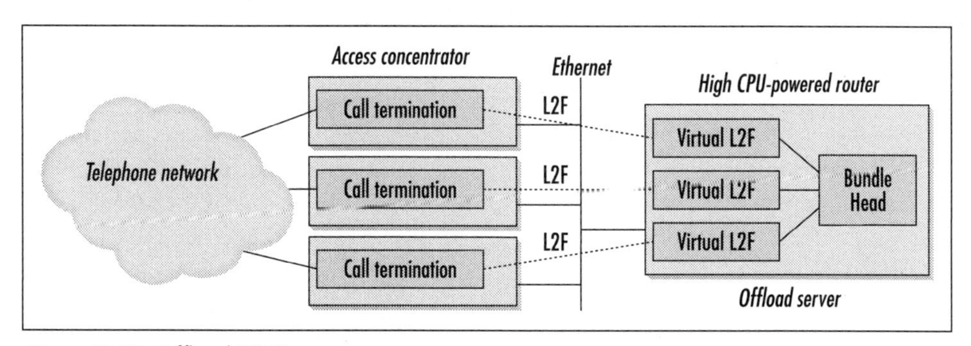

Figure B-22. Offload MMP servers

T1 Performance Monitoring

Populus vult decipi, decipiatur.
(The people want to be deceived,
let them be deceived.)
—Cardinal Carlo Caraffa

Equipment that terminates T1 circuits is often designed to collect data on the performance of the link. WAN links frequently enable remote offices to connect to corporate information resources and are a necessary prerequisite to productivity. When WAN links degrade or fail, end users notice quickly and solutions must be found and implemented promptly.

Several standards address performance monitoring of T1 links. This appendix lists the most common counters that are available in end-user devices. Monitoring this data can help you spot trouble before it becomes critical, and time series data may help correlate performance issues with other network changes.

CSU/DSU functionality runs over a wide range of price points and functionality sets. Inexpensive CSU/DSUs don't bother to collect this data at all and may not even have indicator lights for all of it. High-end CSU/DSUs may collect the data for a short time period so it is available to local network administrators. For network administrators, the best CSU/DSUs are often the integrated units sold by router vendors, because they collect performance data for long periods and expose it to network management applications by the router SNMP agent. As you read this chapter, keep in mind that not all equipment will maintain all the counters. Refer to the equipment vendor's documentation for details.

Collecting Performance Data

To allow carriers to monitor T1 link performance, a message-based protocol is defined for use on the facilities data link. Formatting of the messages is based on LAPD. Performance reports are sent once per second and have the structure shown in Figure C-1. Each byte is transmitted right-to-left (bit 1 to bit 8), with the bytes transmitted in order downward.

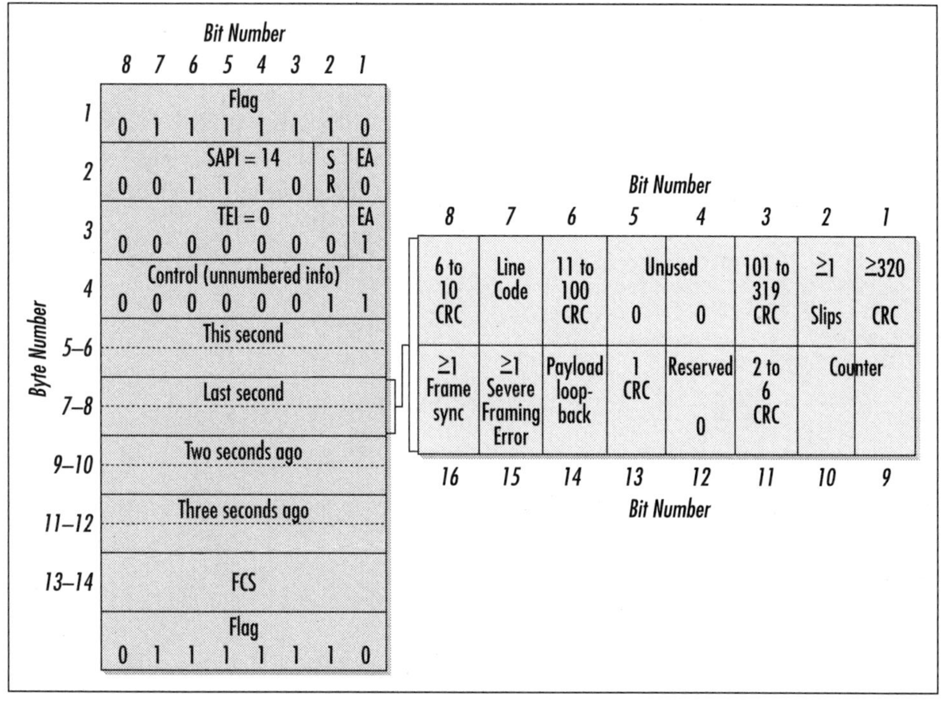

Figure C-1. Performance report message format

The format of the performance report message (PRM) should look familiar—it is based on LAPD, which is in turn derived from HDLC. Like HDLC, the message begins and ends with a unique flag sequence—0111 1110—to mark the beginning and end of the LAPD frame, and it uses a frame check sequence to validate data integrity. Like any other HDLC transmissions, the performance reports are subject to the zero stuffing procedure described in Chapter 8.

LAPD was designed to multiplex several logical information channels over a single physical channel. Logical addresses are defined by the *Service Access Point Identifier* (SAPI) and *Terminal Endpoint Identifier* (TEI). For T1 performance reports, The SAPI is always 14 and the TEI is always 0.* Messages from the carrier set the

* Other SAPI numbers are used for different purposes. Path and signal testing are done over the facilities data link with messages on SAPI 15, and an idle code may also be transmitted on SAPI 15.

command/response (C/R) bit to 1, and user messages have the C/R bit set to 0. After the addressing information, the extended addressing (EA) bit is set to 1 to indicate the end of the address information. Because performance reports are not acknowledged, they are transmitted as unnumbered information. To guard against data loss, each PRM includes data for the previous 4 seconds. Four consecutive seconds are reported on in each message through the use of four information elements, which describe statistics for a 1-second interval.

Statistics Reported in the PRM

Each 1-second interval has a 2-byte information element assigned to it. An information element is shown in Figure C-1 in detail. Six bits report the number of CRC errors on the link in six tiers. Other bits report on framing errors, line code violations, and frame slips. Table C-1 shows the use of each bit in a 1-second information element.

Table C-1. PRM information element bit interpretation

Bit number	Interpretation when set to 1 (except where noted)
13	One CRC error.
11	2 to 5 CRC errors.
8	6 to 10 CRC errors.
6	11 to 100 CRC errors.
3	101 to 319 CRC errors.
1	320 or more CRC errors.
15	At least one severe framing error resulting in a loss of the framing pattern. Mutually exclusive with setting bit 16.
16	Error in framing pattern bit that does not result in a loss of the framing pattern. Set for framing problems that do not cause frame-synchronization loss; mutually exclusive with bit 15.
7	At least one line code violation.
2	At least one slip.
14	Set when payload loopback is activated.
9, 10	Modulo 4 counter for seconds. Bit 10 is the more significant bit.
4, 5	Set to zero; under study for synchronization.
12	Reserved and set to zero.

CRC errors

CRC errors are generated when a received superframe fails the CRC check. Approximately 333 frames per second pass through a T1 circuit. Depending on how the DTE interprets a CRC failure, the tiers above 100 CRC errors could correspond to a data loss of one-third, and it may also result in the signal being declared out of frame. One of the drawbacks to using the CRC over the entire superframe is that even a single bit error will result in a CRC failure. If one bit per superframe is corrupted, the error rate is only about 2×10^{-4}. While that may be a fairly high data rate for a modern digital transport facility, it is nowhere near the 0.33 that might be implied by looking at the CRC error rate in isolation.

Framing errors

Bits 15 and 16 are set in response to framing errors, but they are mutually exclusive. Bit 16 is the milder of the two error conditions. It indicates that framing bits were received in error, but there was no loss of the framing signal. Bit 15 is set when the framing signal is lost. Unlike CRC errors, these two bits indicate that at least one framing error occurred, but do not include any indication of the severity.

Line code errors

T1 standards define line code errors as both bipolar violations and long zero strings, but some T1 equipment may report only bipolar violations as line code errors. (Naturally, intentional bipolar violations inserted by the line code, as in B8ZS, are exempt.) In theory, the number of line code errors can indicate whether the line code is mismatched. B8ZS data fed to an AMI device will cause that device to record a large number of bipolar violations. Likewise, AMI data fed to a B8ZS device may occasionally have strings of more than seven zeros and thus lead to a number of excessive zero string events. However, only the former situation is detected by most equipment. No indication of severity is possible; bit 7 only indicates the presence of at least one line code violation.

Slip events

A timing problem may lead to a slip event, in which a full frame of 193 bits is replicated or deleted. When bit 2 is set, at least one slip has occurred. It is likely to be a controlled slip, because an uncontrolled slip would result in a change of frame alignment and a loss of the framing signal.

Near- and Far-End Statistics

T1 equipment maintains two different sets of statistics. *Near-end* counters collect data on what the local T1 terminating equipment receives. Any CRC errors detected by the local CSU/DSU will increment the near-end CRC error counter and any counters that depend on it. Near-end statistics generate performance reports for the remote end.

T1 lines are bidirectional, however, and it is not unheard of for a line to have trouble in one direction only. Some types of T1 equipment can also maintain *far-end* counters to monitor the outgoing path. Far-end counters are maintained by monitoring incoming performance reports from the remote end. To get far-end counters that mean anything, the remote equipment must support use of the PRM.

Far-end counters are not really troubleshooting tools. Near-end statistics can provide highly granular resolution because they are updated in real time according to the events on the line. The PRM updates far-end statistics only once per second, so they are inappropriately delayed for many types of troubleshooting. When troubleshooting, it is best to obtain the statistics at both ends from real-time access to the near-end counters stored in the equipment rather than relying on PRMs. Figure C-2 shows both near- and far-end performance statistics.

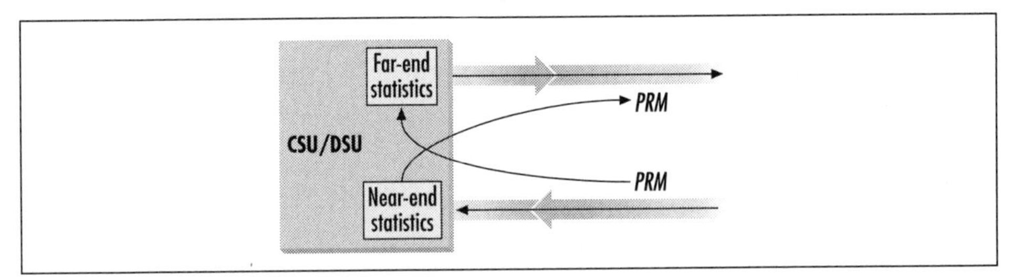

Figure C-2. Near- and far-end performance statistics

An Overview of the Monitoring Process

Many transient conditions can affect the performance of a T1, if only for short periods of time. Effective monitoring requires that low-level details be suppressed to allow administrators to focus on solving high-level problems. ANSI T1.231 outlines one common approach, which underpins SNMP monitoring of T1 lines. A rough overview of the process is shown in Figure C-3.

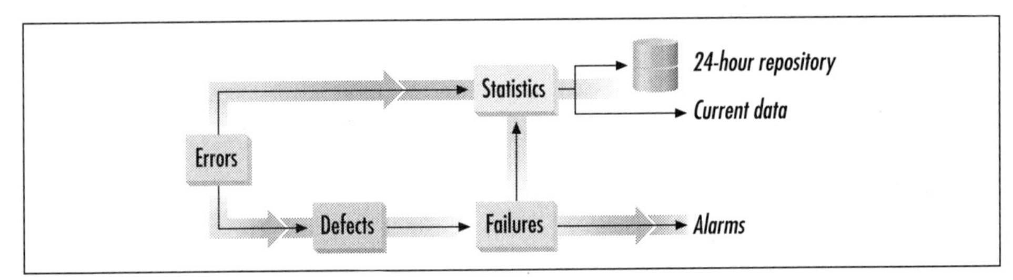

Figure C-3. Relationship between errors, defects, and failures

Errors are the building blocks of the T1.231 approach.* An error is simply an unexpected transient condition. Although the specification forbids errors, they occur over short enough time spans that there is little performance impact from a single isolated error.

* New documents use the term anomaly instead of error. I find the former to be needlessly arcane, so I use the term error throughout this appendix.

Defects are the result of persistent errors. Lone errors may be caused by any number of random events anywhere along the path of the T1, but defects persist for longer periods, so there may be a deterministic, resolvable cause. Defects may also lead to performance degradations.

Persistent defects lead to *failures*. Failures occur only when the line is down. No data can be transmitted, so immediate attention is required. Failures are pager material, but defects are usually not.

Performance data is the final component of the monitoring architecture. Carriers often guarantee certain levels of performance, but it is often up to the user to notice statistics falling below the guaranteed levels and demand a refund.

Failures, Alarms, and Signaling

Failures are the highest-priority events because they indicate drastic problems. Most failures are simply extended presences of the associated defect. The only exception is that so-called *path failures* (frame loss and alarm indication), which are associated with the logical data path between endpoints, have mutually exclusive signaling and only one can be declared at any point. Here is a list of failures, the conditions of which are detailed in Table C-2:

Loss of signal (LOS) failure
: Signal loss is the most basic failure. Signal loss initially triggers the LOS defect. If the defect persists, failure is declared. LOS failure is cleared when a signal with an appropriate ones density is received at the network interface. The time to clear the failure depends on the equipment. Generally, it is only a few seconds, though it is permitted to be as many as 20.

Loss of frame (LOF) failure
: Signal loss corresponds to the most basic transmission on a T1. Performing the T1 framing appropriately is one step up from moving electrons from one end to the other. If either aspect breaks down, the line will fail.

  LOF failure is declared when the out of frame (OOF) condition persists. (OOF is described later in this appendix.) Like the signal loss alarm, it may take up to 20 seconds to clear the LOF failure.

  LOF is not declared when the alarm indication signal (AIS) is present. AIS is a higher-priority alarm, and the signaling bits must be used for only one of the two.

Alarm indication signal (AIS) failure
: A persistent AIS defect causes AIS failure to be declared. Because AIS has a higher priority than LOF, AIS can be declared immediately if no framing signal is detected. Like the two previous signals, it may take up to 20 seconds to clear an AIS failure.

When many devices are placed in the loopback mode, the data is looped back toward its source and the AIS signal is sent on in place of the data to alert equipment further along the path that a device along the way is in loopback mode. This is a configurable option that is frequently activated on smart jacks.

If you determine that neither of the CSU/DSUs at the end locations are in loopback mode, ask the carrier to trace the source of the AIS signal for you. If it is a piece of equipment along the signal path, the carrier can usually deactivate the loopback by using a network maintenance command.

Remote alarm indication (RAI) failure

When the incoming link on a T1 fails, the CSU/DSU transmits a code in the outbound direction to inform the other end that the link has failed. RAI was previously referred to as the *yellow alarm*, and some documents refer to it as the *far end alarm*. RAI is declared when an incoming RAI signal is detected, and it is cleared when the RAI signal is no longer received.

Table C-2. T1 failure conditions

Failure	Declaration condition	Clearance condition	T1.231 reference
LOS	2–3 continuous seconds with LOS defect	At most, 20 continuous LOS-free seconds[a]	6.2.1.1.1
LOF	2–3 continuous seconds; or OOF defect (except when AIS present)	At most, 20 continuous OOF-free seconds; or AIS failure declared (priority rule)	6.2.1.2.1
AIS	2–3 continuous seconds AIS defect; or AIS defect when LOF is declared	At most, 20 continuous AIS defect free seconds	6.2.1.2.2
RAI (receive)	Detection of RAI signal	No RAI signal detected	6.2.2.2.1
RAI (send)	LOS, LOF, or AIS failure declared; transmitted for a minimum of 1 second	Immediately on clearing of underlying failure[b]	6.4.1

[a] Typically, an alarm is cleared in much faster than 20 seconds. Some equipment, however, incorporates legacy functions and may take up to 20 seconds to clear an alarm.

[b] RAI is sent as long as the underlying cause persists. If the underlying causal alarm is subject to a 20-second waiting period, RAI persists throughout the waiting period.

With alarms stacked in priority and some being mutually exclusive, it is worth considering how equipment typically determines whether an alarm should be declared. Figure C-4 shows the logic a T1 receiver typically uses on an ESF link. When pulses are detected on the line but no framing sequence is detected, if the content is substantially all ones, AIS is declared. Otherwise, the OOF or LOF alarm will be declared. To prevent noisy lines from corrupting a great deal of user data, the terminating equipment typically requires that the CRC check pass for 24 superframes before declaring that the signal is in frame.

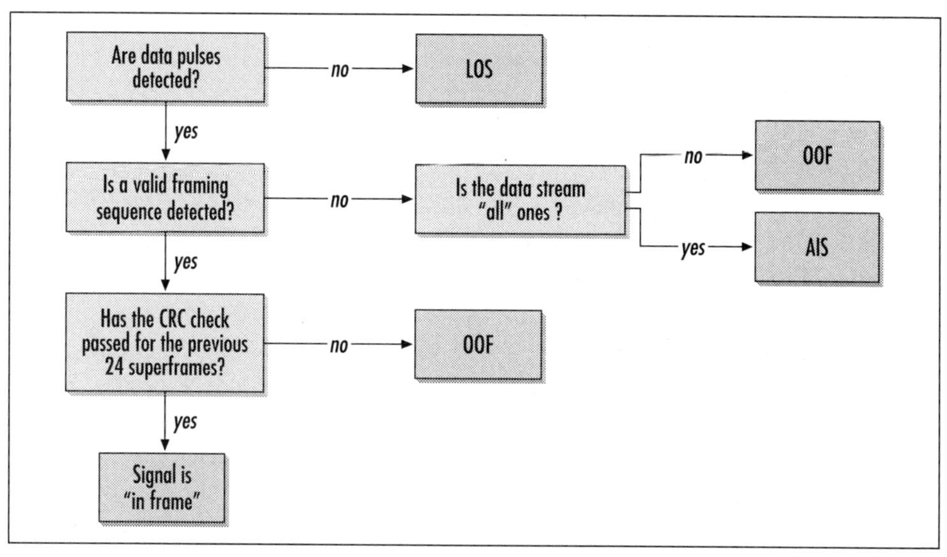

Figure C-4. Typical alarm declaration procedure for ESF T1

Performance Defects

Defects alert administrators to potential performance issues. Defects may grow into full-blown failures. Unlike path failures, path defects are not mutually exclusive and can be declared simultaneously. Here is a list:

Loss of signal (LOS) defect

When a period of 100 to 250 unit intervals (bit times) has passed with no pulses, the LOS defect is declared. This defect is cleared when a comparable time interval has passed with the incoming signal meeting the pulse density requirements. Generally, this is taken to mean a comparable interval with a pulse density of at least 12.5%.

Out of frame (OOF) defect

Both the superframe (SF) format and the extend superframe (ESF) format incorporate framing bits. Incorrect framing bits are called *frame bit errors*. A single frame bit error is not enough to declare an out of frame condition. Most equipment uses an *n* out of *m* density requirement, where *n* framing bits out of *m* consecutive framing bits must be in error to lose the framing. Typical values lead to an error density requirement of between 40–60% frame bit errors over short periods of time. See Table C-3 for details. OOF is terminated when the signal is reframed, which requires a sequence of *m* clean frame bits.

As a practical matter, most ESF equipment avoids clearing the OOF condition until the CRC is good for several superframes. Typically, 24 error-free superframes are required, but this number is not specified in any standard.

AIS signals cause an OOF defect because AIS is an unframed, all-ones signal. However, the AIS defect triggers an AIS alarm, which overrides the LOF alarm.

Severely errored framing (SEF) defect

Signals that are drastically out of frame lead to a severely errored framing defect. Terminating equipment maintains a monitoring window for frame bits. Two frame bit errors in the window cause a severely errored framing defect to be declared. On SF-framed links, the window is 0.75 ms and only the terminating frame (F_t) bits are examined. ESF-framed links use a 3-ms window and examine only the ESF frame synchronization bits.

Out of frame signals almost always cause a severely errored framing defect. For clarity, the rest of this appendix uses "out of frame" where "severely errored framing defect" is used in the specification. The two nearly always go hand-in-hand, and I find "out of frame" to be a more descriptive term.

Alarm indication signal (AIS) defect

When an unframed all-ones signal is received for an interval between 3 and 75 ms, the AIS defect is declared.* It is cleared after the same interval.

Because an AIS signal is unframed, it leads to an out of frame defect as well as severely errored framing.

Table C-3. T1 defect conditions

Defect	Declaration condition	Clearance condition	T1.231 reference
LOS	100 to 250 bit times with no pulses	100 to 250 bit times with at least 12.5% pulse density (12 to 32 pulses, depending on window size)	6.1.2.1.1
OOF	n frame bit errors out of m consecutive framing bits; typical values are (n=2, m=4), (n=2, m=5), and (n=3, m=5)	m consecutive error free framing bits	6.1.2.2.1
SEF	2 frame bit errors in the monitoring window SF: 0.75 ms, only F_t bits used ESF: 3 ms, only frame bits	Framed signal with at most one frame bit error per window	6.1.2.2.2
AIS	3–75 ms of an incoming unframed all-ones sequence	Properly framed signal received; or signal with ones density less than 99.9%	6.1.2.2.3

* Customer-side AIS signals are not all ones. Technically, this requirement is the reception of a signal that has a ones density of 99.9%.

Errors

Errors are the building blocks of the whole system. Here is a list of errors, which are summarized in Table C-4:

Bipolar violation (BPV)

When used as an error counter, BPVs refer only to unintentional BPVs. BPVs inserted by the B8ZS line code are not counted as errors. Thus, on AMI-coded links, any occurrence of two consecutive identical-polarity pulses is counted. B8ZS-coded links, on the other hand, count only BPVs that are not part of the zero-substitution code.

Excessive zeros (EXZ)

Line codes also specify a maximum number of consecutive zeros: 15 for AMI-coded links and 7 for B8ZS. Longer strings of consecutive zeros indicate a problem and are counted as an EXZ event. Not all equipment can detect and report EXZ events.

Line code violation (LCV; occasionally CV-L)

Both BPVs and EXZs are line code problems. For reporting purposes, they are combined for a total number of line code violations.

Path code violation (PCV; also CV-P)

Path coding is an error in the logical path from end to end. For ESF-framed links, it is a count of the CRC errors. SF-framed links have no check for correct transmission along a path, so the number of framing bit errors is used as a proxy.

Controlled slip (CS)

A controlled slip is the deletion or duplication of a DS1 frame, as described in Chapter 5. Although controlled slips may cause problems with the higher-level data stream, they do not cause any physical layer issues. In particular, the framing bits are maintained, so no out of frame defect occurs.

Frame bit error (FE)

Frame bit errors are declared only for the six framing bits in the ESF frame. On SF-framed links, the terminal framing bits (F_t) are always counted for frame bit errors; signaling framing bits (F_s) may also be used.

Table C-4. Basic error events

Error	Description	T1.231 reference
Bipolar violation (BPV)	Unintentional violations of bipolar line code	6.1.1.1.1
Excessive zeros (EXZ)	AMI: greater than 15 zeros B8ZS: greater than 7 zeros	6.1.1.1.2

Table C-4. Basic error events (continued)

Error	Description	T1.231 reference
Line code violation (LCV; CV-L)	BPV + EXZ	6.5.1.1
Path code violation (PCV; CV-P)	SF: framing bit errors ESF: CRC errors	6.5.2.1
Controlled slip (CS)	Replication or deletion of 192-bit superframe of data; framing is maintained, so OOF is not declared	6.1.1.2.3
Frame bit error (FE)	Unexpected framing bit ESF: applies to 6 framing bits only SF: Always applies to F_t bits; may also apply to F_s bits	6.1.1.2.2

Performance Data

In addition to errors, monitoring performance data allows administrators to spot line degradations before they turn into stressful line-down situations. Most performance data is reported as the number of seconds for which a particular condition occurred. Here is a list of problems you might encounter, which are summarized in Table C-5:

Line errored seconds (LES; also ES-L)
: Line errors include BPVs, EXZs, and LOS defects. The number of seconds with at least one of any of those errors is a line errored second.

Controlled slip seconds (CSS)
: A controlled slip second is a second during which at least one controlled slip occurred. Controlled slips can be accurately measured only by CSU/DSUs.

Errored seconds (ES)
: As defined for the path (as opposed to the line), an errored second is a second with at least one error off the following smorgasbord:

- Path code violations
- Controlled slips
- Out of frame signals (severely errored framing defects)
- AIS

An errored second for the path may also be abbreviated as ES-P. Early drafts of T1.231 that were used as the basis for the SNMP monitoring of T1 links used a different definition, which is discussed in Appendix D.

Errored second type B (ESB)
: An errored second type B is a parameter that measures data corruption. It is a 1-second interval with between 2 and 319 CRC errors, but without severely errored framing problems or AIS. Because framing is functional and AIS is not present, the path appears to be functional.

Severely errored seconds (SES)

Severely errored seconds are 1-second intervals with serious problems. Most CSU/DSUs that keep statistics keep records of the number of severely errored seconds. Some of the errors that cause this are:

- Out of frame signals (one or more severely errored framing defects)
- One or more AIS defects
- 320 or more CRC errors (ESF only)
- If only the F_t framing bits are measured, four or more frame bit errors. When both the F_t and F_s framing bits are measured, eight or more framing bit errors. (SF only)

Severely errored is an appropriate description for a second with 320 or more CRC errors. At 8,000 frames per second and 24 frames per superframe, only 333 superframes are transmitted per second. T1 does not discard severely errored data, but it is usually treated as suspect by data-communications equipment. If the receiver discards all superframes failing the CRC check, 96% of the data is lost!

SEF/AIS second (SAS-P)

An SEF/AIS is a 1-second interval with at least one SEF event or one AIS event. This counter contains the number of 1-second intervals in which the line is unavailable due to framing and testing conditions.

Unavailable second (UAS)

Ten consecutive severely errored seconds remove the path from use. The path becomes usable after 10 consecutive seconds with no severely errored seconds. While the line is unavailable, all other parameter counts are frozen because they are intended to measure trouble with the line when it is in service.

Table C-5. Performance data

Name	Description	T1.231 reference
Line errored second (LES; ES-L)	1-second interval with at least BPV, EXZ, or LOS	6.5.1.2
Controlled slip second (CSS)	1-second interval with at least one controlled slip	6.5.2.9
Errored second (ES; ES-P)	1-second interval with a controlled slip, SEF defect, or AIS defect, plus: ESF: CRC errors SF: framing errors	6.5.2.2
Errored second type B; also bursty errored second (BES)	(ESF only) 2–319 CRCs, no SEF, no AIS	6.5.2.4

Table C-5. Performance data (continued)

Name	Description	T1.231 reference
Severely errored second (SES)	SEF, AIS, plus: ESF: >=320 CRC errors SF: 8 frame bit errors in both the F_t and F_s bits, or 4 bit errors in the F_t bits	6.5.2.5
SEF/AIS second (SAS-P)	1-second interval with either SEF or AIS events	6.5.2.6
Unavailable second (UAS)	10 consecutive severely errored seconds begin an unavailable period, which can be ended only by 10 consecutive seconds with no severely errored seconds. All 1-second intervals between are considered unavailable. Performance counters are not incremented during unavailable time.	6.5.2.10

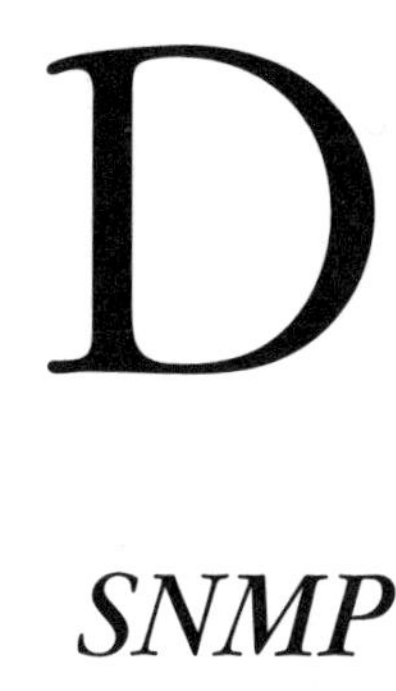

SNMP

Every thought is an afterthought.
—Hannah Arendt

Managing far-flung networks requires a helping hand. Most network administrators, when given the tasks of running large networks with many leased lines, turn to SNMP, the Simple Network Management Protocol. This appendix presents an overview of two MIBs that may be useful to network administrators with extensive T1 deployments.

RFC 2495: DS1 MIB

RFC 2495 defines the MIB for DS1, DS2, E1, and E2 interfaces. This section will consider only the DS1 portion of RFC 2495.

Differences from T1.231

RFC 2495 was written at the time when a draft revision of T1.231 was working through the ANSI standards process. Some of the definitions in RFC 2495 are based on early drafts of T1.231 and thus are not fully consistent with T1.231. These differences are important because data communications equipment vendors implement the RFC, but telecommunications equipment vendors implement the ANSI specifications. Here are some differences:

- RFC 2495 defines the out of frame (OOF) defect in exactly the same way that ANSI T1.231 defines the severely errored framing (SEF) defect. In practice, this is only a minor difference because SEF defects are nearly always accompanied by ANSI-defined OOF defects.

- The definition of a severely errored framing second (SEFS) in RFC 2495 is equivalent to the ANSI definition of a severely errored framing/AIS second (SAS-P).
- RFC 2495 uses the term "bursty errored second" to refer to ANSI's errored second type B.
- As RFC 2495 defines an errored second, bipolar violations are also an underlying cause. ANSI defines an errored second on an SF-framed link as a 1-second interval that contains framing bit errors, controlled slips, severely errored framing (identical to the RFC's out of frame defect), and AIS defects. RFC 2495 adds bipolar violations to the list of triggers. There is no difference between the RFC 2495 and ANSI definitions of an errored second on an ESF-framed link.
- The RFC 2495 definition of a severely errored second for SF-framed links does not include AIS defects, but does label a second severely errored if 1,544 or more line code violations occur. Both definitions are identical for ESF-framed links.
- RFC 2495 defines degraded minutes (DMs), which measure when the bit error rate is higher than expected for modern digital facilities (10^{-6}).
- RFC 2495 specifies that defects must be present for between 2 and 10 seconds before declaring the corresponding failure alarm, not the 2 to 3 second interval in T1.231.
- RFC 2495 includes a loopback pseudofailure mode so that network managers may determine whether or not lines are in service. Although loopback modes do not correspond to hardware or software failures, they do remove the line from service.

Table D-1 summarizes the differences between RFC 2495 and ANSI T1.231. For obvious reasons, this appendix uses RFC terminology. Details on what constitutes certain errors are also presented in the discussion of statistical tables in the DS1 MIB.

Table D-1. Differences between RFC 2495 and ANSI T1.231

RFC 2495	ANSI T1.231
Out of frame defect	Severely errored framing defect
Severely errored framing second	Severely errored framing/AIS second
Bursty errored second	Errored second type B
Bipolar violations also make an interval count as an errored second on SF-framed links	Errored seconds caused only by severely errored framing, controlled slips, frame synchronization bit errors, and AIS defects
On SF-framed links, severely errored second is a result of framing errors, out of frame defects, or more than 1,544 line code violations	On SF-framed links, a severely errored second is the result of framing errors, severely errored framing, or AIS

Table D-1. Differences between RFC 2495 and ANSI T1.231 (continued)

RFC 2495	ANSI T1.231
Degraded minutes measure time periods over which the bit error rate is between 10^{-3} and 10^{-6}	No analogous feature
2–10 second window for defects to grow into failures	2–3 second window
Loopback pseudofailure mode	No analogous feature

The Structure of the MIB

Figure D-1 illustrates the organization of the DS1 MIB. The top branch is collapsed to *.iso.org.dod.internet* (an OID of .1.3.6.1). The DS1 MIB falls under the *transmission* section of MIB-2, at *mgmt.mib-2.transmission.ds1* (2.1.10.18).

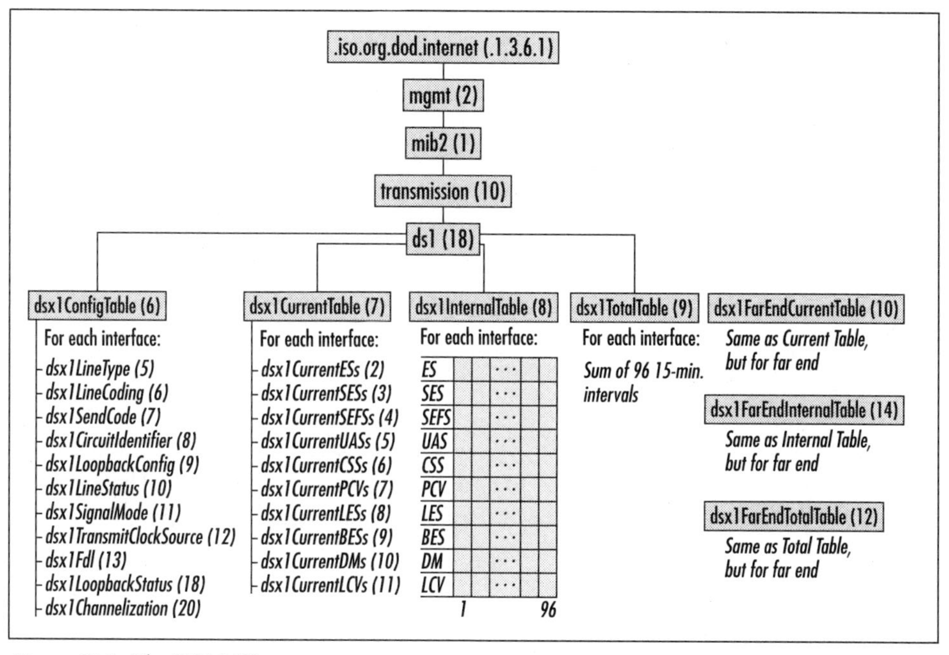

Figure D-1. The DS1 MIB

At a high level, the DS1 MIB is simple. Several tables hold configuration and statistical data about each DS1. Performance data is available in three ways: the current 15-minute interval, a data structure containing 15-minute interval data for the past 24 hours, and totals for the past 24 hours. Additionally, data is available from both the near side and the far side. Implementation of the near side is mandatory, while the far-side data must be collected based on the performance reports sent over the facilities data link. Each of these branches will be described in detail, with particular attention paid to useful objects in the branch.

The configuration table

Configuration for each DS1 is held in the configuration table, *ds1.dsx1ConfigTable*. Each row in the configuration table contains configuration information for one DS1. One of the items in the row is the DS1's interface index in the MIB-2 interfaces table. All objects in this section begin with *ds1.dsx1ConfigTable.dsx1ConfigEntry* (18.6.1). Unless otherwise noted, the data type for these objects is an enumerated data type defined for that object in the RFC. The objects are:

dsx1LineType
: T1 lines are *dsx1ESF* (2) for ESF-framed links or *dsx1D4* (3) for SF-framed links.

dsx1LineCoding
: The most common settings are *dsx1B8ZS* (2) and *dsx1AMI* (5).

dsx1SendCode
: CSU/DSUs can use the facilities data link to send commands to the remote end. Many CSUs can also generate and send test patterns. The enumerated type defined for this object will report whether the CSU/DSU is sending loopback commands or test patterns.

dsx1CircuitIdentifier (string, maximum length 255)
: Telecommunications carriers always ask for the circuit ID when troubleshooting. This object keeps it handy for you.

dsx1LoopbackConfig
: Under most conditions, this object is set to *dsx1NoLoop* (1), which indicates that no loopback is configured. Network management stations can write to this object to configure an interface for loopback.

dsx1LineStatus (integer bit-map, range 1 to 131071)
: This object contains status information for the line. It returns an integer with information including alarms, failure modes, and loopback status. The bit positions in Figure D-2 are the various statuses that may be reported. Multiple status indications may be returned by adding the appropriate contributions.

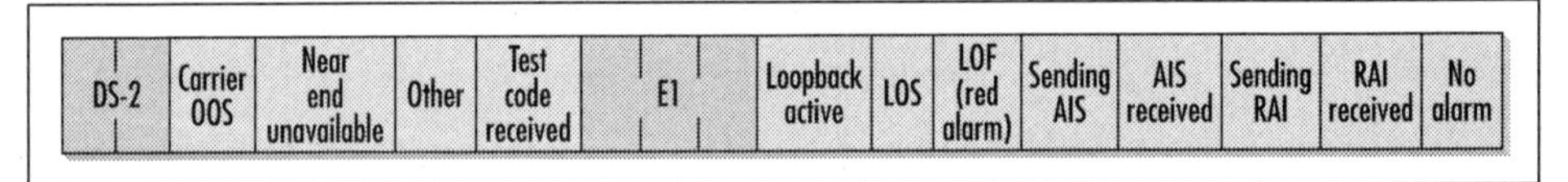

Figure D-2. dsx1LineStatus codes

dsx1SignalMode
: Using T1 for channelized data transport requires sending signaling information to set up circuits through the telephone network. This object returns the signaling mode in use on an interface. Unchannelized T1 is set to *none* (1). Channelized T1 is set to *robbedBit* (2). If the T1 is the underlying physical transport for an ISDN PRI, it is set to *messageOriented* (4).

dsx1TransmitClockSource

This object reports the source of the transmit clock. In most installations it is set to *loopTiming* (1), which corresponds to recovering the transmit clock from the received data. Other choices include *localTiming* (2) for a self-generated clock and *throughTiming* (3) when the receive clock from another interface on the router is used.

dsx1Fdl

If the interface does not support use of the facilities data link, this object is set to *dsx1FdlNone* (8). Two major standards for using the facilities data link exist: AT&T publication 54016 and ANSI T1.403. These are assigned the values, *dsx1Att54016* (4) and *dsx1AnsiT1403* (2), respectively. The *dsx1Fdl* object represents the sum of the capabilities of the device, so it is often set to 6.

dsx1LoopbackStatus (integer bit-map, range 1–127)

When a loopback mode is activated, this object reports the active loopback types. Like the *dsx1LineStatus* object, this object is a bit map in which the values of the individual bits in Figure D-3 correspond to the various loopback types that can be returned.

Figure D-3. dsx1LoopbackStatus

dsx1Channelization

If the T1 is unchannelized, this object is set to *disabled* (1). When it is broken out into individual DS0s, it is set to *enabledDs0* (2).

The current statistical table

Statistics are gathered from each interface at 15-minute intervals. These intervals may, but need not, line up with quarter hours on the wall. The current statistical table holds the numbers being collected in the current 15-minute interval. All entries are Gauge32 counters, which range from 0 to 4,294,967,295.

During a 15-minute window, the current statistical data is available from the *dsx1CurrentTable*. All entries in the table begin with *ds1.dsx1CurrentTable.dsx1CurrentEntry* (18.7.1). Here are the entries:

dsx1CurrentESs

This is the number of errored seconds (ESs) in the current 15-minute interval. An errored second is a 1-second interval with at least one of the following:

- Path code violation
- OOF defect

- At least one controlled slip
- At least one AIS defect
- Bipolar violation (SF-framed links only)

dsx1CurrentSESs

This object measures the number of severely errored seconds (SESs) in a 15-minute interval. SESs are 1-second intervals that contain any one of the following errors.

For ESF-framed links:

- 320 or more path code violations
- At least one OOF defect
- At least one AIS defect

For SF-framed links:

- Framing bit errors
- At least one OOF defect
- 1,544 or more line code violations

dsx1CurrentSEFSs

A severely errored framing second (SEFS) is a 1-second interval with either OOF defects or AIS defects. ANSI standards refer to severely errored framing seconds as SEF/AIS seconds.

dsx1CurrentUASs

This counter measures the number of unavailable seconds (UASs). Unavailability commences with 10 continuous SESs or the declaration of a failure state. During a period of unavailability, this is the only counter that is incremented.

Because unavailability is defined to start at the beginning of a 10-second stretch of SESs SNMP agents must either lag statistical reporting by 10 seconds (to the onset of unavailability) or alter line statistics to reflect a line being declared unavailable.

dsx1CurrentCSSs

This object counts controlled slip seconds (CSSs), or seconds in which at least one controlled slip occurs. Excessive controlled slips may be due to timing problems and should be investigated promptly.

dsx1CurrentPCVs

Path code violations (PCVs) are problems with the data channel riding on the T1 physical layer. Framing bit errors cause path code violations on SF-framed links. On ESF-framed links, CRC failures also lead to path code violations.

dsx1CurrentLESs

Line errored seconds (LESs) are 1-second intervals in which one or more line code violations occur.

dsx1CurrentBESs

Bursty errored seconds (BESs) apply only to ESF links. In a bursty errored second, framing is maintained, but between 2 and 319 CRC checks fail.

dsx1CurrentDMs

Degraded minutes (DMs) are minutes in which the functionality of the line is degraded, but still usable; these correspond to 60-second intervals in which the error rate is estimated to be greater than 10^{-6} but less than 10^{-3}.

Degraded minutes are calculated only from usable time, which is defined as seconds that are not unavailable seconds or severely errored seconds. Usable seconds are grouped into consecutive 60-second blocks to form minutes. Degraded minutes are 60-second blocks with an estimated error rate exceeding 10^{-6}.

dsx1CurrentLCVs

Bipolar violations and long zero strings indicate problems with the line code. Both are grouped together under the generic name, line code violations (LCVs).

The interval table

Historical data for the most recent 24-hour period is maintained in the *dsx1IntervalTable*. Figure D-4 shows the logical structure of data in the interval table.

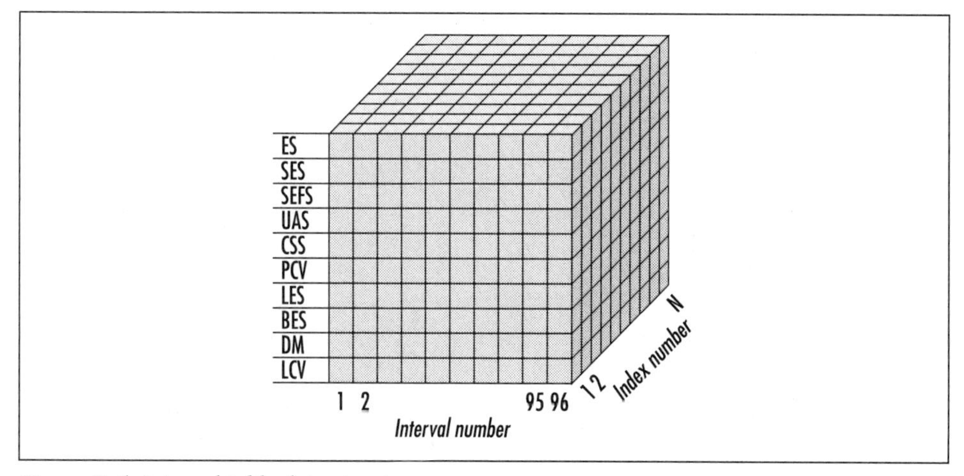

Figure D-4. Interval table data structure

Each DS1 interface is given a unique identifying number, which can be cross-referenced to the MIB-2 interface index number. Time is divided into 15-minute intervals; 96 intervals comprise one 24-hour day. Performance data for the 96 preceding 15-minute intervals is available in the SNMP analog of a three-dimensional array.

At the conclusion of each 15-minute interval, the 96th entry is pushed out of the table, each interval in the table is shifted down one, and the current interval statistics are copied to interval 1.

The total table

To make it easier to obtain total counts for the previous 24-hour period, the DS1 MIB also includes the *dsx1TotalTable*. Instead of performing 96 SNMP queries on the appropriate objects in the interval table and adding, network managers can consult the appropriate cell in the total table. Figure D-5 shows how the total table relates to the data in the interval table.

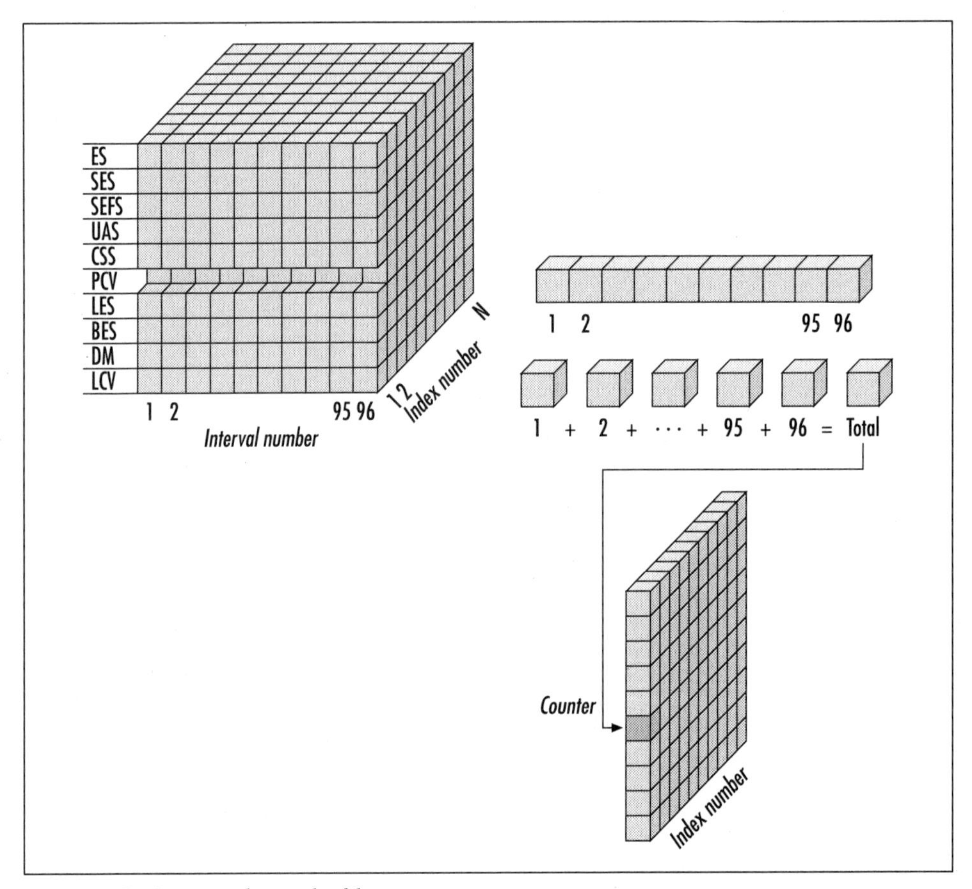

Figure D-5. Creating the total table

For the purposes of performing the addition, each invalid 15-minute interval is considered to have a zero count. To offer a feel for acceptable numbers in the total table, Table D-2 shows the performance thresholds used by the U.S. Defense Information Systems Agency, the organization that maintains the data network used by the U.S. military and intelligence community, to evaluate circuit performance.

Table D-2. DISA performance threshold guidelines

Parameter	Threshold (24-hour period)
Change of frame alignment[a]	511 (major) 17 (minor)
Frame slip[a]	255 (major) 4 (minor)
Errored seconds	864
Severely errored seconds	225

[a] Multiple events per 10-second period are counted as a single event toward the threshold.

Guidelines for Statistical Data

The Defense Information Systems Agency (DISA) publishes a set of guidelines for the performance of leased lines used in the Defense Information Infrastructure (DII). Currently, these guidelines are published in DISA Circular 300-175-9, "DII Operating-Maintenance Electrical Performance Standards." The circular is available online at *http://www.disa.mil/pubs/circulars/dc3001759.html.*

The far end tables

If the line supports it, statistics are also collected for the far end of the DS1. Far-end statistics are gathered from performance reports over the facilities data link. Far-end statistics are not available on SF-framed T1s because they do not support the facilities data link channel. Implementation of far-end tables is optional.

DS1 trap

RFC 2495 defines one trap for DS1 interfaces. When the line status changes, a *dsx1LineStatusTrap* can be sent to the network management station. MIB-2 traps such as *linkUp* and *linkDown* may also be sent, but they are specified in other RFCs.

RFC 2115: Frame Relay DTE MIB

A frame relay DTE device can terminate a large number of virtual circuits on each physical interface. Any MIB hoping to be useful to frame relay network administrators must report on the physical interfaces, the associated logical connections, and the mapping between the two.

Structure

Tables in the frame relay MIB report on the LMI configuration, the status of each virtual circuit, and errors observed on each interface. The MIB is shown in Figure D-6.

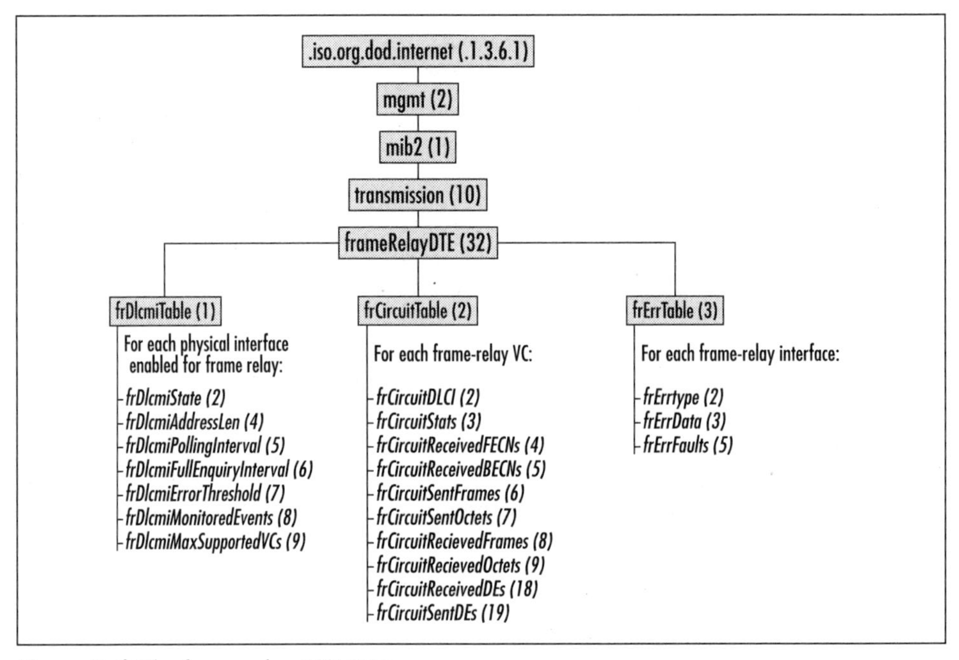

Figure D-6. The frame relay DTE MIB

RFC 2115 also defines the DLCI textual convention (data type) as an integer ranging from 0 to 8,388,607. In practice, most networks use DLCIs ranging from 0 to 1,023.

DLCMI Table

The first table contains data about the Data Link Connection Management Interface (DLCMI). For each frame relay physical interface, it reports on LMI parameters. Objects in this table begin with *frameRelayDTE.frDlcmiTable.frDlcmiEntry* (32.1.1). Here is a list:

frDlcmiState (RFC-defined enumerated type)
: This object reports the LMI type in use on the interface. The most common values are *itut933a* (5) for the ITU Q.933 Annex A specification and *ansiT1617D1994* (6) for the ANSI T1.617a-1994 Annex D specification.

frDlcmiAddressLen (RFC-defined enumerated type)
: This object reports the length of the address field in bytes, although, it is an enumerated type. Its possible values are *twoOctets* (2), *threeOctets* (3), and *fourOctets* (4).

frDlcmiPollingInterval (integer, range 5–30)
This object reports the value of the T391 timer.

frDlcmiFullEnquiryInterval (integer, range 1–255)
This object reports the value of the N391 counter.

frDlcmiErrorThreshold (integer, range 1–10)
This object reports the value of the N392 counter.

frDlcmiMonitoredEvents (integer, range 1–10)
This object reports the value of the N393 counter.

frDlcmiMaxSupportedVCs (DLCI)
This object reports the maximum number of virtual circuits allowed on the interface. Network management stations may attempt to set the value of this object, but the frame relay network usually imposes a cap. If the attempted new value exceeds the old value, an error is returned.

Circuit table

Frame relay physical interfaces may carry several logical connections. Each virtual circuit has an entry in the circuit table. LMI connections are circuits on DLCI 0 and thus have entries in the circuit table. All entries in this section begin with the prefix *frameRelayDTE.frCircuitTable.frCircuitEntry* (32.2.1) and are as follows:

frCircuitDlci (DLCI)
Circuits are multiplexed over physical connections by assigning each virtual circuit a DLCI. In addition to a physical-interface index, each entry in the circuit table includes the DLCI value.

frCircuitState (RFC-defined enumerated type)
Circuits are either *active* (2) or *inactive* (3).

frCircuitReceivedFECNs and frCircuitReceivedBECNs (Counter32)
These two objects count the FECNs and BECNs received on a circuit and provide some indication of congestion on the circuit. Counting FECNs and BECNs can assist in monitoring capacity and ensuring that service levels are not degrading.

frCircuitSentFrames, frCircuitReceivedFrames, frCircuitSentOctets, and frCircuitReceivedOctets (Counter32)
These objects may be used to monitor the traffic on a per–virtual circuit basis. Monitoring traffic patterns may assist in selecting appropriate committed information rates from the service providers.

frCircuitReceivedDEs and frCircuitSentDEs (Counter32)
These objects monitor the number of frames sent and received on the circuit that have the discard eligible bit set. In conjunction with the previous objects, they can assist in evaluating whether or not the purchased set of services is appropriate.

Error table

Errors are placed into a table so that network managers may analyze them. Each interface occupies one row in the table, with the row describing the last error seen on the interface. Errors are timestamped so they can be correlated to clock time. Each error results in up to 1,600 bytes of the offending packet that is being stored in the table as *frErrData* for further analysis.

frErrType (RFC-defined enumerated type)
: Nine different specific errors are enumerated with a tenth "catch-all" code. The list of error types is shown in Table D-3.

frErrFaults (Counter32)
: This object tracks the number of times the interface has gone down since it was initially brought up.

Table D-3. frErrType enumeration

Error	Code	Description
unknownError	1	Error not described by other codes
receiveShort	2	Similar to a runt frame on an Ethernet, a *receiveShort* error corresponds to a frame that was too short to contain a full address field; if RFC 1490 encapsulation is used, a missing or incomplete protocol identifier also leads to this error
receiveLong	3	Incoming frame exceeded maximum frame length for the interface
illegalAddress	4	Received frame did not match the address format configured on the interface
unknownAddress	5	Inbound frame received on inactive or disabled DLCI
dlcmiProtoErr	6	Unintelligible LMI frame
dlcmiUnknownIE	7	LMI frame with an invalid information element
dlcmiSequenceErr	8	Unexpected sequence number in LMI frame
dlcmiUnknownRpt	9	LMI frame with an invalid report type IE
noErrorSinceReset	10	No errors since interface was initialized

Traps

To prevent traps from overwhelming network capacity or the network management station, the frame relay MIB includes the rate-limiting mechanism, *frameRelayDTE.frameRelayTrapControl.frTrapMaxRate* (32.4.2), which specifies the number of milliseconds that must elapse between sending traps. It ranges up to one hour; setting it to zero disables rate control. By default, rate control is disabled.

Only one trap is defined. When the status of a frame relay virtual circuit changes, a *frDLCIStatusChange* trap may be sent to the network management station. Virtual circuits may change status by being switched between the inactive and active states or they may be created and destroyed by other mechanisms.

PPP MIBs

To aid in the management of devices that terminate PPP sessions, several MIBs were defined to allow SNMP management stations to access PPP data. Four RFCs were published, each describing a subset of the PPP MIB. This appendix considers the two major RFCs, which provide the object definitions for the Link Control Protocol MIB and the IP Control Protocol MIB. In addition to the two major RFCs, there are bridging and security MIBs.

Taken together, the PPP MIBs define large groups. Broadly speaking, the major groups presented in this appendix are the Link Group, the LQR Group, LQR Extensions, and the IP Group. The first three are defined by RFC 1471, the LCP MIB. RFC 1473 defines the IPCP MIB.

RFC 1471: Link Control Protocol MIB

RFC 1471 creates the PPP MIB in the *transmission* group, putting it on par with both the DS1 and frame relay DTE MIBs. In addition to defining the branch off of the *transmission* group, it also defines the Link Group, the LQR Group, the Tests Group, and the IP Group.

PPP Link Group

The Link Group reports information on LCP and the lowest layers of the PPP suite. It is composed of two tables: the link status table and the link configuration table. The Link Group is shown in Figure D-7.

Link status table

The link status table is used to report on the operational parameters of a link and its current status. All entries in the table begin with the prefix *ppp.pppLink.pppLinkStatusTable.pppLinkStatusEntry* (23.1.1.1), and rows in the table are indexed by the main interface index. All the entries in the table are read-only. They are:

pppLinkStatusPhysicalIndex (Integer, range 0–2147483647)

Operating systems may present multiple logical interfaces to higher-layer programs when running PPP. This object links the PPP logical interface to the lower-layer physical interface on which PPP is running.

pppLinkStatusBadAddresses (Counter)

This counter reports the number of PPP frames received on the link with a bad address field in the HDLC header. Under most circumstances, the address field should be set to 0xFF. Any frames with incorrect address fields increase the general error count on the interface.

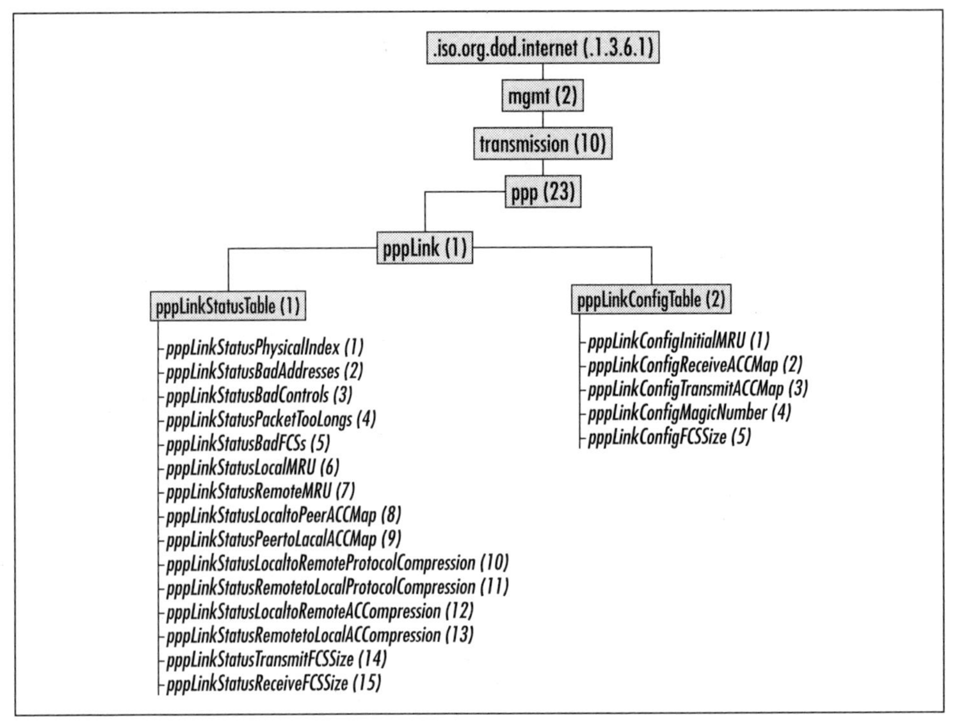

Figure D-7. Link Group of the PPP LCP MIB

pppLinkStatusBadControls (Counter)

This counter reports the number of PPP frames received on the link with a bad control field in the HDLC header. Under most circumstances, the control field should be set to 0x03. Any frames with incorrect control fields increase the general error count on the interface.

pppLinkStatusPacketTooLongs (Counter)

This counter reports the number of frames that were discarded because they exceeded the MRU on the link. Some PPP implementations are flexible regarding the MRU, and implementations should always be able to receive frames of 1,500 bytes. This counter counts only the number of frames that are discarded for being too long; any frames that increment this counter will also increment the general error count on the interface.

pppLinkStatusBadFCSs (Counter)

This counter reports the number of frames discarded because the calculated frame check sequence did not match the transmitted frame check sequence. Any frames that increment this counter will also increment the general error count on the interface.

pppLinkStatusLocalMRU (integer)

When the PPP link has reached the Open state, this object is set to the value of the MRU for frames transmitted by the remote end to the local end. If the PPP link has not reached the Open state, the value of this object is undefined.

pppLinkStatusRemoteMRU (integer)

When the PPP link has reached the Open state, this object is set to the value of the MRU for frames transmitted by the local end to the remote end. If the PPP link has not reached the Open state, the value of this object is undefined.

pppLinkStatusLocalToPeerACCMap (string, length 4)

This object contains the asynchronous control character map used by the local end for transmitted frames. If the PPP link has not reached the Open state, the value of this object is undefined.

pppLinkStatusPeerToLocalACCMap (string, length 4)

This object contains the asynchronous control character map used by the remote end for frames it transmits to the local end. If the PPP link has not reached the Open state, the value of this object is undefined.

pppLinkStatusLocalToRemoteProtocolCompression (enumerated)

This object is set to *enabled* (1) or *disabled* (2) to indicate whether outbound frames are transmitted using protocol field compression. If the PPP link has not reached the Open state, the value of this object is undefined.

pppLinkStatusRemoteToLocalProtocolCompression (enumerated)

This object is set to *enabled* (1) or *disabled* (2) to indicate whether inbound frames are transmitted using protocol field compression. If the PPP link has not reached the Open state, the value of this object is undefined.

pppLinkStatusLocalToRemoteACCompression (enumerated)

This object is set to *enabled* (1) or *disabled* (2) to indicate whether inbound frames are transmitted using address and control field compression. If the PPP link has not reached the Open state, the value of this object is undefined.

pppLinkStatusRemoteToLocalACCompression (enumerated)

This object is set to *enabled* (1) or *disabled* (2) to indicate whether inbound frames are transmitted using address and control field compression. If the PPP link has not reached the Open state, the value of this object is undefined.

pppLinkStatusTransmitFcsSize (integer, range 0–128)

This object is set to the size of the FCS field, in bits, on outbound frames on the link. Normally, a 4-byte FCS is used and the value of this object is 32. If the PPP link has not reached the Open state, the value of this object is undefined.

pppLinkStatusReceiveFcsSize (integer, range 0–128)

This object is set to the size of the FCS field, in bits, on inbound frames on the link. Normally, a 2-byte FCS is used and the value of this object is 16. If the PPP link has not reached the Open state, the value of this object is undefined.

Link configuration table

The link configuration table monitors and sets the values of configuration parameters for the link. All entries in the table begin with the prefix *ppp.pppLink.pppLinkConfigTable.pppLinkConfigEntry* (23.1.2.1), and rows in the table are indexed by the main interface index. All the entries in the table are writable so administrators can configure the underlying parameters. When the network management station changes a value, it goes into effect only when PPP is reinitialized on the link. Here is a list of objects:

pppLinkConfigInitialMRU (integer, range 0–2147483647)
: This object is set to the initial MRU value advertised to the peer when PPP negotiations commence. Its default value is 1,500.

pppLinkConfigReceiveACCMap (string, length 4)
: This object is a 32-bit string that is equivalent to the asynchronous control character map required by the local end to decode any escaped characters in the incoming data stream. Its default value is all ones (0xFFFFFFFF). The remote node may also negotiate to escape additional characters, so the ACCM applied to the incoming data stream is derived from both this map and the transmit map on the remote end.

pppLinkConfigTransmitACCMap (string, length 4)
: This object is a 32-bit string that is equivalent to the asynchronous control character map required by the local end to send any special characters through its transmitter. Its default value is all ones (0xFFFFFFFF). The remote node may also negotiate to escape additional characters, so the ACCM applied to the outgoing data stream is derived from both this map and the transmit map on the remote end.

pppLinkConfigMagicNumber (enumerated)
: This object is set to *false* (1) or *true* (2) depending on whether magic number negotiation is enabled on the PPP interface. Its default value is *false*, though PPP implementations must comply with any negotiations attempted by the remote peer.

pppLinkConfigFcsSize (integer, range 0–128)
: This object is set to the number of bits in the FCS that the local node attempts to negotiate with its peer. Its default value is 16, which reflects the default 2-byte FCS used in PPP.

RFC 1471: LQR Group

Implementation of the LQR Group is mandatory for all PPP implementations that also implement Link Quality Monitoring. Like the Link Group, it is split into a status table and a configuration table. The LQR Group is shown in Figure D-8.

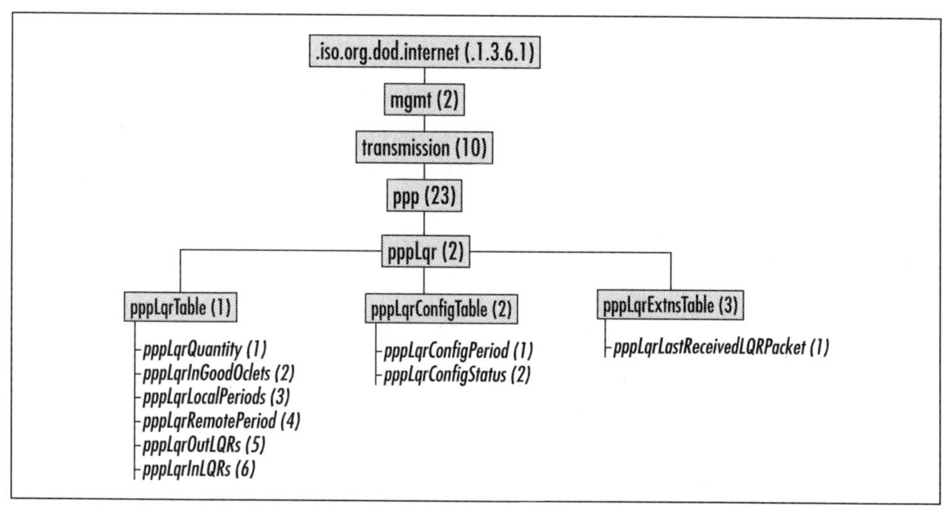

Figure D-8. LQR Group of the PPP LCP MIB

LQR table

The LQR table provides access to operational parameters and statistics for the local PPP implementation. All entries in the table begin with the prefix *ppp.pppLqr.pppLqrTable.pppLqrEntry* (23.2.1.1), and rows in the table are indexed by the main interface index. All the entries in the table are read-only because they provide access to statistics. Here is a list of the entries:

pppLqrQuality (enumerated integer)
: This object is set to *good* (1), *bad* (2), or *not-determined* (3). Link Quality Monitoring determines the quality of the link based on the number of link quality reports received. This object does not report on the policy used to determine link quality. Some implementations may measure quality but refrain from making a determination, which is why the third option may be used.

pppLqrInGoodOctets (Counter)
: This object reports the value of the *InGoodOctets* counter. The counter attempts to measure the number of bytes received in good frames, so frames that are discarded or have errors cannot cause the *InGoodOctets* counter to rise.

pppLqrLocalPeriod (integer, range 1–2,147,483,648)
: This object reports the amount of time, measured in hundredths of a second, between the transmission of link quality reports sent by the local end.

pppLqrRemotePeriod (integer, range 1–2,147,483,648)
: This object reports the amount of time, measured in hundredths of a second, between the transmission of link quality reports sent by the remote end.

pppLqrOutLQRs (Counter)

This object reports the value of the *OutLQRs* counter. When LCP attempts to establish a link, the value of this counter is set to zero, and it is only reset when LCP terminates the link. While the link is active, this counter is incremented by one for each quality report transmitted by the local end.

pppLqrInLQRs (Counter)

This object reports the value of the *InLQRs* counter. When LCP attempts to establish a link, the value of this counter is set to zero, and it is only reset when LCP terminates the link. While the link is active, this counter is incremented by one for each quality report received by the local end.

LQR configuration table

The LQR configuration table allows a network management station to alter the configuration settings for Link Quality Monitoring. All entries in the table begin with the prefix *ppp.pppLqr.pppLqrConfigTable.pppLqrConfigEntry* (23.2.2.1), and rows in the table are indexed by the main interface index. All of the entries in the table are read-write to allow administrative access. Changes to the objects in this table do not go into effect until the PPP implementation is restarted. The entries are:

pppLqrConfigPeriod (integer, range 0–2,147,483,647)

This object is set to the value the local end will use for the LQR reporting period. By default, it is set to zero.

pppLqrConfigStatus (enumerated integer)

This object is either *enabled* (1) or *disabled* (2), depending on whether LQR negotiation should be attempted with the peer. The RFC does not specify exactly how an implementation should react to a management station setting a row to *disabled*; an agent may immediately remove the entry from the table, or it may keep the entry around while noting that it has been disabled.

LQR extensions (optional)

This table is intended to allow network management stations to use SNMP to assist in making the policy determination of whether a link is good or bad. Only one object is in the table, and it begins with the prefix *ppp.pppLqr.pppLqrExtnsTable.pppLqrExtnsEntry* (23.2.3.1). Rows in the table are indexed by the main interface index. Here is a description of the object:

pppLqrExtnsLastReceivedLqrPacket (string, length 68)

This object holds the data field, minus any framing and link layer headers. It begins with the magic number and ends with the *PeerOutOctets* counter.

RFC 1471: Test Group

The Test Group allows network managers to write to objects and cause the PPP implementation to perform network troubleshooting tests. There are two objects in the group, each of which corresponds to a test; the Test Group is shown in Figure D-9.

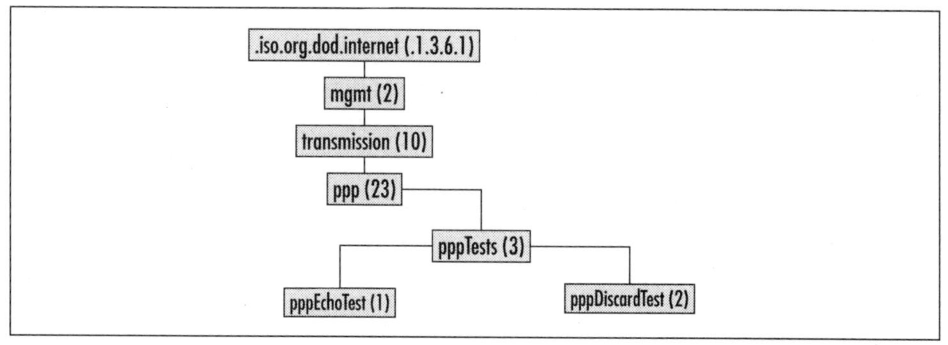

Figure D-9. Test Group of the PPP LCP MIB

The results of the test are reported through the interface extensions MIB, specified in RFC 1229. The particular object of interest is the *ifExtnsTestResult* object, which is present in a table of test results. *ifExtnsTestResult* can be set to a number of values, but the two most important values are *success* (2) and *failed* (7). The fully qualified path to *ifExtnsTestResult* is in the *experimental* branch, rather than the *mgmt* branch, at *experimental.ifExtensions.ifExtnsTestTable.ifExtnsTestEntry* (3.6.2.1.5). Here are descriptions of the objects:

pppEchoTest
: Invoking this object causes a PPP implementation to send out an LCP Echo-Request. If the Echo-Reply is received successfully, the related *ifExtnsTestResult* will be set to *success.*

pppDiscardTest
: Invoking this object causes a PPP implementation to send out an LCP Discard-Request. If the Discard-Request can be successfully transmitted, *ifExtnsTestResult* will be set to *success*; the definition of "transmission failure" is implementation-dependent.

RFC 1473: IP Group

The IP Group enables network administrators to view the IP settings associated with each PPP link and configure IPCP-related parameters. The IP Group is defined in RFC 1473, and is shown in Figure D-10.

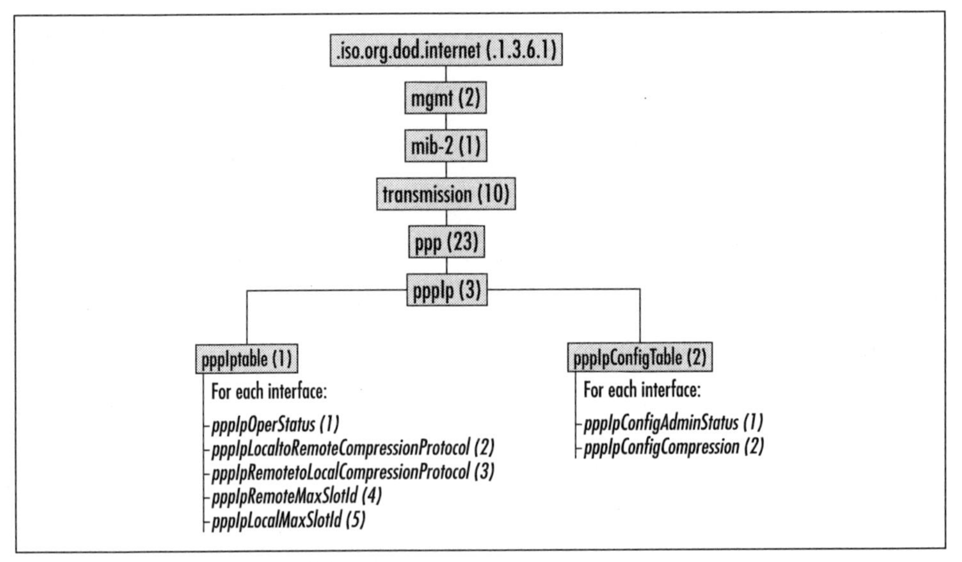

Figure D-10. PPP IP Group of the PPP LCP MIB

IP table

The IP table is used to report the current IP configuration of each PPP link. All entries in the table begin with the prefix *ppp.pppIp.pppIpTable.pppIpEntry* (23.3.1.1), and rows in the table are indexed by the main interface index. All the entries in the table are read-only, and are listed here:

pppIpOperStatus (enumerated integer)
: This object reports on whether IPCP has reached the Open state. If it has, this object will be *opened* (1); otherwise, it will be *not-opened* (2).

pppIpLocalToRemoteCompressionProtocol (enumerated integer)
: This object shows the IP compression protocol used by the local end when sending data to the remote end. Currently, only Van Jacobson compression is supported, so this object takes only the values of *none* (1) and *vj-tcp* (2). This object is defined only when the IPCP state engine has reached the Open state.

pppIpRemoteToLocalCompressionProtocol (enumerated integer)
: This object shows the IP compression protocol used by the remote end when sending data to the local end. Currently, only Van Jacobson compression is supported, so this object takes only the values of *none* (1) and *vj-tcp* (2). This object is defined only when the IPCP state engine has reached the Open state.

pppIpRemoteMaxSlotId (integer, range 0–255)
: Van Jacobson compression associates header information with slots. More slots can hold header information for a greater number of simultaneous connections, but they require more memory and more processing cycles on both

ends. This object reports on the maximum number of slots advertised by the remote end. When Van Jacobson compression is not in use, this object takes the value of zero. Like the other compression-related objects, it is meaningful only when the IPCP state engine has reached the Open state.

pppIpLocalMaxSlotId (integer, range 0–255)
: This object reports on the maximum number of slots advertised by the local end. When Van Jacobson compression is not in use, this object takes the value zero. Like the other compression-related objects, it is meaningful only when the IPCP state engine has reached the Open state.

IP configuration table

The IP configuration table allows a network management station to alter the configuration settings for the IPCP configuration. All entries in the table begin with the prefix *ppp.pppIp.pppIpConfigTable.pppIpConfigEntry* (23.3.2.1), and rows in the table are indexed by the main interface index. All the entries in the table are read-write to allow administrative access. Here is a list of entries:

pppIpConfigAdminStatus (enumerated integer)
: Network administrators may use this object to change the administrative state of the IPCP state engine. Setting it to *open* (1) will send an Open event to lower PPP layers, which will attempt to begin negotiation. Setting it to *closed* (2) will send a Close event to lower layers, which will tear down the IPCP connection.

pppIpConfigCompression (enumerated integer)
: This object may be set to *none* (1) or *vj-tcp* (2). When it is set to *none*, IPCP will not attempt to negotiate any IP header compression. Any other value will cause IPCP to negotiate for that type of IP packet compression. Currently, only Van Jacobson header compression is defined. Changes to the objects in this table do not go into effect until the PPP implementation is restarted.

Cable Pinouts and Serial Information

The most useful piece of learning for the uses of life is to unlearn what is untrue.
—Antisthenes

The telco's responsibilities end at the demarc, and you are responsible for supplying a device to interface with their network. In many cases, the device is a modem; in others, it is a CSU/DSU. Most likely, your equipment will connect to the interface device using serial communications. This appendix contains a short introduction to serial communication basics, plus pinouts for the common interfaces used with T1 circuits.

Introduction to Serial Communications

In the most general sense, serial communications transfer a data stream between two devices one bit at a time. Serial links have, by definition, two devices, and two common setups are illustrated in Figure E-1.

The *data terminal equipment* (DTE) is the device where data is generated or consumed by users. The canonical example of a DTE is a dumb terminal, which displays data it receives and sends data in the form of user keypresses. Serial links also have devices called *data communications equipment* (DCEs), which are responsible for sending the bitstream to another location. In the dumb terminal setup, the DCE is a modem or terminal server. Routers often use high-speed serial links to connect to devices, which interface with the telephone network. On T1 links, a serial link is used between the router and the CSU/DSU. The router is the serial DTE, and the CSU/DSU is the serial DCE.

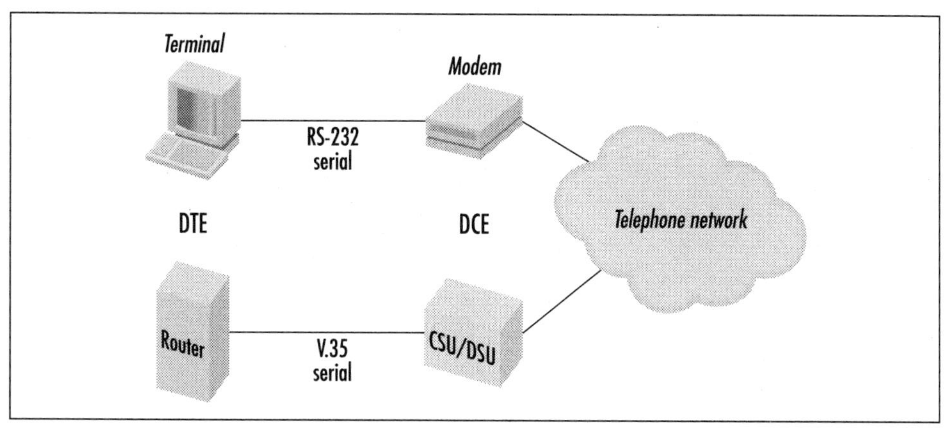

Figure E-1. Two common serial links

Serial interfaces are cheap, relatively easy to construct, and found everywhere:

- RS-232 interfaces are used on Intel-based microcomputers to connect modems and mice.
- RS-422, the big brother of RS-232, runs at higher speeds with greater resistance to interference. It was adopted by Apple for serial-based AppleTalk networks.
- V.35 is commonly used as a medium-speed serial interface between routers and CSU/DSUs.

Serial Line Signaling

Serial lines transmit data as a varying voltage. In order to make sense of the voltage, receivers must know when to measure the voltage. Two common types of serial cabling exist: *common-ground* cables and *differential* cables.

Common-ground cabling, which is paired with *unbalanced* transmission methods, uses one ground line as the reference point. Incoming voltages on the signal line are measured using the ground wire as a reference. For ones, the voltage is high, and for zeros, it is the same as the ground wire. Common-ground cable is cheap and easy to construct, but it is prone to electromagnetic interference because certain types of interference can resemble the voltage going high. Figure E-2 illustrates this problem.

Twisting wires together helps to alleviate the interference on common-ground circuits; because the interference tends to affect all wires somewhat equally, the induced voltage will appear on both the signaling lead and the ground lead at the same time and will tend to cancel itself out. However, interference-induced currents can still cause problems, so differential transmission is used in demanding circumstances.

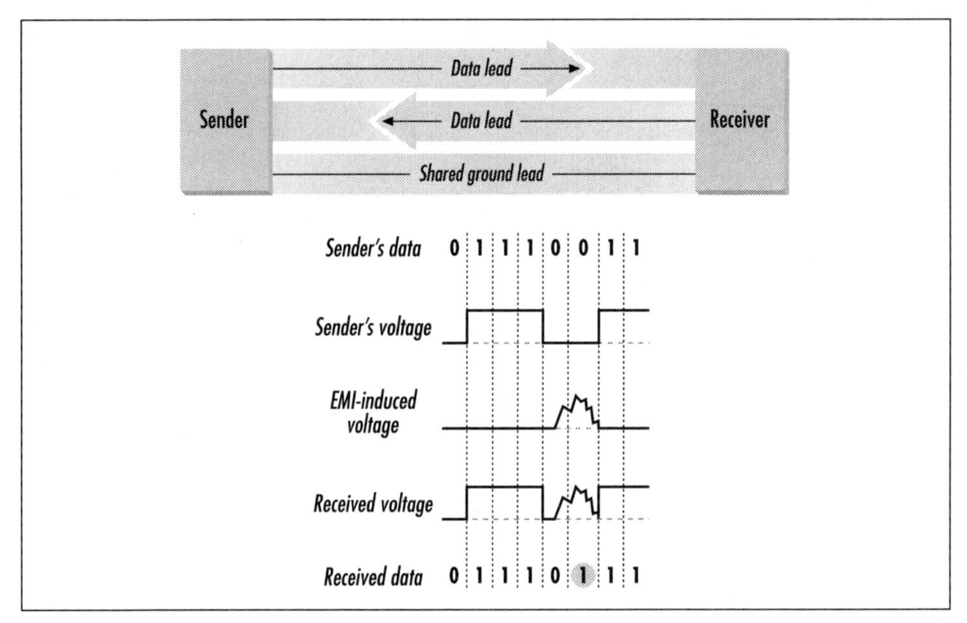

Figure E-2. Interference and common-ground wiring

Differential cables use two wires for each signal. Balanced transmission is paired with differential cables. As in unbalanced transmission, the signal is transmitted as the difference between the voltage on two leads. However, balanced transmission dedicates two wires to each circuit and splits the intended voltage difference between the two leads. Stray electromagnetic interference is likely to affect both wires used in a differential pair. By subtracting one voltage from the other, any stray interference is usually canceled out. Figure E-3 illustrates this procedure. The sender's voltage is transmitted on line A with a corresponding negative pulse on line B. Interference is absorbed by both lines, but when the voltage difference is taken, it is far too small to affect the operation of the circuit.

Differential transmission has far better noise immunity and performs better at high speeds than common-ground transmission. (Twisted-pair Ethernet uses differential transmission.) Its chief disadvantages are increased equipment and cabling costs. Two wires are required to transmit one signal, and the equipment at both ends must be prepared to add voltages together to reconstruct the signal. As a cost compromise, V.35 uses a blend of common-ground and differential circuits.

One final note on cabling: higher-speed cables must be built to tighter tolerances, which often requires equipment to which most end-users do not have access. The effectiveness of a custom cable often depends on the speed at which it must operate. If you need to use long cables, buy them from a supplier who can ensure that they comply with any relevant standards.

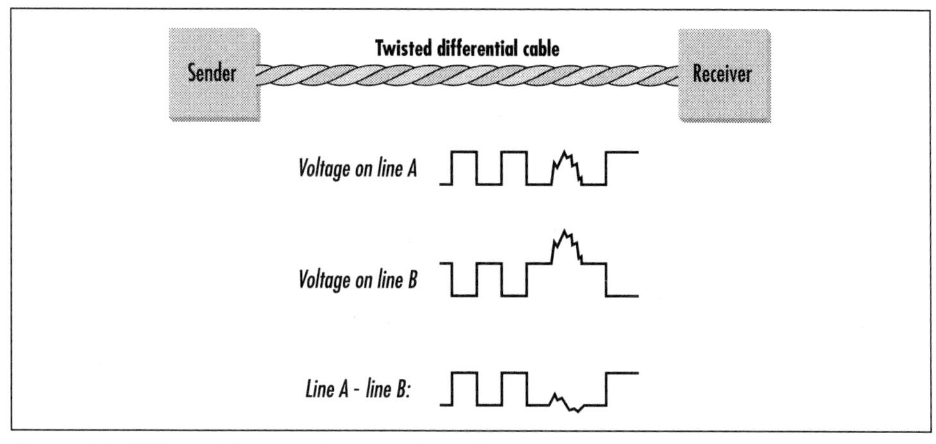

Figure E-3. Differential transmission and interference

Common Serial Signals

ITU Recommendation V.24 defines the circuit names used in many ITU-specified serial communication systems. V.28 defines the corresponding electrical characteristics. V.24 circuit identifiers have been adopted by many manufacturers, with an occasional modification. Common signal names used on most types of serial links are listed in Table E-1. When differential transmission is used, each line is marked with a polarity, such as RX+ and RX– for the two wires used to receive data.

Table E-1. Serial signals and their purposes

Name	Common abbreviations	Function
Carrier Detect Data Carrier Detect Carrier Sense	CD, DCD, CS	Used by the DCE to indicate that it has established a link and can transmit data to the remote end of the circuit
Transmit Data Send Data	TX, TD, SD	Used when sending user data to the DCE
Transmit Clock Transmit Timing	TCLK, TXC, TC, TT	Used by the DCE to provide timing to the DTE
External Clock Serial Clock Transmit External	SCTE	Used by the DTE to supply timing to the DCE
Receive Data	RX, RD	Used when receiving data from the DCE
Receive Clock Receive Timing	RCLK, RXC, RC, RT	Used when receiving a clock signal from the DCE
Clear to Send	CTS	Used by the DCE to indicate that it is okay for the DTE to send data
Request to Send	RTS	Used by the DTE to check whether data transmission is allowed

High-Speed Serial: V.35

V.35 is an old specification that has been obsolete since 1988. Most "V.35" interfaces are actually V.36, V.37, or V.38 interfaces. V.35 is fairly sparse—it does not even specify the form factor of the interface. ISO 2593 contains the specifications for the blocky, long-pinned Winchester connector. However, the industry refers to such interfaces and cables as V.35, so I will too in order to maintain consistency.

To save space, V.35 uses common-ground transmission on signaling leads, but the data and clock signals use differential signaling. Connectors have 34 possible pins, which are lettered A through NN. (G, I, O, and Q are not used.) When used for data-communications applications, connectors typically only have 17 or 18 pins. To ensure that the cables are inserted correctly, the screw attachment mechanism allows connectors to be attached only one way. Figure E-4 shows male and female V.35 connectors.

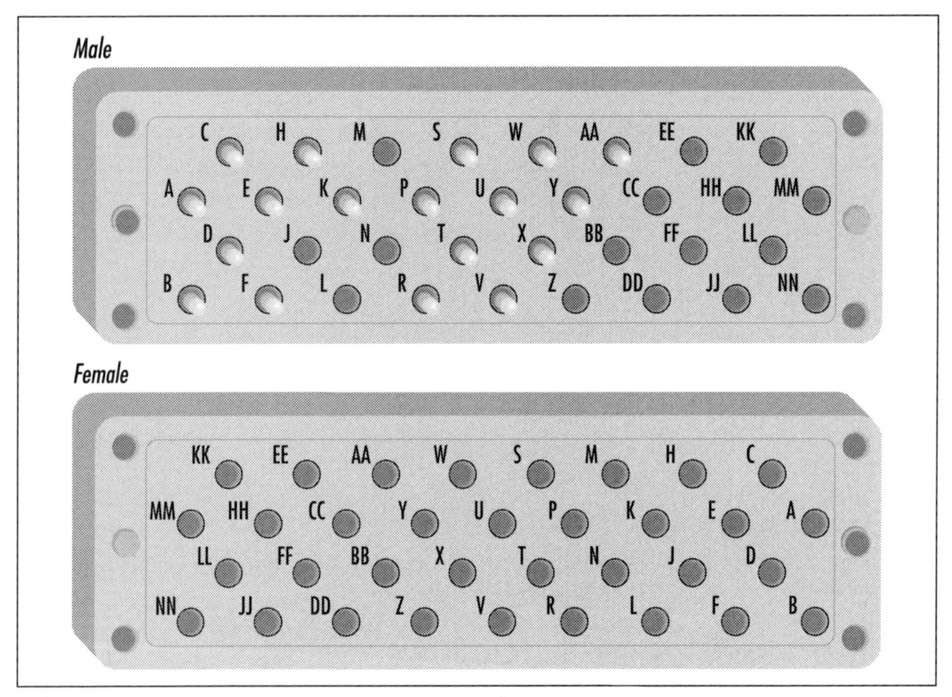

Figure E-4. Male and female V.35 connectors

Each pin is given a purpose and is assigned to a particular serial circuit. Figure E-5 shows the circuit assignments by abbreviation. More detail on the circuits is in Table E-2.

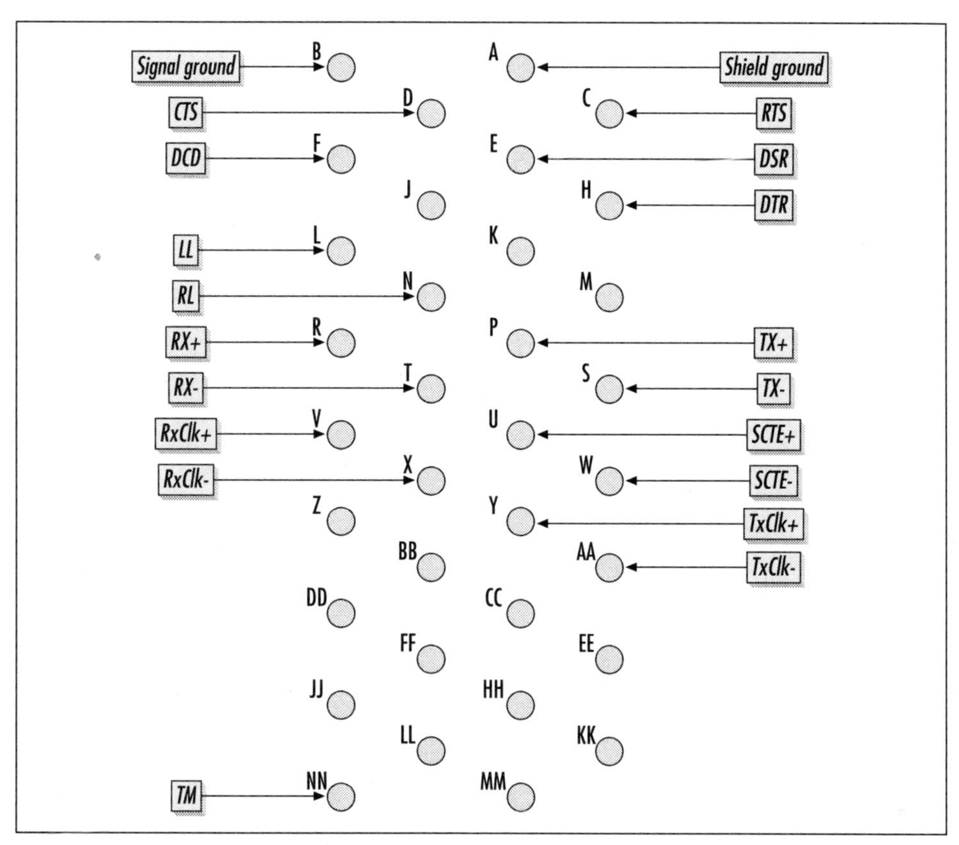

Figure E-5. Circuit identifiers

Table E-2. V.35 pin assignments

Pin	V.24 circuit	Circuit name	Circuit abbrev.	Circuit type	Circuit direction
A	101	Protective Ground Shield	PGND SHD	Ground	N/A
B	102	Signal Ground	GND or SGND	Ground for common-ground circuits	N/A
C	105	Request to Send	RTS	Common-ground control	To DCE
D	106	Clear to Send	CTS	Common-ground control	From DCE
E	107	Data Set Ready	DSR	Common-ground control	From DCE
F	109	Received Line Signal Detector Carrier Detect	RLSD DCD	Common-ground control	From DCE
H	108.2	Data Terminal Ready	DTR	Common-ground control	To DCE
L	141	Local Loopback	LL	Common-ground control	To DCE
N	140	Remote Loopback	RL	Common-ground control	To DCE

Table E-2. V.35 pin assignments (continued)

Pin	V.24 circuit	Circuit name	Circuit abbrev.	Circuit type	Circuit direction
P	103	Send Data A	TD A TX+, SD+	Differential Data	To DCE
R	104	Receive Data A	RD A RX+, RD+	Differential data	From DCE
S	103	Send Data B	TD B TX–, SD–	Differential data	To DCE
T	104	Receive Data B	RD A RX–, RD–	Differential data	From DCE
U	113	Serial Clock Transmit External A	SCTE A	Differential timing	To DCE
V	115	Serial Clock Receive A Receive Timing	SCR A RT+ RxClk+	Differential timing	From DCE
W	113	Serial Clock Transmit External B	SCTE B	Differential timing	To DCE
X	115	Serial Clock Receive A Receive Timing	SCR B RT– RxClk–	Differential timing	From DCE
Y	114	Serial Clock Transmit A Send Timing	SCT A ST+ TxClk+	Differential timing	From DCE
AA	144	Serial Clock Transmit B	SCT B ST– TxClk–	Differential timing	From DCE
NN	142	Test Mode	TM	Common-ground control	From DCE

Even though Pin NN, the Test Mode signal, can occasionally be used, it is not frequently pinned out. Pin J, the Received Line Signal Detector (RLSD) is not used and not pinned on data communications equipment. Pins M, Z, BB, CC, DD, EE, FF, and MM are reserved for future standards and are not pinned. Pins HH, KK, JJ, and LL are reserved for country-specific features, but are not pinned out.

Common V.35 Cabling Problems

In T1 circuits, V.35 runs at 1.5 Mbps. Do not be tempted to save money by making your own cable according to a manufacturer pinout. V.35 needs to run fast within tight electrical tolerances. Hand-built cable will not work. Any money saved by scrimping on cable will haunt you when the cable turns out to have a slightly flaky lead or improper grounding.

Custom-built V.35 cables will not work. Do not try this at home!

Long V.35 connections may cause problems if the ground level at the two ends of the connection is too different. If a voltage differential exists between a third-floor router and a basement CSU/DSU, control signals may not be interpreted correctly.

RJ-48X

In the 1970s, AT&T developed a system of codes for the different types of jacks used at customer premises. These jack codes were called *Universal Service Order Codes* (USOCs), and jack specifications were known as *Registered Jacks* (RJs). The USOC lives on in Part 68 of the FCC rules, which has adopted parts of the USOCs and the RJ system.

Common designations are RJ-11, which is a modular telephone jack found on all modems, and RJ-45, which is the jack used by Category 5 UTP Ethernet wiring. Each jack has a number to designate its type, as well as an optional letter. Some of the letters are listed in Table E-3.

Table E-3. Meaning of letters in the USOC

Letter	Meaning
C	Surface- or flush-mounted jack
W	Wall-mounted jack
S	Single-line jack
M	Multiline jack
X	Complex jack

T1s are terminated by RJ-48X jacks, which are complex jacks because they incorporate shorting bars for automatic loopback. The RJ-48 pinout is given in Table E-4.

Table E-4. RJ-48X pinout

Pin number	Usage
1	RX ring
2	RX tip
3	Grounding
4	TX ring
5	TX tip
6	Grounding

Table E-4. RJ-48X pinout (continued)

Pin number	Usage
7	Unused
8	Unused

Pins 3 and 6 are used for grounding by connecting the leads to the metal shields on the cable. Regulatory approval depends on the presence of the shield grounding.

Regulatory approval requires that the grounding pins be used. Without a properly grounded cable, your installation cannot achieve regulatory approval. Put another way, do not make these cables at home!

Pins are numbered in Figure E-6. Transmission is based on a tip and ring scheme, so leads in Figure E-6 are labeled with Ts and Rs.

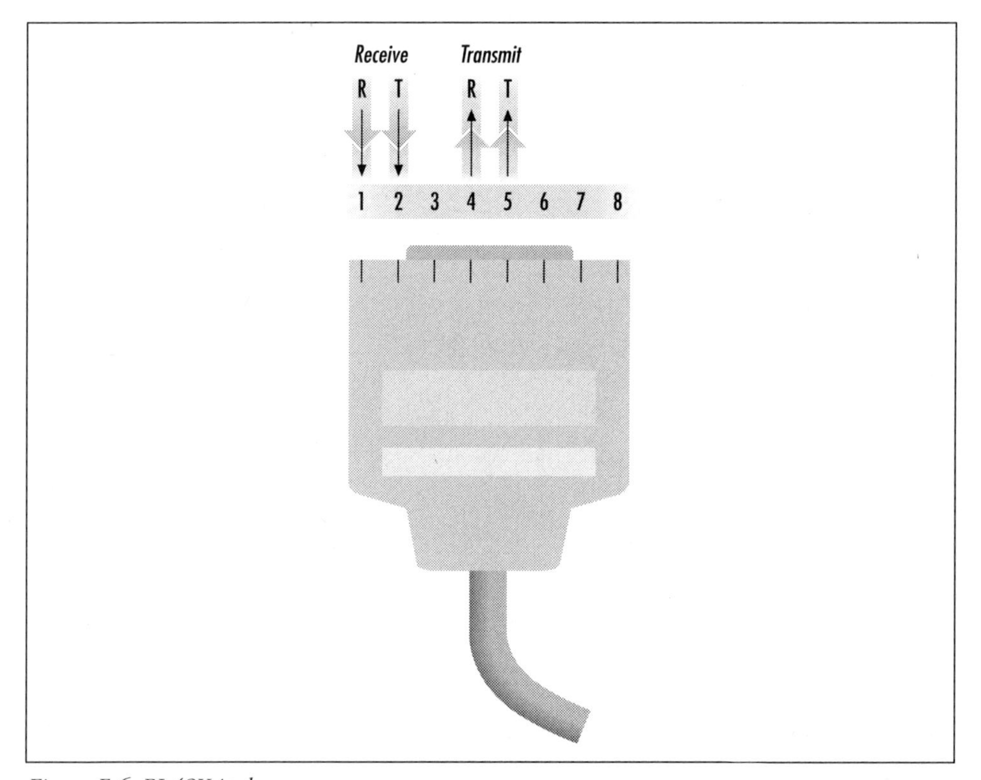

Figure E-6. RJ-48X jack

Shorting bars create an automatic loopback, as Figure E-7 shows. When the plug is removed from the jack, pin 1 is connected to pin 4 with a shorting bar, and pin 2 is connected to pin 5. The receive pair from the network is electrically connected to the transmit pair, and any pulses are redirected back to the network. To ensure that any transmitted pulses from the network are reconstituted by the first repeater, the round-trip loss from the last inbound repeater to the jack and back to the first outbound repeater must be within the loss guidelines for a single path between two repeaters. Normally, one path between repeaters is engineered to a 30 dB loss. Half the loss corresponds to 15 dB, which is why end spans may have a maximum loss of only 15 dB.

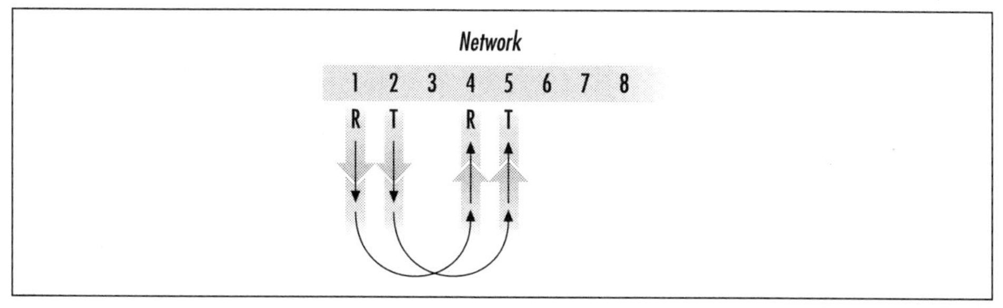

Figure E-7. Shorting bars

DA-15

Some T1 equipment uses DA-15 connectors and requires a DA-15 to RJ-48 cable, which is available from the equipment manufacturer. The female DA-15 pinout is shown in Figure E-8, and its pinout is in Table E-5.

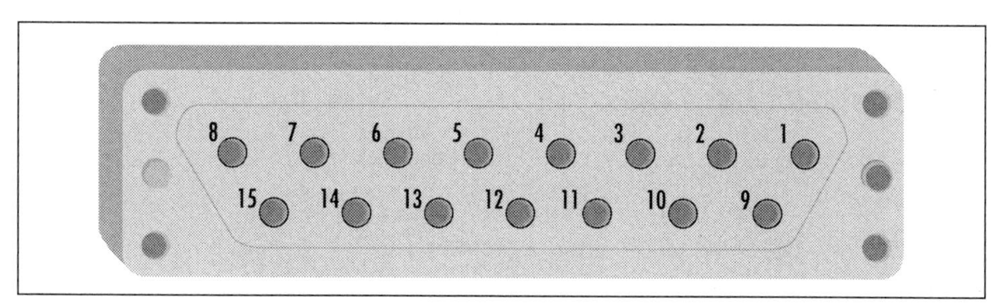

Figure E-8. Female DA-15 connector

Table E-5. DA-15 pinout for T1

Pin numbers	Purpose
1	Transmit to telco, ring
9	Transmit to telco, tip
3	Receive from telco, ring
11	Receive from telco, tip
2 and 4	Grounding

Reference

The wonderful thing about standards is that there are so many to choose from.
—Admiral Grace Hopper or Andrew Tannenbaum?*

Any work offering complete coverage of telecommunications standards is likely to be extremely boring. This book has attempted to show how standards are used in common practice without an unnecessary plodding focus on the low-level details. There may be times, however, when network engineers need to sit down with the specification and fully understand it. For that reason, this appendix lists the standards you might encounter when working with T-carrier systems and the common link layers.

Standards Bodies

Several bodies have worked on the standardization efforts for the technologies described in this book:

- The American National Standards Institute (ANSI), primarily through Committee T1
- The International Telecommunications Union—Telecommunication Standardization Sector (ITU-T)
- The Frame Relay Forum (FRF)

* It appears the saying even applies to itself! Andrew Tannenbaum wrote this in both editions of *Computer Networks*, the first of which was published in 1981. However, some evidence indicates that Admiral Hopper first said this in the late 1970s, and the saying is attributed to her by *The Unix-Haters Handbook*. Tim Salo posed the question on the IETF mailing list in September 1994, and he received a variety of responses, which are summarized at major mailing list archives. (Among many other locations, his summary is archived at *http://mlarchive.ima.com/ietf/1994/1820.html*.)

- The International Organization for Standardization (ISO)
- The Internet Engineering Task Force (IETF) and the Internet Assigned Numbers Authority (IANA)
- Telcordia Technologies (formerly Bellcore)

Each of these bodies produces standardization documents, but differences in philosophy and organization lead to differences in the way standards may be obtained.

ANSI Committee T1

Alliance for Telecommunications Industry Solutions
1200 G Street, NW, Suite 500
Washington, DC 20005
http://www.t1.org

ANSI delegates standardization work to experts by accrediting an expert group to carry out the work. Committee T1 is accredited by ANSI to produce technical specifications for U.S. telecommunications networks. Committee T1 standards are easy to spot. They are numbered starting with *T1*, followed by a period and a document number, such as *T1.403*. ANSI standards are copyrighted and must be purchased.

International Telecommunications Union

ITU
Place des Nations
CH-1211 Geneva 20
Switzerland
http://www.itu.org

The ITU develops recommendations for telecommunications through the work of its telecommunications sector, ITU-T. Related work is carried out in other sectors, so the "-T" is used to indicate the telecommunications standardization efforts. When only one ITU sector is involved, most documentation drops the "-T" in favor of the simpler "ITU."

The ITU's standardization work is divided into several series, and the resulting standards are labeled with the series letter and a document number. The most common series of wide-area networking standards are shown in Table F-1.

Table F-1. Common ITU standards series for the data network engineer

Series letter	ITU description	Contents
Q	Switching and signaling	Frame formats and management protocols
I	Integrated Services Digital Network (ISDN)	The ITU vision of the communications network of the future

Table F-1. Common ITU standards series for the data network engineer (continued)

Series letter	ITU description	Contents
X	Data networks and open system communications	Digital packet networking
V	Data communication over the telephone network	Digital transmission over analog facilities
G	Transmission systems and media, digital systems and networks	Network interfaces and timing

ITU recommendations must be purchased. Many recommendations are inter-linked, which means that understanding a topic often requires reading several recommendations. Large companies, especially telecommunications-equipment vendors and carriers, often obtain site licenses. If you work for a major vendor or carrier, you may have access to most, if not all, recommendations.

Frame Relay Forum

http://www.frforum.com
frf@frforum.com

ITU recommendations are detailed documents that can be a challenge to implement fully. The Frame Relay Forum (FRF) publishes implementation agreements (IAs). IAs are interpretations of ITU recommendations; essentially, they are agreements by FRF members to make some of the complex ITU functionality optional. Such an approach allows for a phased rollout of complex features contained in existing standards. Compatibility testing performed by interoperability labs is based on the IA, which covers that particular functionality. IAs may be downloaded from the FRF web site for no charge.

International Organization for Standardization

International Organization for Standardization (ISO)
1, rue de Varembé, Case postale 56
CH-1211 Geneva 20
Switzerland
http://www.iso.ch
central@iso.ch

The ISO is composed of representatives from 130 national standards bodies.* Information technology specifications are carried out with the International Electrotechnical Commission (IEC) and are published as standards from the Joint Technical

* ISO is not an acronym. It is derived from the Greek *isos*, meaning equal. Equality is certainly an appropriate goal for a standardization organization. It also avoids the need for the ISO to have different acronyms in each of its official languages.

Committee (JTC). ISO/IEC JTC specifications are numbered and dated. For example, the most recent HDLC specification is ISO/IEC 3309:1993, which gives both a specification number (3309) and a revision date (1993).

Internet Engineering Task Force

http://www.ietf.org

The IETF is different from most other standards bodies. In keeping with the functional anarchy of the Internet, the IETF has no fixed membership, and leadership is by example and technical contribution.

IETF standards are called *Requests for Comments* (RFCs). RFCs are numbered in chronological order. Not all RFCs are standards; the RFC series is a living history of the Internet, in which the best current practice information and descriptions of the current state of networking are mixed. Many RFCs dated April 1 are worth reading for the humor value.*

RFCs are frequently updated to cope with changing technology. Unlike many other standards documents, RFCs are never revised and republished with the same numerical identifier. Each update to a protocol is published in a new RFC. New protocol versions are published in new RFCs that "obsolete" (i.e., supersede) older RFCs. The IETF maintains a list of RFCs that includes status information for each one. Manually checking the IETF list can be a chore, though. The RFC editor has made a search engine available that will pull up any related RFCs, including updates to preexisting RFCs, at *http://www.rfc-editor.org/rfcsearch.html*.

Internet Assigned Numbers Authority

http://www.iana.org

Devising standard protocol specifications is only half the battle in creating interoperable network protocols. A plethora of numerical identifiers are required to keep the network functioning.

All protocols may have option codes. Transport-layer protocols must be numbered and identified in the network layer header. Different network layers may have numerical identifiers in link layer headers. To take a concrete example from this book, consider the numerical identifiers in PPP. Interoperability requires that two implementations agree on PPP protocol numbers, LCP codes, LCP option numbers, as well as codes and option numbers for each network protocol

* One of the classic April 1 RFCs is RFC 1149, "Standard for the transmission of IP datagrams on avian carriers." An implementation was first made available as this book was being reviewed, and one of the reviewers helpfully included a link to the field trial: *http://www.blug.linux.no/rfc1149/*.

transmitted across the link. Coordination of numerical assignments is handled by the Internet Assigned Numbers Authority (IANA). IANA is headquartered at the Information Sciences Institute at the University of Southern California.

Telcordia (Formerly Bellcore)

http://www.telcordia.com

Telcordia was formed in the 1984 AT&T divestiture. It was originally known as Bell Communications Research (Bellcore), the shared research arm of the Regional Bell Operating Companies. SAIC subsequently acquired Bellcore and renamed it Telcordia.

Bellcore was responsible for a great deal of the signaling specification that occurred in the post-divestiture telecommunications era. Although Bellcore technical reports were, in theory, only reports, products and services were often developed around them, and they have the force of standards.

Physical Layer Standards

Much of this book is about physical networks. Several standards apply to the physical-transmission medium, many of which were originally derived from AT&T technical publications.

DS0 (DDS)

Although 56/64-kbps leased lines, also known as DDS leased lines, were not a major component of this book, much of the information presented about T1 can be applied to them. These two standards define the DS0:

AT&T Technical Publication 62310
: Offering low-speed (56/64 kbps) leased lines was one of AT&T's original data services and is defined in this publication.

ANSI T1.410
: Carrier to Customer Metallic Interface—Digital Data at 64 kbit/s and Subrates"

  ANSI followed up the previous publication with its own additions to the initial AT&T engineering work.

DS1 Standards

AT&T laid the foundations for T1 networking. As a result, two AT&T technical publications lay the foundation for subsequent engineering standards:

AT&T Technical Publication 62411
: This publication defines the SF (D4) format and T1 line characteristics.

AT&T Technical Publication 54016

This publication extended T1 technology by defining ESF framing and the first use of the FDL. The FDL specification in T1.403 is more useful because it is based on periodic messaging rather than this publication's command/response system.

ANSI T1.403

"Network and Customer Installation Interfaces—DS1 Electrical Interface"

This is the modern T1 line specification. In addition to including the technical details from the previous two publications, it defined a new FDL that was more useful for polled monitoring.

ANSI T1.231

"Digital Hierarchy—Layer 1 In-Service Digital Transmission Performance Monitoring"

T1.231 describes the error conditions that may occur on T1 and T3 circuits and what equipment must do to monitor for error conditions. Appendix C is an introduction to the T1-specific parts of T1.231.

Timing

Like many other endeavors, the secret to WAN success is timing. Several specifications set limits on timing performance, which must be achieved by digital networks:

ANSI T1.101

"Synchronization Interface Standards for Digital Networks"

ANSI has standardized the different strata of network clocks, as well as their relative performance information. That information can be found in this specification.

ITU Recommendation G.821

"Error performance of an international digital connection operating at a bit rate below the primary rate and forming part of an integrated services digital network"

This recommendation defines the acceptable error rates for international leased-line connections operating at less than primary rate (T1 or E1) speeds. Performance targets are, however, independent of the physical network medium used.

ITU Recommendation G.826

"Error performance parameters and objectives for international, constant bit rate digital paths at or above the primary rate"

This recommendation is an equivalent to G.821, but for circuits operating at speeds above the primary rate.

DISA circular 300-175-9 (8 June 1998)

"DII Operating-Maintenance Electrical Performance Standards"

DISA, the Defense Information Systems Agency, maintains the *Defense Information Infrastructure* (DII), a data network used by the U.S. Department of Defense. This circular, which is available from *http://www.disa.mil/pubs/*, discusses the performance requirements for leased lines used in the DII.

Serial

Once the telco line is terminated by the CSU/DSU, a serial link is used to connect the CSU/DSU to the customer router. Several specifications define the characteristics of that serial link:

ITU Recommendation V.24

"List of definitions for interchange circuits between data terminal equipment (DTE) and data circuit-terminating equipment (DCE)"

This recommendation defines the generic circuits used on a serial interface of any type. It forms the basis for a great deal of the ITU's work on serial communications.

ITU Recommendation V.28

"Electrical characteristics for unbalanced double-current interchange circuits"

V.28 is the electrical counterpart to V.24. It defines the electrical characteristics of a standard serial circuit.

ITU Recommendation V.35

"Data transmission at 48 kilobits per second using 60–108 kHz group band circuits"

V.35 is a label attached to most devices that terminate leased lines and have serial interfaces for attachments to routers. V.35 is obsolete; a good number of devices labeled "V.35" are actually V.36, V.37, or V.38.

ITU Recommendation V.36

"Modems for synchronous data transmission using 60–108 kHz group band circuits"

ITU Recommendation V.37

"Synchronous data transmission at a data signaling rate higher than 72 kbit/s using 60–108 kHz group band circuits"

ITU Recommendation V.38

"A 48/56/64 kbit/s data circuit-terminating equipment standardized for use on digital point-to-point leased circuits"

ITU Recommendation V.54

"Loop test devices for modems"

V.54 defines the different types of loopback test modes incorporated into CSU/DSUs.

ISO/IEC 2593

"Information technology—Telecommunications and information exchange between systems—34-pole DTE/DCE interface connector mateability dimensions and contact number assignments"

This standard defines a standard 34-pin metal connector designed for use in serial-communication applications. It was adopted by V.35 and its successors and is commonly used for serial interfaces that connect to telecommunications devices. Unlike many ISO standards, this standard is available in English only.

Channelized T1 Signaling

Signaling was an important part of the early T1 specifications. Two major standards were used:

EIA/TIA 464

"Requirements for Private Branch Exchange (PBX) Switching Equipment"

Private Branch Exchanges (PBXs) are used to connect customer telephones to the telco's trunk lines. Frequently, PBXs incorporate T1 interfaces in order to serve high-capacity telephone sites. T1 interfaces must incorporate the inter-switch T1 signaling so that the PBX and the CO switch can communicate.

Bellcore (Telcordia) TR-NWT-000057

"Functional Criteria for Digital Loop Carrier System"

This technical reference is a component of the Transport Systems Generic Requirements. It also describes T1 signaling systems.

ISDN

A great number of ISDN standards have been written. ISDN is often pejoratively referred to as "Interesting Services Doing Nothing" or "IS it Done Now?" because the ongoing standardization process can be time consuming. Here is a list of standards:

ITU Recommendation Q.921

"ISDN user-network interface—Data link layer specification"

Frame relay shares several features with ISDN. The frame relay data link layer incorporates several features from the ISDN data link layer specification.

ITU Recommendation Q.930

"Recommendation Q.930 (03/93)—Digital Subscriber Signalling System No. 1 (DSS 1)—ISDN user-network interface layer 3—General aspects"

ISDN signaling is based on its own protocol stack. Q.921 is the data link layer, and Q.930 defines the basic network layer parameters.

ITU Recommendation Q.931

"Recommendation Q.931 (05/98)—ISDN user-network interface layer 3 specification for basic call control"

Call-control signaling on the D channel is specified in Q.931. It is baroque, complex, and confusing. You should dare to tread in these waters only for a specific purpose. Before embarking on a journey through these woods, make sure to tell a good friend or two so they can call for help if you fail to show up at your intended destination on time.

ANSI T1.408

"ISDN Primary Rate—Customer Installation Metallic Interfaces, Layer 1 Specification"

This is an old specification that has since been incorporated into recent revisions of T1.403.

Frame Relay

Frame relay standardization work was carried out in parallel by the ITU and ANSI. Standards documents often refer to frame relay services as *frame mode bearer services*. Here is a list of standards:

ITU Recommendation X.25

"Interface between Data Terminal Equipment (DTE) and Data Circuit-terminating Equipment (DCE) for terminals operating in the packet mode and connected to public data networks by dedicated circuit"

X.25 was one of the first specifications for public data networks. It incorporated hop-by-hop flow control and error correction, which contributed to the high latency of packets through X.25 networks. X.25 is still widely used for low-data-rate connectivity for point-of-sale terminals and similar applications.

ITU Recommendation Q.921

"ISDN user-network interface—Data link layer specification"

Frame relay shares several features with ISDN. The frame relay data link layer incorporates several features from the ISDN data link layer specification.

ITU Recommendation Q.922

"ISDN data link layer specification for frame mode bearer services"

Q.922 specifies the frame format used for frame relay services. It incorporates several features from Q.921 by reference, which is why it is helpful to have a copy of each.

ANSI T1.618

"Digital Subscriber Signaling System No. 1 (DSS1)—Core Aspects of Frame Protocol for Use with Frame Relay Bearer Service"

ITU Recommendation Q.933

"Integrated services digital network (ISDN) digital subscriber signaling system No. 1 (DSS 1)—Signaling specifications for frame mode switched and permanent virtual connection control and status monitoring"

The title is a bit wordy, but Q.933 is just the ITU's LMI specification. Sections of Q.933 refer to ISO/IEC TR 9577.

ANSI T1.617

"Digital Subscriber System No. 1 (DSS1) Signaling Specification for Frame Relay Bearer Service"

T1.617, and its later follow-up, T1.617a, describe a similar but slightly different version of the LMI.

FRF.1.1 and FRF.1.2

"PVC User-to-Network Interface (UNI) Implementation Agreement"

Q.933 and T1.617 are both complex standards. To make deployment easier, FRF.1 specified a streamlined set of options so that frame relay could be deployed faster.

ISO/IEC TR 9577

"Information technology—Protocol Identification in the network layer"

This technical report is available at no charge from the ISO. It describes how to identify both ITU-standardized and ISO-standardized protocols in a number of other protocols. It was adopted by the IETF for use with RFC-1490 encapsulation. Unlike most other ISO and ITU standards, TR 9577 is available at no charge.

ITU Recommendation X.263

"Information technology—Protocol identification in the Network Layer"

This ITU recommendation contains the same text as the ISO/IEC TR 9577.

RFC 2427 (updated version of RFC 1490)

"Multiprotocol Interconnect over Frame Relay"

RFC 1490 was the first IETF standard for multiprotocol encapsulation over frame relay links. It defines the multiprotocol encapsulation header for different network protocols, plus the encapsulation format for bridged packets.

RFC 2427 was an evolutionary update that clarified some minor points and removed the specification of Inverse ARP, among other things. A complete list of changes from RFC 1490 is present in the final section of the RFC.

RFC 2390

"Inverse Address Resolution Protocol"

frame relay networks may bring new DLCIs into existence dynamically. New DLCIs are reported in the LMI PVC status message with the "new" bit set. These announcements, however, do not include any higher-layer information. Inverse ARP allows a frame relay device to request the IP address associated with a given DLCI.

ITU Recommendation T.50

"Recommendation T.50 (09/92)—Information technology—7-bit coded character set for information interchange"

T.50 describes character sets used by other standards. Parts of it are used by both LMI Annex A and LMI Annex D.

ITU Recommendation V.4

"Recommendation V.4 (11/88)—General structure of signals of International alphabet No. 5 code for character oriented data transmission over public telephone networks"

International Alphabet 5 is used by the LMI Annex A. This recommendation specifies IA5.

HDLC

The basis of HDLC is laid out in a series of ISO/IEC joint specifications. Bill ("Chops") Westfield initially described Cisco's multiprotocol specifications in an email to a public mailing list in 1990. It was subsequently quoted several times on other mailing lists and is readily available from web search engines. Two standards form the basis of the portions of HDLC relevant to data communications:

ISO/IEC 3309

"Information technology—Telecommunications and information exchange between systems—High-level data link control (HDLC) procedures—Frame structure"

This document describes the basic HDLC frame structure and the frame check sequence, but does not describe how the address and control fields are used.

ISO/IEC 4335

"Information technology—Telecommunications and information exchange between systems—High-level data link control (HDLC) procedures—Elements of procedures"

This specification describes how the various frame types use the control field.

Ethernet Type Codes

Cisco HDLC makes use of Ethernet type codes to identify higher-layer protocols. The IEEE is the official keeper of Ethernet type codes; their list of "EtherType Field Public Assignments" can be found at *http://standards.ieee.org/regauth/ethertype/type-pub.html*. IANA also maintains a list of assigned Ethernet types at *http://www.iana.org/assignments/ethernet-numbers/*.

PPP

PPP has been standardized through the IETF process, so an excellent history of the evolution of the protocol has been captured through the RFC series. Open development has also made a number of good references readily available. I particularly recommend *PPP Design, Implementation, and Debugging*, by James Carlson.

Motivation, Background, and History

Documents listed in this section are of interest because they offer insight into the way the protocol evolved and the challenges facing protocol designers at the time. None of the RFCs in this section describe present standards, nor should they be used to guide any implementation efforts.

RFC 1055

"A Nonstandard for Transmission of IP Datagrams Over Serial Lines: SLIP"

Despite the title, SLIP was the first standard method of transmitting IP over serial lines. It lacked framing, flow control, and the ability to send anything other than IP, but it was a start.

RFC 1144

"Compressing TCP/IP Headers for Low-Speed Serial Links"

This RFC is unusual in that it is available in both ASCII text and PostScript, with the latter format having more detailed and prettier diagrams. In 1990, state of the art modems were far slower than they are today, and time-shared computing was a necessity for researchers needing even moderate power. One of SLIP's many problems is that it transmitted full headers for both TCP and IP. Interactive connections frequently send packets with 1 or 2 bytes of data. LAN media can easily handle packets with 40-plus header bytes for 1 data byte, but a 2,400-baud modem makes the procedure excruciating. Van Jacobson, the networking wizard responsible for writing *tcpdump* and for helping us to understand the behavior of TCP/IP in many other contexts, developed a procedure for compressing the TCP and IP headers down to only a handful of bits. Not surprisingly, the procedure is known as Van Jacobson header compression, or simply Van Jacobson compression. When used with SLIP, it is called Compressed SLIP, or CSLIP.

Van Jacobson's work on SLIP compression was funded in part by federal research dollars. As he said in a Usenet post in 1988, "Research to make slow things go fast sometimes makes fast things go faster." That post is archived at *http://wwwhost.ots.utexas.edu/ethernet/enet-misc/load/vanj-enet-posting.html.*

RFC 1547

"Requirements for an Internet Standard Point-to-Point Protocol"

Before PPP was specified, a set of design requirements was drawn up. These requirements from 1989 were initially published in RFC 1547 in 1993 for completeness. RFC 1547 concludes with a discussion of the various serial-line protocols available at the time and their weaknesses.

Current State-of-the-Art PPP

Several RFCs define the PPP suite. The list in this appendix is complete for common features as of this writing, but more features may be added to PPP in the future. Up-to-date listings of all the officially assigned PPP protocol numbers may be retrieved from the IANA web site at *http://www.isi.edu/in-notes/iana/assignments/ppp-numbers.*

RFC 1661

"The Point-to-Point Protocol (PPP)"

This RFC is the foundation for all modern PPP implementations. It describes the encapsulation format, the option-negotiation state engine, and LCP.

RFC 1662

"PPP in HDLC-like Framing"

This RFC specifies the link layer framing for PPP. Adaptations laid out in this document allow PPP to run across nearly any serial link. This RFC borrows some of the common HDLC header components, as well as the standard HDLC stuffing procedure for bit-synchronous links. It also defines the octet-stuffed interface used on asynchronous links and discusses the use of PPP across asynchronous-to-synchronous converters.

RFC 1663

"PPP Reliable Transmission"

This RFC describes how to use the ITU's LAPB to create a reliable transmission layer under PPP. While interesting in theory, reliable delivery can play havoc with the TCP flow-control algorithms, which infer network congestion from packet loss.

ISO/IEC 7776

"Information technology—Telecommunications and information exchange between systems—High-level data link control procedures—Description of the X.25 LAPB-compatible DTE data link procedures"

LAPB is the link layer for ISDN. Unlike many other link layers developed by the networking industry, LAPB is reliable. Damaged or lost frames are automatically retransmitted. When PPP is used with reliable transmission, the PPP layer rides on LAPB rather than the standard modified HDLC layer.

Network Control Protocols

Network Control Protocols have been defined for most common network protocols. Table F-2 lists the standardized PPP NCPs at the time this book was written.

All NCPs have similar structure. They each inherit the option-negotiation state engine from RFC 1661 and negotiate options in a similar fashion to LCP. Each NCP standard defines the PPP protocol number for encapsulating the higher-layer protocol, the PPP protocol number for its related NCP, and the options used by that NCP to bring up the network layer in question.

Table F-2. Standards for PPP NCPs

Network protocol	Related NCP RFCs
IP	1332: The PPP Internet Protocol Control Protocol (IPCP)
	1877: PPP Internet Protocol Control Protocol Extensions for Name Server Addresses
	2290: Mobile-IPv4 Configuration Option for PPP IPCP
IPX	1552: The PPP Internetworking Packet Exchange Control Protocol (IPXCP)
AppleTalk	1378: The PPP AppleTalk Control Protocol (ATCP)
NetBIOS	2097: The PPP NetBIOS Frames Control Protocol (NBFCP)
SNA	2043: The PPP SNA Control Protocol (SNACP)
DECnet	1376: The PPP DECnet Phase IV Control Protocol (DNCP)
802.1d bridging	2878: PPP Bridging Control Protocol (BCP)
OSI	1377: The PPP OSI Network Layer Control Protocol (OSINLCP)

Multilink PPP

MP is specified in one RFC, and not a particularly long one at that. RFC 1717 was the predecessor to the current specification, RFC 1990:

RFC 1990

"The PPP Multilink Protocol (MP)"

This standard defines MP. It includes information on MP encapsulation, fragmentation, reassembly, and detecting fragment loss and negotiating MP with LCP.

Several extensions to RFC 1990 may be used for Multi-Chassis MP. Nortel's method has been specified in an informational RFC. Two of the layer-two protocols used are also specified, though only L2TP was an open, publicly developed standard:

RFC 2701

"Nortel Networks Multilink Multinode PPP Bundle Discovery Protocol"

This is Nortel's MMP specification. It is discussed in detail in Appendix B.

RFC 2661

"Layer Two Tunneling Protocol 'L2TP'"

L2TP is not discussed in detail in this book. It is used by the Nortel MMP specification for fragment tunneling.

RFC 2341

"Cisco Layer Two Forwarding (Protocol) 'L2F'"

Cisco's MMP code uses L2F for fragment forwarding. L2F is also not discussed in this book. L2F was a precursor to L2TP.

Other protocols I did not cover were the Ascend (Lucent) Stacks and Cisco's Stack Group Bidding Protocol (SGBP). Neither is publicly documented.

Two informational RFCs describe additional approaches to using multiple links. MP+ is specific to Ascend/Lucent. BAP and BACP form the basis of the call-steering approach described in Appendix B:

RFC 1934

"Ascend's Multilink Protocol Plus (MP+)"

RFC 1934 is only an informational RFC. It contains errors that make an interoperable implementation impossible. It does not offer any advantages over a standard MMP implementation, so it was not discussed in this book.

RFC 2125

"Bandwidth Allocation Protocol (BAP) & Bandwidth Allocation Control Protocol (BACP)"

BAP and BACP are the basis for the call-steering approach to terminating MP sessions on multiple devices.

ITU Recommendation E.164

"The international public telecommunication numbering plan"

E.164 is the ITU's standard for international telephone numbers. E.164 numbers are used in many places in multilink protocols.

Other PPP Extensions

RFC 1570

"PPP LCP Extensions"

This RFC specified the first set of extensions to LCP. It describes two new LCP codes for the Identification and Time-Remaining frames. It also defines several new LCP options for alternative frame check sequences, the Self-Describing-Padding option, the use of the Callback option for shifting dial-up costs, and the seldom used Compound-Frames option for encapsulating multiple PPP frames in a single link layer frame.

RFC 2153

"PPP Vendor Extensions"

This RFC defines the procedure by which vendors are supposed to implement proprietary options to PPP. In practice, most vendors simply take an unused option without bothering to ask IANA for an option number.

RFC 1989

"PPP Link Quality Monitoring"

PPP offers an open framework for monitoring link quality. RFC 1989 defines the only standardized link quality monitoring protocol.

RFC 1962

"The PPP Compression Control Protocol (CCP)"

PPP can compress packets before sending them across links. CCP is best used when the connection is paid for on traffic volume or has limited transmission capacity.

RFC 1968

"The PPP Encryption Control Protocol (ECP)"

PPP can also be used to encrypt packets before transmitting them across the link. ECP is used only after the packets are compressed with CCP. The PPP reliable delivery option is often used to ensure sequential loss-free delivery because retransmission requires key resynchronization.

PPP authentication

This book did not discuss PPP authentication. For completeness, however, relevant standards are listed here for interested readers:

RFC 1334

"PPP Authentication Protocols"

This was the first RFC to describe authentication on PPP links. It remains the standard for the Password Authentication Protocol (PAP). PAP is the lowest-common authentication denominator. Because it is so simple, it is widely supported.

RFC 1994

"PPP Challenge Handshake Authentication Protocol (CHAP)"

CHAP is a far stronger authentication protocol that eliminates the major weakness of PAP—transmission of secret authentication data in the clear. RFC 1994 is its modern incarnation.

RFC 2433

"Microsoft PPP CHAP Extensions"

Microsoft Windows dial-up systems use a variant of CHAP, called MS-CHAP or CHAP-80, to authenticate dial-up links. It is described in this RFC.

RFC 2759

"Microsoft PPP CHAP Extensions, Version 2"

Updates to MS-CHAP are described in this document.

RFC 2284

"PPP Extensible Authentication Protocol (EAP)"

Rather than issue a new RFC each time a new authentication method is devised, the EAP allows the use of several different authentication methods with an underlying homogenous protocol.

SNMP

SNMP by itself is a wide topic. Only a few T1-specific RFCs were presented in detail in this book. For interested readers, this reference list includes the earlier documents that laid out the specifications for SNMP itself.

SNMP Foundations

RFC 1157

"Simple Network Management Protocol (SNMP)"

This is the first version of the SNMP wire protocol. It includes the ill-fated "security" based on community strings.

RFC 1155

"Structure and Identification of Management Information for TCP/IP-based Internets"

SNMP lays out data in a highly structured format, referred to as the Structure of Management Information (SMI). This RFC defines the basic data types used by SNMP MIBs.

ISO 8824

"Information technology—Open Systems Interconnection—Specification of Abstract Syntax Notation One (ASN.1)"

ASN.1 is a rigorous syntax for describing structured data. ASN.1 is the language of SNMP because the MIBs are defined using ASN.1 syntax.

RFC 1213

"Management Information Base for Network Management of TCP/IP-based Internets: MIB-II"

Basic system information is contained in MIB-2. Among other things, MIB-2 contains the *interfaces* table to which the DS1 and frame relay DTE MIBs cross-index.

SNMP Version 2

SNMPv2 provided several enhancements over the initial SNMP standards. It made tables easier to work with, expanded the data types available to MIB authors, and improved the efficiency of the protocol. Its major failing was in not improving the "security" model of SNMPv1.

RFC 2578

"Structure of Management Information Version 2 (SMIv2)"

This RFC updated the SMI for use with SNMPv2.

RFC 2579

"Textual Conventions for SMIv2"

This RFC defines "textual conventions," which are data types, for use in SNMPv2-specific MIBs.

RFC 1905

"Protocol Operations for Version 2 of the Simple Network Management Protocol (SNMPv2)"

SNMPv2 added some protocol operations to the basic wire protocol. This document lays out the new protocol operations.

SNMP Version 3

SNMPv3 was designed to fill in the major holes remaining in SNMPv2 (most notably, the security model from earlier versions of SNMP2, which provided no security at all).

RFC 2570

"Introduction to Version 3 of the Internet-standard Network Management Framework"

This RFC introduces SNMP Version 3. The protocol is complex and several RFCs are required for the full specification.

RFC 2571

"An Architecture for Describing SNMP Management Frameworks"

The overall SNMPv3 architecture is described in this RFC. It describes the new security mechanisms so sorely needed in earlier versions of SNMP, such as user-based access control to individual objects in the MIB.

RFC 2572

"Message Processing and Dispatching for the Simple Network Management Protocol (SNMP)

This RFC describes the composition of SNMPv3 messages and how they are processed. SNMPv3 allows for multiple message processing models within the SNMP framework; this RFC contains the specifications that allow this coexistence.

RFC 2574

"The User-Based Security Model for Version 3 of the Simple Network Management Protocol (SNMPv3)"

SNMPv3 provides message-level security against modification and eavesdropping. RFC 2574 describes the threats to SNMP data and how SNMPv3 defends against the identified threats.

RFC 2575

"View-based Access Control Model for the Simple Network Management Protocol (SNMP)"

View-based access control allows administrators to create user-specific access controls on MIB objects.

RFC 2576

"Coexistence Between Version 1, Version 2, and Version 3 of the Internet-standard Network Management Framework"

To make sense of all the SNMP data floating around, network management stations must be "multilingual"—they must be able to interpret all existing versions of SNMP. Processes that collect data and expose it to SNMP may also need to be multilingual if they must present a backward-compatible interface to a network management station.

SNMP MIBs Related to T1

RFC 2495

"Definitions of Managed Objects for the DS1, E1, DS2, and E2 Interface Types"

Appendix D described only the DS1 interface MIB. The monitoring data used by this MIB was based on an early draft of T1.231. It was not revised as T1.231 was, so the definitions of some quantities are slightly different.

RFC 2115

"Management Information Base for Frame Relay DTEs Using SMIv2"

Several frame relay MIBs exist, some of which are more appropriate to frame relay switching devices. This MIB is appropriate for use at the edges of frame relay devices. Several carriers insist on this MIB solely for the purpose of reporting DLCI status changes with SNMP traps.

RFC 1471

"The Definitions of Managed Objects for the Link Control Protocol of the Point-to-Point Protocol"

This MIB reports on the statuses of various LCP parameters used on PPP links. For a complete list, see the LCP MIB exposition in Appendix D.

RFC 1472

"The Definitions of Managed Objects for the Security Protocols of the Point-to-Point Protocol"

This RFC defines the PPP Security Group, which can be used to report authentication data for PPP links. It was not discussed in this book because authentication was not discussed.

RFC 1473

"The Definitions of Managed Objects for the IP Network Control Protocol of the Point-to-Point Protocol."

Each PPP link that has been configured for use with IP may include data on the configuration of each link based on this MIB, which defines the IP Group of the PPP MIB. It is comparatively simple and is essentially a way of monitoring the compression protocols in use on each PPP link.

Glossary

When visiting another country, the first task is to learn the language. Telcos are a foreign realm for many data networkers. They have different customs, attitudes, and a different language. This glossary should help you talk the talk and look up terms quickly when necessary.

AIS

Alarm indication signal. AIS is also known as the *blue alarm*. When no incoming signal is detected, a CSU/DSU transmits an unframed all-ones pattern to the network to maintain synchronization and announce its presence to the network.

ACCM

Asynchronous control character map.

AMI

Alternate Mark Inversion. An encoding scheme, also known as *bipolar return to zero*, in which one bits (marks) are transmitted as pulses and zeros (spaces) are indicated by the absence of a pulse. Pulses alternate in polarity.

ANI

Automatic Number Identification. A similar service to CLID, which transmits the number of the calling party.

asynchronous

Data transmission that relies on the endpoints to send explicit stop and start instructions. While easier to do than synchronous transmission, the overhead of start and stop messages makes it impractical for high-speed links.

B8ZS

Bipolar with 8 Zero Substitution. An encoding scheme that replaces eight zeros with intentional bipolar violations. Eight zeros are replaced with a code that has two intentional bipolar violations. The substitution takes the form 000V10V1, where V is a pulse of the same polarity as the previous pulse. B8ZS enables clear channel transmission because it ensures adequate pulse density even when long strings of zeros must be transmitted.

BER
: Bit Error Rate. The fraction of bits in error on a particular circuit. If 100,000 bits are transmitted and 1,000 are received in error, the BER is .01, or 1%. BER is often expressed in scientific notation, where 1 bit error per 1,000 bits is equal to 1×10^{-4} or 1E-4.

BERT
: Bit Error Rate Test. A test that measures the BER.

blue alarm
: Older term for AIS. See AIS.

BPV
: Bipolar violation. In AMI encoding schemes, successive ones are supposed to be sent with alternating polarities. When this pattern is broken, it is known as a bipolar violation. Bipolar violations are the result of line noise, unless expressly inserted by a clear-channel line code such as B8ZS.

bit robbing
: For signaling purposes, T1 links may rob the least-significant bit out of every time slot in every sixth frame. The lost bit is called the *robbed bit*, which is an appropriate term for data network engineers because the robbed bit takes away 12.5% of the capacity.

carrier
: Another term for telco.

CCC
: Clear channel capability. Lines that do not need to rob bits from the data stream for signaling or timing purposes are said to be clear channel lines. This term is used to describe 64k DDS lines and higher levels of the T-carrier hierarchy that employ 64k channels.

CLID
: Calling Line Identification; also referred to as "Caller ID." It is the transmission of the caller's telephone number to the receiver.

CPE
: Customer premises equipment. Hardware purchased by customers and used for the purpose of connecting to the network. In the context of T1, it means the CSU/DSU.

CS

Controlled slip. Replication or deletion of the 192-bit data frame, usually due to buffer overrun or underrun caused by timing mismatch between the relatively inaccurate CPE clock and the much more accurate CO clock.

CSU

Channel Service Unit. The piece of your CSU/DSU that talks to the telco network, understands framing and line coding, and provides electrical isolation of your network from the telco network.

D4

(After the Western Electric channel bank of the same name.) A mistaken identifier sometimes used to refer to the superframe (SF) format on T1 links. SF existed long before the D4 channel bank.

DACS

Digital Access and Cross-connect System. Also called a DCS.

DCS

Digital Cross-connect System. CO equipment that connects incoming digital lines to outgoing digital lines. DCSs are the foundation of data transmission over the telco network.

demarc

Short for demarcation point. To the telco, anything beyond the demarc is a customer problem. To the customer, anything beyond the demarc is the telco's problem.

DL

Data link. In this book, I have referred to the ESF facilities data link as the FDL, though some standards may refer to it as the DL.

DNIS

Dialed Number Information Service. A channelized T1 service that transmits the dialed number to the recipient.

DS0

Digital Stream, level zero. A single 64-kbps channel that can be used to transmit a sequence of pulses across the telephone network.

DS1

Digital Stream, level 1. 24 DS0s are combined into a DS1, which supplies 1.536-Mbps connectivity, with 8 kbps framing and signaling overhead for a total speed of 1.544 Mbps.

DSU

Data Service Unit. The part of the CSU/DSU that interfaces with our familiar world of routers, switches, and packets. It has a serial port to interface with compatible data equipment.

E&M

Ear and Mouth signaling. A method switches use to tell neighboring switches which channels are being used for telephone calls. E&M signaling has both wink-start and immediate-start variants.

ES

Errored second. A 1-second interval with path code violations, controlled slips, out of frame errors, or AIS defects.

ESF

Extended superframe. ESF is a newer frame format that allows for error detection across an entire span and includes an embedded data link that can be used for signaling and performance reporting. Now the default framing format on nearly all new T1s.

extended demarc

The telco will put the demarc where it is convenient. To make it useful, an extended demarc must be constructed so the jack is near the customer DTE.

EXZ

Excessive zeros. An error recorded when the line code has had too many zeros and has lost synchronization. This is >15 for AMI and >7 for B8ZS. Most equipment does not track EXZ events.

FDL

Facilities data link. On ESF-framed T1s, the FDL provides a 4-kbps channel that can be used for control, signaling, and performance reporting.

FE

Frame bit error. A single framing bit that does not have the expected value. Only framing bits are considered on ESF; SF framing bits are always considered, and SF terminal bits may be considered.

FXO

Foreign exchange office. The CO-side counterpart to a foreign exchange station (FXS).

FXS

Foreign exchange station. A telephone connected through a PBX, or the PBX itself.

glare

The deadlock that results when both ends of a line attempt to use it at the same time. It is amusing to think of telephone switches staring at each other, but the word glare was probably chosen for a different reason.

HDLC

A common ISO-standardized link layer protocol, and the grandfather of most subsequent link layer protocols, including all varieties of the ITU's Link Access Procedure and the IETF's PPP.

ISDN

Integrated Services Digital Network.

ITU

International Telecommunications Union. The ITU is responsible for international telecommunications standardization through several of its component bodies.

ITU-T

The ITU organ responsible for telephone standardization. ITU-T was formerly known as the CCITT (Consultative Committee on International Telephone and Telegraph).

LAPB

Link Access Procedure, Balanced. The ISO specification for the X.25 defines LAPB as its own network layer. LAPB may be used as part of an X.25 network or to provide a reliable lower layer for PPP (see RFC 1663).

LAPD

Link Access Procedure, D Channel. An ITU specification used in ISDN for control signaling that was adopted for sending performance reports over the T1 FDL. Because telcos use performance reports to monitor service levels, this LAPD (unlike the more familiar LAPD—the Los Angeles Police Department) might actually help enforce your rights.

LBO

Line build out. Circuitry that attenuates the signals from the CSU/DSU to the appropriate amplitude for the nearest line repeater.

LOF

Loss of frame. When a T1 CSU/DSU is unable to synchronize framing patterns with the remote end for 2.5 seconds, LOF is declared. LOF is sometimes called the *red alarm*, after the color of the light on an old channel bank.

LOS

Loss of signal. When no incoming pulses are received by a T1 CSU/DSU for 100 to 250 bit times, LOS is declared. Even if only zeros were transmitted as data, some framing bits should result in pulses on the line during that time. LOS indicator lights are often red as well. If LOS persists, LOF will eventually be declared because there is no incoming signal with which to synchronize.

NFAS

Network Facilities Associated Signaling. A means of sharing a single ISDN PRI D channel among several PRI lines. NFAS also allows for a backup D channel in case of the failure of the primary D channel.

NI

Network interface. Used to refer to the RJ-48 jack on the CSU/DSU to connect to the telco network.

NRZI
Non-Return to Zero Inverted. A line code in which zeros change the voltage from the previous bit and ones do not.

OOF
Out of frame. When frame synchronization is lost, an OOF event is recorded. If OOF persists, LOF is declared. Some T1 CSU/DSUs may have a light that blinks briefly on OOF events. OOF is cleared when frame synchronization is regained.

PBX
Private branch exchange. A customer-owned switch that creates a private telephone network at the customer location.

PDH
Plesiochronous digital hierarchy. Lower-speed digital lines are clocked with similar but not identical clocks. Successively higher speeds in the PDH are achieved by time-division multiplexing of signals from the lower levels. In North America, PDH is synonymous with the T-carrier hierarchy. Europe's PDH is the E-carrier hierarchy.

plesiochronous
Network in which timing is almost the same. It uses independent clocks that are within tight tolerances of each other.

POTS
Plain old telephone service. Used in North America to refer to traditional analog telephone service, as opposed to any fancier digital telephony services such as ISDN.

provisioning
A term telcos often use to mean "installing and configuring." It may also refer to configuration that can be done at the CO switch to set up the telephone network to recognize your circuit.

PSTN
Public switched telephone network. A fancy name for the telephone network, used frequently in job interviews.

PVC
Permanent virtual circuit. A dedicated logical path through a frame relay network.

RAI
Remote alarm indication. Also known as *yellow alarm*, after the color of the indicator on Western Electric equipment. When a CSU/DSU enters the red alarm state, an RAI is transmitted in the outgoing direction. The RAI signals to the remote end that the local end is unable to synchronize framing patterns.

red alarm
: See LOS and LOF.

RJ
: Registered jack. A code used with the Universal Service Order Code to specify a standard jack pinout and connector. Common designations are RJ-11 (plain telephone jacks) and RJ-45 (Ethernet).

SEF
: Severely errored framing. Refers to a certain density of frame bit errors, which almost always causes an out of frame defect.

SES
: Severely errored second. One-second interval that is likely to have garbage on data links because framing was lost, AIS was detected, or 320 or more frames failed the CRC check.

SF
: Superframe. The original multiframe format for T1, which was composed when 12 frames were bundled together. Errors were detected solely by any resulting bipolar violations. No error check was available over the entire T1 path. All the framing bits were used for maintaining frame synchronization, so no end-to-end communications were possible outside of the data channel. Full-bandwidth data could cause a false yellow alarm unless certain bits were constrained from being zero, so data transmission capability was limited. All these limitations spurred the development of the ESF format, and SF has largely been supplanted by ESF.

signaling
: Procedures used for communicating control information.

synchronous
: When used to refer to data transmission, it means that data is streamed, with bits transmitted at regular intervals. Frequently, it also implies that a separate clocking signal is sent with the data.

T1
: DS1 delivered over a 4-wire copper interface.

telco
: A company that sells telecommunications services to the public; synonymous with *carrier*. Also known pejoratively to many as "the phone company."

UAS
: Unavailable second. Used in measuring the performance of T1 lines; carriers will often guarantee a certain uptime. Uptime is the number of seconds in the measuring period minus the number of unavailable seconds. Different monitoring specifications define this quantity in different ways.

UI

Unit interval. Synonymous with bit time. The time slot for a single bit.

USOC

Universal Service Order Code. System of codes used by AT&T that make up the Registered Jack system (see RJ).

VJ (also VJC)

Van Jacobson (Compression), named for its inventor, is an ingenious method of compressing TCP/IP headers on slow serial links. Its formal specification is RFC 1144.

yellow alarm

Also known as RAI; see RAI.

ZBTSI

Zero Byte Time Slot Interchange. ZBTSI is a method of clear channel coding that is no longer widely used in T-carrier facilities.

Index

Numbers

0.0.0.0 address (IPCP), 113
255.255.255.255 (discovery protocol), 189

A

A channels, 31
AARP (AppleTalk Address Resolution Protocol), 71
ABM (Asynchronous Balanced Mode), 64, 75
ACCM (asynchronous control character map)
  asynchronous lines and, 77
  bundling maintenance and, 181
  LCP configuration option code, 104
  PPP configuration option, 114
ACFC (Address and Control Field Compression), 104, 114, 182
acknowledgments, frames and, 66
Ack-Rcvd (Ack-Received), 87
Ack-Sent, 87
actions and events in LCP, 88–91
Acton, Lord (quote), 76
Address and Control Field Compression (ACFC), 104, 114, 182
address byte (HDLC frame component), 66
address field (PPP), 83
Address Resolution Protocol (ARP), 189
AIS (alarm indication signal)
  defect, 200
  failure condition table, 198
  keepalive transmission, 49
  LOF and, 197
  OOF and, 200
  path failure, 197
  performance data and, 202
  physical layer diagnostic table, 140
  purpose of, 26
  T1 defect condition, 200
alarm signals
  failures and, 197–200
  superframe and, 26
  transmission durations, 30
A-law standard, 19
all-ones test, bit error rate and, 26
all-zeros test, bit error rate and, 26
Alternate Mark Inversion (see AMI)
alternating pulse polarity, 19
American Bell, 2
American National Standards Institute (see ANSI)
American Telephone and Telegraph (see AT&T)
AMI (Alternate Mark Inversion)
  BPV (bipolar violation) and, 201
  CSU/DSU configuration options, 53
  encoding alternative, 21

We'd like to hear your suggestions for improving our indexes. Send email to *index@oreilly.com.*

AMI (*continued*)
excessive zeros (EXZ) and, 201
line code errors and data, 195
troubleshooting, 143
AMI encoding, 18, 25
Amiel, Henri Frederique (quote), 115
analog loopback, troubleshooting, 133
analog signals
aynchronous systems and, 32
beginnings (1876 through 1950), 2
early problems, 4
FDM as natural fit, 6
analog systems, conversions, 151
ANI (Automatic Number Identification), 160, 183
anomaly (see error checking/detection)
ANSI (American National Standards Institute)
Codeset 5, 128
frame relay frame format and, 125
LMI type, 123
Q.933 differences, 129
standardization efforts, 12
ANSI Committee T1, 237
ANSI T1.101, 241
ANSI T1.231, 196, 241
ANSI T1.403, 12, 25, 241
ANSI T1.408, 244
ANSI T1.410, 240
ANSI T1.617, 245
ANSI T1.617a Annex D, 128
ANSI T1.618, 245
Antisthenes (quote), 226
AppleTalk protocols
AARP, 71
PPP standards, 249
RS-422 and, 227
Arendt, Hannah (quote), 205
ARM (Asynchronous Response Mode), 64
ARP (Address Resolution Protocol), 189
Ascend "Stack" protocols, 186, 250
ASCII, 20, 100
ASN.1, 253
Asynchronous Balanced Mode (ABM), 64, 75
asynchronous communications, 6, 32, 77
asynchronous control character map (see ACCM)
Asynchronous Response Mode (ARM), 64
AT&T (American Telephone & Telegraph)
antitrust suit against, 3
beginnings, 2
DDS offering, 17
Electronic Switching System (ESS), 4
extended superframe development, 28
jack codes developed, 233
PRS and, 32
Telcordia and, 240
(see also Bell Labs)
AT&T Technical Publication 54016, 241
AT&T Technical Publication 62310, 240
AT&T Technical Publication 62411, 240
AUI (attachment unit interface), 60
authentication
bundling maintenance and, 183
configuring for cT1, 161
during initialization, 81
PAP and CHAP, 114
PPP, 78, 82, 83, 251
timing of, 78
authentication protocol, 104
Automatic Number Identification (ANI), 160, 183

B

B8ZS (Bipolar with Eight Zero Substitution)
BPVs (bipolar violations) and, 201
capabilities of, 19
clear channel capability and, 21
CSU/DSU configuration options, 53
EXZ (excessive zeros) and, 201
line code errors and, 195
line code from telco, 59
ordering ISDN, 164
troubleshooting, 143
B8ZS encoding, 21, 25
B (bearer) channels
as subchannels, 162
configuring ISDN PRI, 166
SF-framed links and, 31
backhoe fade, 10
backward explicit congestion notification (BECN) bit, 119
BACP (Bandwidth Allocation Control Protocol), 185
balanced transmissions, 228
Bandwidth Allocation Control Protocol (BACP), 185

bandwidth, exceeding supply, 118
BCP (Bridging Control Protocol), 249
BECN (backward explicit congestion notification) bit, 119
Bell, Alexander Graham, 2
Bell Labs, 2, 3, 4
Bell Telephone Company, 2
Bellcore (Telcordia), 240
Bellcore (Telcordia) TR-NWT-000057, 243
BER (bit error rate), 25
BES (bursty errored seconds), 203, 211
bipolar return to zero (see AMI)
bipolar violation (see BPV)
Bipolar with Eight Zero Substitution (see B8ZS)
biscuits, 13
bit error rate (BER), 25
bit order, 65, 67, 68
bit robbing technique (see robbed-bit signaling)
bit stuffing, 53, 70
bits
  BECN and FECN, 119
  PRM information element, 194
  in superframe, 26
  used for CRC, 29
BITS (building-integrated timing supply), 36
black hole effect, 123
blue alarm (superframe alarm condition), 26
BPV (bipolar violation)
  all-zeros test, 26
  basic error event, 201
  captured in dsx1CurrentTable, 211
  code word containing, 21
  errors, 201
  LES and, 202
  line errors and, 27, 195
  troubleshooting, 143
Bridging Control Protocol (BCP), 249
broadcast, frame as in cHDLC, 71
buffers, overflow and underflow conditions, 41
building-integrated timing supply (BITS), 36
bundle heads, 184, 186, 189
bundling
  channels into frames, 26
  defined, 18
  maintenance, 181–184
  MP architecture and, 168
Burke, Edmund (quote), 22
burst duration/size, frame relay switch and, 118
bursty errored seconds (BES), 203, 211

C

C channels, 31
Cabanis, Pierre J. G. (quote), 44
cabling
  common V.35 problems, 232
  common-ground and differential, 227
  ensuring compatibility, 12
  grounding, 234
  local loopback tests for, 134
  physical layer standards, 242
  pre-connection tasks, 60
  short haul LBO and, 47
  tolerances and, 228
  top T1 trouble spots, 137
call completion, E&M signaling and, 155
call initiation, E&M and, 154
call termination, MMP and, 186
call type, voice and data, 161
Caller ID, 160, 183
Calling Line Identification (CLID), 160
Caraffa, Cardinal Carlo (quote), 192
Carlson, James, 77
Carrier Detect (CD), 48, 229
Carrier Detect (DCD), 231
Carrier Sense (CS), 229
carriers, 5, 116, 134
CAS (channel associated signaling), 149, 153
Category 5 UTP, 58, 140
CCC (Clear Channel Capability), 21
CCITT (Consultative Committee on International Telephone and Telegraph), 11, 123
CCP (Compression Control Protocol), 108, 251
CCS (common channel signaling), 149, 163
CD (Carrier Detect), 48, 229
central office (see CO)
Challenge Handshake Authentication Protocol (CHAP), 114, 252
Change of Frame Alignment (COFA) slips, 41

channel associated signaling (CAS), 149, 153
channel banks, as termination devices, 26
channel data rate (CSU/DSU option), 53
channel map (CSU/DSU option), 53
channel order (CSU/DSU option), 53
Channel Service Unit (see CSU/DSU)
channelized T1 (cT1), 149, 150–162, 243
channels
  bundling into frames, 26
  distinguishing start/end points, 23
  as DS0s, 18
  LAPD and, 66
  signaling methods and, 29
CHAP (Challenge Handshake Authentication Protocol), 114, 252
cHDLC (Cisco HDLC), 66, 70–75, 126
CIR (committed information rate), 117, 118
circuit ID sticker, 59
circuits
  circuit identification information, 59
  circuit-switched networks, 18
  documentation on usage, 53
  MIB table, 215
  recommended documentation, 62
  requirements for timing source, 37
  turn-up process for, 61
  V.24 definitions, 229
Cisco
  Gang of Four, 123
  HDLC standards and, 246
  Stack Group Bidding Protocol, 186
  web site for Ethernet type codes, 247
Cisco HDLC (cHDLC), 66, 70–75, 126
Cisco IOS, debug trace and, 137
clarity, testing, 25
Clear Channel Capability (CCC), 21
Clear to Send (CS), 48
Clear to Send (CTS), 48, 229, 231
CLID (Calling Line Identification), 160
clock source
  avoiding slips, 43
  CSU/DSU configuration option, 52, 53
  settings, 53
clocking
  at data ports, 37–41
  at the network interface, 36
  FDL code words and, 30
  internal/inverted setting, 52, 54
  synchronous communications and, 6, 32
  top T1 trouble spots, 137
  (see also timing)
Close event, 89
Closed (LCP engine state), 85
closing flags, opening flags and, 65
Closing (LCP engine state), 86
CO (central office)
  delivery technology, 8
  end spans and, 12, 14
  switches and ISDN signaling, 163
code rejection messages, 100
code violation line (LCV), 202
code violation path (CV-P), 201
code words, 21, 29
Code-Reject, 98, 100, 168
Code-Reject (RXJ+) event, 90
Code-Reject (RXJ-) event, 90
Codeset 0, 129
Codeset 5, 128
COFA (Change of Frame Alignment) slips, 41
collisions, magic numbers and, 107
combined nodes (in ABM environment), 64
command/response (C/R) bit, 118, 193
committed information rate (CIR), 117, 118
common channel signaling (CCS), 149, 163
common-ground transmission, 227, 228, 230
communication channels, FDL as, 29
compatibility for cables, 12
Compressed SLIP, 247
compression, 77, 114, 182
Compression Control Protocol (CCP), 108, 251
computerization, transistors and, 4
Comte, Auguste (quote), 17
configuration option negotiation, 85
configurations
  channelized T1, 161
  Cisco HDLC, 74
  common CSU/DSU settings, 53
  examined for compability, 93
  FT1 guidelines, 52
  ISDN, 164, 165
  LBO, 47
  MIB table, 208–209
  PPP options, 112
  pre-connection tasks, 60
  top T1 trouble spots, 137
Configure-Ack, 98, 100

Configure-Nak
bundling maintenance and, 181
LCP configuration messages and, 99
LCP type code, 98
magic numbers and, 107
RFC 1332 and, 110
Configure-Reject
LCP configuration messages and, 99
LCP type code, 98
RFC 1332 and, 110
scn action, 90
Configure-Request
address negotiation and, 113
bundling maintenance and, 181
LCP configuration messages and, 99, 100
LCP type code, 98
magic numbers and, 107
options example, 103
congestion notification (FECN/BECN), 119
connectors
DA-15, 235
V.35 interface, 10, 230
Consultative Committee on International Telephone and Telegraph (CCITT), 11, 123
control bits, T1-carrier hierarchy, 23
control byte (HDLC frame component), 66
control characters for transparency, 70
controlled slip seconds (CSS), 203, 210
controlled slips, 41, 202
conversions, analog and digital, 151
copper wiring
bipolar violations and, 27
data streams and, 16
end spans and, 8
internal timing and, 37
scenario, 14
counters, 123, 195, 209
CPI and telephony costs, 4
C/R (command/response) bit, 118, 193
CRC (cyclic redundancy check)
bits used for, 29
errors, 194
performance data, 201, 202
crosstalk
defined and all-ones test, 26
line build out and, 45
NEXT problems, 57
RJ-48 and, 60
shielded cable and, 58
T1 considerations, 14
troubleshooting, 143
CS (Carrier Sense), 229
CS (Clear to Send), 48
CS (controlled slip)
basic error event, 202
characteristics, 41
error, 201
ES and, 202
performance data, 202
troubleshooting, 143
CSS (controlled slip seconds), 203, 210
CSU/DSU (Channel Service Unit/Data Service Unit)
as serial DCE, 226
collecting performance data, 192
CS and, 202
digital or analog loopback, 133
end spans and, 12
FDL as communication channel, 29
from router with V.35, 10
interfaces as bridges, 35
ISP responsibilities, 16
loopback mode and AIS, 198
permission when connecting, 61
role clarification, 44–55
superframe alarm conditions, 26
synchronization process, 24
telco network and, 12
timing and synchronization, 20
timing sources, 33
V.35 and, 227
cT1 (channelized T1), 149, 150–162, 243
CTS (Clear to Send), 48, 229, 231
current statistical table (MIB), 209–211
customer, as frame relay DTE, 116
CV-L (code violation-line), 201, 202
CV-P (code violation-path), 201, 202
cycles, technology and, 3
cyclic redundancy check (see CRC)

D

D channels
as subchannels, 162
configuring ISDN PRI, 166
ESF-framed links and, 31
D1 channel banks, 26
D2 channel banks, 26

D3 channel banks, 26
D4 channel banks, 26
D5 channel banks, 28
DA-15 connectors, 12, 235
DACS (digital access and cross-connect systems), 15
Data Carrier Detect (DCD), 48, 229
data communications equipment (see DCE)
data corruption, 198
data fields, 66
data inversion, 53
data link connection identifier (DLCI), 59, 117, 214
Data Link Connection Management Interface (DLCMI), 214–215
data ports, 37–41
Data Service Unit (see CSU/DSU)
Data Set Ready (DSR), 231
data streams, 16, 18, 35
data terminal equipment (see DTE)
Data Terminal Ready (DTR), 231
data transmission, 19, 20, 36
data types (MIB objects), 208
dB (decibel), 14, 15
DCD (Carrier Detect), 231
DCD (Data Carrier Detect), 48, 229
DCE (data communications equipment)
  characteristics, 226
  CSU/DSU as, 10
  frame relay, 116, 123
DCS (digital cross-connect system), 15
DDS (Digital Dataphone Service), 17, 240
DE (discard eligible) bit, 119
dead phase, PPP link states and, 82
debugging (see troubleshooting)
decibel (dB), 14, 15
DECnet protocol, 71, 249
default settings
  CSU/DSU configuration options, 53
  LBO, 47
  N391, 124
defects, 197, 199
Defense Information Infrastructure (DII), 213
Defense Information Systems Agency (DISA), 213
definitions, v.24 circuit names, 229
Degraded Minutes (DM), 211
DELETE control characters, data transparency and, 70
Demand dialing (configuration option), 114
demarc
  end span, 12
  extended demarc, 13, 57, 143
  installation, 57, 60
density requirement (n out of m), 199
deviation (jitter and wander), 35
Dialed Number Information Service (DNIS), 160
dial-up connections (see modems)
differential cables
  performance, 228
  serial line cabling and, 228
differential signaling, data and clock signals, 230
digital access and cross-connect systems (DACS), 15
digital cross-connect system (DCS), 15
Digital Dataphone Service (DDS), 17, 240
digital encoding (mu-law and A-law), 19
Digital Equipment Corporation (Gang of Four), 123
digital loopback, troubleshooting, 133
digital networks, TDM and, 5
digital signals, 6, 8
Digital Subscriber Line (DSL), 17
digital systems, conversions, 151
digital transmission, 17–21
digitalization, 4, 18
DII (Defense Information Infrastructure), 213
DISA Circular 300-175-9, 213, 242
DISA (Defense Information Systems Agency), 213
DISC (Disconnect) command, 68
discard eligible (DE) bit, 119
Discard-Request, 98, 102
Disconnect (DISC) command, 68
Disconnected Mode (DM) command, 68
discovery messages, format of, 189
discovery protocol, 189
Disraeli, Benjamin (quote), 63
distortion, 4, 14
divestiture, RBOC and AT&T split, 6
DLCI (data link connection identifier), 59, 117, 214

DLCMI (Data Link Connection Management Interface), 214–215
DLCs and PVC maximum, 127
DM (Degraded Minutes), 211
DM (Disconnected Mode) command, 68
DNIS (Dialed Number Information Service), 160
DNS servers, 111
documentation
  circuits in use, 53
  clock settings, 53
  post connection, 62
Down event, 78, 82, 88
drain wires (in ISTP), 58
dropping links, bundling maintenance and, 183
DS0 interface
  as digital stream, 17
  configuring for cT1, 161
  physical layer standards, 240
  T-carrier comparison table, 22
DS1 interface
  as network side, 45
  CS and frame, 201
  physical layer standards, 240
  RFC 2495, 205–213
  T-carrier comparision table, 22
DS2 interface, 22, 205–213
DS3 interface, 22
DS4 interface, 22
DSL (Digital Subscriber Line), 17
DSR (Data Set Ready), 231
DSU (see CSU/DSU)
DSX label (local side), 45, 47
dsx1Channelization (MIB object), 209
dsx1CircuitIdentifier (MIB object), 208
dsx1CurrentBESs (MIB table entry), 211
dsx1CurrentCSSs (MIB table entry), 210
dsx1CurrentDMs (MIB table entry), 211
dsx1CurrentESs (MIB table entry), 209
dsx1CurrentLCVs (MIB table entry), 211
dsx1CurrentLESs (MIB table entry), 211
dsx1CurrentPCVs (MIB table entry), 210
dsx1CurrentSEFSs (MIB table entry), 210
dsx1CurrentSESs (MIB table entry), 210
dsx1CurrentTable (MIB table entry), 209
dsx1CurrentUASs (MIB table entry), 210
dsx1IntervalTable (MIB interval table), 211
dsx1LineCoding (MIB object), 208
dsx1LineStatus (MIB object), 208
dsx1LineStatusTrap (DS1 trap), 213
dsx1LineType (MIB object), 208
dsx1LoopbackConfig (MIB object), 208
dsx1LoopbackStatus (MIB object), 209
dsx1SendCode (MIB object), 208
dsx1SignalMode (MIB object), 208
dsx1TotalTable (MIB total table), 212
dsx1TransmitClockSource (MIB object), 209
DTE (data terminal equipment)
  characteristics, 226
  clocking, 52, 53, 54
  CRC errors and, 194
  frame relay, 116, 124
  pre-connection tasks, 60
  router as, 10
DTMF (dual-tone multifrequency), 156
DTR (Data Terminal Ready), 231
dual-tone multifrequency (DTMF), 156
duty cycle for pulses, 18

E

E1 interface, 18, 205–213
E2 interface, 205–213
EA (extended address) bit, 118, 193
Echo-Reply
  LCP link integrity messages and, 102
  LCP type code, 98
  magic numbers and, 107
Echo-Request
  bundling and, 168
  LCP link integrity messages and, 102
  LCP type code, 98
  magic numbers and, 107
Echo-Response, 168
ECP (Encryption Control Protocol), 251
ED (Endpoint Discriminator)
  bundling maintenance and, 183
  class values table, 180
  LCP option code, 108
EIA/TIA 464, 243
Electronic Switching System (ESS), 4
E&M signaling
  call completion, 155
  call initiation, 154
  idle state, 154
  wink start, 155
Emerson, Ralph Waldo (quote), 56

encapsulation
  cHDLC configuration option, 74
  configuring frame relay, 130
  configuring ISDN PRI, 166
  enabling PPP option, 112
  LCP and, 85
  MP, 169–170
  PPP, 77, 83
  with RFC 1490, 120–122
encoding schemes, 18
Encryption Control Protocol (ECP), 251
end spans
  CSU/DSU to CO, 12
  network outages and, 10
  to CO, 14
  transmission losses in, 46
Endpoint Discriminator (see ED)
equalization, 45, 47
error checking/detection
  alternating pulse polarity and, 19
  CRC, 194
  defects as, 197
  flags and, 65
  framing, 195
  in QRS tests, 25
  on leased-line circuits, 20
  line and path, 27
  T1 performance monitoring, 201
  timer expiration, 124
  (see also troubleshooting)
error table (MIB), 216
Error Threshold Counter (N392), 124, 125
errored second type B (ESB), 202
errored seconds (ES), 202, 203, 209
errored seconds-line (ES-L), 202, 203
errored seconds-path (ES-P), 203
ErrType, table of potential values, 216
ES (errored seconds), 202, 203, 209
ESB (errored second type B), 202
ESF (extended superframe)
  CSU/DSU configuration options, 53
  errors and, 200, 201, 203
  FE and, 201
  features of, 27–30
  line framing from telco, 59
  OOF defect and, 199
ES-L (errored seconds-line), 202, 203
ES-P (errored seconds-path), 202, 203
ESS (Electronic Switching System), 4
establishment phase, PPP link states and, 82
Ethernet, 10, 70, 140, 247
Ethernet 802.2 LLC, 65
EtherTalk protocol, 71
Ethertype, OUI and, 121
events, 88–91, 197
excessive zeros (EXZ), 201, 202
Exchange Information (XID) command, 68
explicit timing, 35, 38
extended address (EA) bit, 118, 193
extended demarc, 13, 57, 143
extended superframe (see ESF)
extensibility, PPP and, 76
external timing, 52
EXZ (excessive zeros), 201, 202

F

facilities data link (FDL), 29, 53
failures, alarms and signaling, 197–200
far end alarm (see RAI)
far end table (MIB), 213
far-end counters, statistics and, 195
FCC (Federal Communications Commision), 61, 151, 233
FCS (frame check sequence), 66, 84
FDL (facilities data link), 29, 53
FDM (frequency division multiplexing), 4, 6
FE (frame bit error), 199, 201
FECN (forward explicit congestion notification) bit, 119
fiber optics, 6, 8
Finagle's Law (Murphy's Law), 132
First Rule, troubleshooting, 136
flags, 65, 118
foreign exchange (FXS) ground start, 156
foreign exchange (FXS) loop start, 158
foreign exchange office (FXO), 156
formats
  discovery messages, 189
  LCP frames, 96–103
  link quality report, 106
  performance report message format, 29
formula for decibels, 15
forward explicit congestion notification (FECN) bit, 119
fractional T1 (FT1), 24, 52

fragmentation
  fragment numbers, 174
  MP, 171–177, 178
frame bit error (FE), 199, 202
frame check sequence (FCS), 66, 84
frame components
  flag, 65
  formats, 109
  frame bit sequences, 28, 29
  frame check sequences, 115
  frame loss, 197
  headers, 63
Frame Reject (FRMR) command, 68
frame relay
  address mapping, 130
  configuring, 130
  maximum number of PVCs, 127
  network overview, 59, 116–118
  problems, 146
  protocol used by T1 lines, 63
  RFC 2115, 213–216
  standards, 244
  switches, 118
Frame Relay Forum (FRF), 118, 123, 238
frames
  as layer 2 construct, 63
  bundling channels into, 26
  defined, 24
  errors in, 195
  HDLC components, 65–70
  LMI formats, 125–130
  MP variations, 169
framing bits
  signaling framing bits, 26
  synchronization and, 28
  T1-carrier hierarchy, 23
  terminal framing bits, 26
framing considerations
  CSU/DSU configuration option, 51, 53
  LOF failure, 197
  n out of m density requirement, 199
  PPP, 77, 83
  RFC 1662, 77
  slip events and, 195
frCircuitDlci (MIB object), 215
frCircuitReceivedBECNs (MIB object), 215
frCircuitReceivedDEs (MIB object), 215
frCircuitReceivedFECNs (MIB object), 215
frCircuitReceivedFrames (MIB object), 215
frCircuitReceivedOctets (MIB object), 215
frCircuitSentDEs (MIB object), 215
frCircuitSentFrames (MIB object), 215
frCircuitSentOctets (MIB object), 215
frCircuitState (MIB object), 215
frDLCIStatusChange (MIB object), 216
frDlcmiAddressLen (MIB object), 214
frDlcmiErrorThreshold (MIB object), 215
frDlcmiFullEnquiryInterval (MIB object), 215
frDlcmiMaxSupportedVCs (MIB object), 215
frDlcmiMonitoredEvents (MIB object), 215
frDlcmiPollingInterval (MIB object), 215
frDlcmiState (MIB object), 214
frequency division multiplexing (FDM), 4, 6
frErrFaults (MIB object), 216
frErrType (MIB object), 216
FRF (Frame Relay Forum), 118, 123, 238
FRF.1 frame relay standards, 123
FRF.1.1 frame relay standards, 245
FRF.1.2 frame relay standards, 118, 245
FRMR (Frame Reject) command, 68
FT1 (fractional T1), 24, 52
FXO (foreign exchange office), 156
FXS (foreign exchange station), 156, 158

G

Gang of Four, 123
Gauge32 counters, 209
glare, 155, 164
global DLCIs, unique identifiers in, 117
GND (Signal Ground), 231
GPS (Global Positioning System), 34
ground start signaling, 156
grounding cables, 234
guidelines for leased lines, 213

H

HDLC (High-level Data Link Control Protocol)
  overview, 63–75
  performance report message and, 193
  PPP foundation in, 76
  problems, 147
  specification example, 238
  standards for, 246

HDLC (*continued*)
U-frame codes, 68
(see also Cisco HDLC)
(see also mod-128)
(see also mod-8)
headers, 117, 126
Heraclitus (quote), 1
hints, 94, 99
Hopper, Grace (quote), 236
hunt groups, multiple lines in, 159

I

IA (implementation agreement), 238
IANA (Internet Assigned Numbers Authority)
Ethernet type codes and, 247
queries to address, 189
standards body, 239
web site for PPP protocol numbers, 248
IBM, SDLC development, 64
identification (LCP type code), 99
identification message, 102
idle state, 154, 156, 158
IE (information element), 126–127, 194
IEC (International Electrotechnical Commission) specifications, 238
IEEE 802, 63
IEEE (Institute of Electrical and Electronics Engineers), 247
IETF (Internet Engineering Task Force), 65, 76, 239
ifExtnsTestResult (PPP Test Group object), 223
I-frame (information frame), 66
implementation agreement (IA), 238
InARP (Inverse ARP), 130
inband signaling, 53, 153
incoming calls, 159, 187
indicators (CSU/DSU table), 48
individually shielded twisted pair (ISTP), 58
inference, from pulse reception, 35
information element (IE), 126–127, 194
information frame (I-frame), 66
InGoodOctets counter (PPP LQR Group), 221
Initial (LCP engine state), 85
initialization, 78, 80, 82
initialize restart counter (irc) action, 90
InLQRs counter (PPP LQR Group), 222
installation
and termination for T1, 57
ordering T1 connectivity, 56
post connection, 62
pre-connection tasks, 60
Institute of Electrical and Electronics Engineers (IEEE), 247
Integrated Services Digital Network (see ISDN)
interfaces (see network interfaces)
interference, 58, 228
internal timing, 37
International Alphabet 5, 128
International Electrotechnical Commission (IEC) specifications, 238
International Organization for Standardization (see ISO)
International Telecommunications Union (see ITU)
Internet, 7, 17
Internet Assigned Numbers Authority (see IANA)
Internet Engineering Task Force (IETF), 65, 76, 239
Internet Protocol (see IP)
Internet Service Provider (see ISP)
Internetwork Packet Exchange (see IPX)
interoperability, 4, 239
interruptions, backhoe fade, 10
interval table (MIB), 211
Inverse ARP (InARP), 130
IP Address (IPCP), 110
IP addresses
assigned by ISP, 59
cHDLC configuration option, 75
configuring for cT1, 161
configuring frame relay, 130
PPP and, 113
pre-connection tasks, 60
IP Addresses (IPCP), 109
IP configuration table (PPP IP Group), 225
IP (Internet Protocol)
cHDLC protocol code, 71
fragmentation contrasted with MP, 178
fragments, 178
NLPID assignments, 120
packets, 80
PPP, 113, 224, 249

PPP initializing links, 78
protocol-rejection messages and, 101
SLARP numbering in, 72
SNAP encapsulation and, 121
stack configuration, 111
state machine diagram, 78
IP Subnet Mask (IPCP), 110
IPCP (IP Control Protocol), 77, 109–112
IPSec security associations, 92
IPv4 routers, 77
IPX (Internetwork Packet Exchange)
cHDLC protocol code, 71
packet initialization, 81
PPP initializing links, 82
PPP standards, 249
protocol-rejection messages and, 101
SNAP encapsulation and, 121
state machine diagram, 78
IPXCP negotiation/packet initialization, 81
irc (initialize restart counter) action, 90
ISDN BRI (Basic Rate Interface), 149, 162
ISDN (Integrated Services Digital Network)
fragmentation-loss rules scenario in, 176
levels of service and, 167
LMI frame formats and, 125
signaling, 30, 66
standards, 243
(see also LAPD)
ISDN PRI (Primary Rate Interface)
features, 162–166
ISDN and, 149, 162
ISO 8824, 253
ISO 33090, 70
ISO (International Organization for Standardization)
bit order in diagrams, 65
HDLC specification, 65
standards body, 238
ISO/IEC 2593, 243
ISO/IEC 3309, 65, 238, 246
ISO/IEC 4335, 246
ISO/IEC 7776, 249
ISO/IEC TR 9577, 120, 245
ISP (Internet Service Provider)
as subsidiaries, 56
assigning IP addresses, 59
calling for help, 135
circuit turn-up process, 61
connection needed, 15
filing affidavits, 61
link diagram, 8
MP and, 167
protocol support, 101
responsibilities for supplying, 16
ISTP (individually shielded twisted pair), 58
ITU (International Telecommunications Union)
bit order in diagrams, 65
HDLC as basis for LAP, 65
LMI type, 123
nomenclature for frame relay, 118
overview, 11
protocols, 30, 120
specifications, 125
standards committee, 237
ITU Recommendation E.164, 250
ITU Recommendation G.821, 241
ITU Recommendation G.826, 241
ITU Recommendation Q.921, 243, 244
ITU Recommendation Q.922, 245
ITU Recommendation Q.930, 244
ITU Recommendation Q.931, 244
ITU Recommendation Q.933, 245
ITU Recommendation Q.933 Annex A, 129
ITU Recommendation T.50, 246
ITU Recommendation V.4, 246
ITU Recommendation V.24, 242
ITU Recommendation V.28, 242
ITU Recommendation V.35, 242
ITU Recommendation V.36, 242
ITU Recommendation V.37, 242
ITU Recommendation V.38, 242
ITU Recommendation V.54, 243
ITU Recommendation X.25, 244
ITU Recommendation X.263, 245
ITU Series V standards, 11
ITU-T (International Telecommunications Union-telecommunications sector), 11, 237

J

jack codes, AT&T development of, 233
jitter, 35
JTC (Joint Technical Committee), 238

K

KA (keepalive)
  functions, 124
  SLARP protocol and, 73, 74
  transmission, 48, 49
kbps (phone company definition), 18
Kettering, C. F. (quote), 8
kronos, 32

L

L2F (Layer Two Forwarding), 191
L2TP (Layer Two Tunneling Protocol), 187
LAN (local area network), 10, 63
LAP (Link Access Procedure), 65, 66, 83
LAPB (Link Access Procedure, Balanced), 108
LAPD (Link Access Procedure, D Channel)
  defined, 30
  LAPF compared with, 118
  performance report message based on, 193
  Q.921 and, 163
  variation on LAPD, 66
LAPF Core (Q.922), 118
LAPM (Link Access Procedure for Modem), 66
Layer Two Forwarding (L2F), 191
Layer Two Tunneling Protocol (L2TP), 187
LBO (line build out)
  common settings, 48
  CSU/DSU configuration option, 51, 53
  estimating during installation, 59
  overview, 44–48
  short-haul and long-haul contrasted, 44
  top T1 trouble spots, 137
  troubleshooting, 141
LCP (Link Control Protocol)
  actions and events, 88–91
  bundling and, 168
  Open command to, 80
  options for MP, 177–180
  PPP and, 77, 78, 82, 85–108
  state engine, 91–96
LCV (line code violation), 201, 202, 211
leased lines
  as expensive option, 116
  authentication and, 83
  DDS connections, 17
  link termination and, 96
  performance guidelines for, 213
LES (line errored seconds), 202, 203, 211
line build out (see LBO)
line code
  CSU/DSU configuration options, 53
  errors, 195
  line speed and, 59
  troubleshooting violations, 143
line code violation (LCV), 201, 202, 211
line equalization, 45, 47
line errored seconds (LES), 202, 203, 211
Link Access Procedure, Balanced (LAPB), 108
Link Access Procedure, D Channel (see LAPD)
Link Access Procedure for Modem (LAPM), 66
Link Access Procedure (LAP), 65, 66, 83
Link Configuration packets (LCP packet class), 85
link configuration table (PPP Link Group), 220
Link Control Protocol (see LCP)
Link Dead (PPP link state), 82
link integrity messages, 102
link integrity verification information element, 126
Link Integrity Verification Polling Timer (T391), 123, 124
link layer
  MP as fragmentation mechanism, 177
  PPP and, 76, 77, 84
  problems in, 145–148
  protocols used in, 63
Link Maintenance packets (LCP packet class), 85
link management
  adding to existing bundle, 182
  LCP state engine and, 91–96
  link configuration, 92
  link encapsulation, 81
  link establishment, 82
  link initialization, 91
  link monitoring, 85
  link quality monitoring, 182
  link termination, 83, 85, 95
  link thrashing, 186
  PPP, 77

steps in initializing, 78, 82
subordinancy in, 113
techniques for multiple, 167
link mode (cHDLC configuration option), 75
Link Quality Monitoring (LQM), 105, 114
link status table (PPP Link Group), 217–219
LL (Local Loopback), 231
LLC (Logical Link Control), 65
LMI Annex A, 123, 129
LMI Annex D, 128
LMI Annex G, 122
LMI (Local Management Interface)
configuring frame relay, 130
features, 122–130
information needed, 59
loading coils, 2
Local Loopback (LL), 231
Local Management Interface (see LMI)
localTiming (dsx1TransmitClockSource), 209
LOF (loss of frame)
CSU/DSU, 48
failure, 197, 198
physical layer diagnostic table, 140
red alarm problems, 139
logical addresses (SAPI and TEI), 193
Logical Link Control (LLC), 65
Long Range Navigation (LORAN), 34
loop timing, 36, 43, 52
loopback considerations
AIS failure and, 198
local or remote, 50, 133, 134
magic numbers and, 77, 106
received clock signal and, 40
remote smart jack and, 29
retention signal, 30
RJ-48X and, 233
testing, 132
loop-start signaling, 158
loopTiming (dsx1TransmitClockSource), 209
LORAN (Long Range Navigation), 34
LOS (loss of signal)
CSU/DSU, 48
defect, 199
failure, 197
failure condition table, 198
LES and, 202
physical layer diagnostic table, 139
red alarm problems and, 139
T1 defect condition, 200
loss of frame (see LOF)
loss of signal (see LOS)
LQM (Link Quality Monitoring), 105, 114
LQR, 222
LQR Group, 221, 222
Lucent (MCMP), 186, 191

M

magic numbers
asynchronous links use of, 77
bundling maintenance and, 181
LCP option code, 104, 106
PPP configuration option, 113
master/slave timing characteristics, 36
MAX access server, 191
Maximum Receive Unit (see MRU)
Maximum Reconstructed Receive Unit (see MRRU)
Maximum Transmission Unit (MTU), 114
M-block connector, 10
MCMP (Multi-Chassis MP), 186, 191
message types (ESF code words), 29
messageOriented (dsx1SignalMode), 208
messages, 99–108, 149, 193
MF (multifrequency), 156
MIB
for PPP, 217–225
RFC 2115, 213–216
SNMP, 254
structure of, 207–213
MIB tables
circuit, 215
configuration, 208–209
current statistical, 209–211
error, 216
far end, 213
interval, 211
total, 212
Microsoft DNS and WINS servers, 111
Mill, John Stuart (quote), 167
MMP (Multi-Chassis MP), 167, 184–191
mod-8, 66, 108
mod-128, 66, 108
modems
as asynchronous communications, 6
asynchronous systems and, 32
compared with DTE, 10
configuring for cT1, 161

modems (*continued*)
establishing paths, 80
ITU standards, 11
LAPM and, 66
Numbered Mode and, 108
power output limitations of, 151
PPP and, 77
Starting state and, 85
Monitored Events Count (N393), 124, 125
monitoring considerations, 122, 196
monopoly, AT&T as regulated, 3
most recent fragment (MRF) number, 174
MP+ (LCP configuration option code), 104
MP (Multilink PPP)
features, 167–184
fragmentation, 177, 178
frames, 169
MRRU and, 104, 108
standards, 249
MRAC-34 connector, 10
MRF (most recent fragment) number, 174
MRRU (Maximum Reconstructed Receive Unit)
bundling maintenance and, 182
LCP option code, 108, 177
multilink LCP configuration option code, 104
MRU (Maximum Receive Unit)
LCP option code, 104
PPP configuration option, 114
MTU (Maximum Transmission Unit), 114
mu-law standard, 19
Multi-Chassis MP (see MCMP for Lucent solution; MMP for nonproprietary solution)
multifrequency (MF), 156
Multilink Endpoint Discriminator (LCP codes), 104
Multilink Multinode Bundle Discovery Protocol (RFC 2701), 186
Multilink PPP (see MP)
multiplexing
frame relay and, 116
functions and DCSs, 15
T-carrier hierarchy and, 22–31
Murphy's Law (quote), 132

N

N391 (Status Polling Counter), 124
N392 (Error Threshold Counter), 124, 125
Naks, hints in, 94
NAT, bringing up circuits, 61
national ISDN-2 (NI-2) specification, 163
navigational systems (GPS and LORAN), 34
NCP (Network Control Protocol)
features, 109–112
network layer protocol phase and, 83
packet initialization, 81
PPP and, 77, 82, 84, 249
PPP initializing links, 78
near-end counters, 195
near-end crosstalk (NEXT), 57, 143
negotiation
configuration option, 85
LCP, 81
of magic numbers, 113
timing of messages and, 101
NetBIOS protocol
DNS and WINS server assignments, 111
PPP standards, 249
protocol-rejection messages and, 101
Network Control Protocol (see NCP)
Network Facilities Associated Signaling (NFAS), 162
Network Interface Unit (NIU), 12
network interfaces
line or payload loopbacks, 133
receive clock inference, 35
RJ-48X standard, 12
top T1 trouble spots, 137
transmit clocking, 36
Network Layer Protocol Identifier (NLPID), 120
network layer protocol phase (PPP link state), 83
network masks, 75
network timing, 36
network-protocol management (PPP), 77
NEXT (near-end crosstalk), 57, 143
NFAS (Network Facilities Associated Signaling), 162
NI (network interface), 12, 133
NI-2 (national ISDN-2) specification, 163
NIU (Network Interface Unit), 12
NLPID (Network Layer Protocol Identifier), 120
noise, analog signal problems, 4
Non-Return to Zero Inverted (NRZI), 70
Normal Response Mode (NRM), 64
Nortel, 186

Northern Telecom (Gang of Four), 123
notification requirement, 61
NRM (Normal Response Mode), 64
NRZI (Non-Return to Zero Inverted), 70
Numbered Mode, 104, 107
Nyquist criterion, 18

O

OAM&P, 7
OCU (Office Channel Unit), 14
off-hook, E&M signaling and, 154
1-in-8 test, 25
one (1)
  as mark, 19
  as voltage pulse, 18
  all-ones test, 26
ones density, 24
one-way calling, ordering, 159
on-hook, E&M signaling and, 154
OOF (out of frame)
  defect, 199
  ES and, 202
  LOF and, 197
  out of frame, 48
  SES and, 203
  T1 defect condition, 200
Open command, 78, 80, 82
Open event, 89
Open state, 87
opening flags, closing flags and, 65
ordering, 159, 164
OSI (Open Systems Interconnect)
  models, 12, 63
  protocols, 120, 249
OUI (Organizationally Unique Identifier), 121
out of frame (see OOF)
outgoing calls, 157, 158
OutLQRs counter (PPP LQR Group), 222

P

packets
  as layer 3 construct, 63
  fragmenting for transmission, 169
  IPXCP negotiation, 81
  LCP classes of, 85
  RFC 1661 and RFC 1662, 77
  trading, 61
PAP (Password Authentication Protocol), 114, 251
Part 68 (USOC jack codes), 233
path code violation (PCV), 201, 202, 210
PBX (Private Branch Exchange), 47, 156, 243
PCM (Pulse Code Modulation), 18
PCV (path code violation), 201, 202, 210
PDH/SDH (plesiochronous digital hierarchy/Synchronous Digital Hierarchy), 32, 33, 34
peer-initiated link initialization, 91
peer-initiated link termination, 95
performance considerations
  defects and degradation, 197
  differential transmission and, 228
  list of defects, 199
  monitoring, 192–204
  reports, 30, 53, 193
performance data
  as monitor component, 197
  collecting, 193–196
  FDL and, 29
  problem list, 202–204
performance report message (PRM), 193, 194–196
permanent virtual circuit (PVC), 116, 117, 127
P/F (poll/final) bit, 66
PFC (Protocol Field Compression)
  bundling maintenance and, 182
  LCP configuration option code, 104
  PPP configuration option, 114
PGND SHD (Protective Ground Shield), 231
phase lock loop (PLL), 41
phase locked (timing/synchronization), 21
physical layer
  initializing, 80
  LAN interface to, 63
  PPP initializing links, 78, 82
  problems in, 139–145
physical layer standards, 240–244
physical topologies (SDLC), 64
PID (Protocol Identifier), 121
pinouts
  female DA-15, 235
  RJ-48X, 233
  smart jack, 13
  V.35 pin assignments, 230–232
PLL (phase lock loop), 41
point of presence (see POP)

Point-to-Point Protocol (see PPP)
poll/final (P/F) bit, 66
polling, 73, 123
Polling Verification Timer (T392), 123, 124
POP (point of presence)
  defined, 16
  discovery operation at, 190
  frame relay setup and, 116
PPP Design, Implementation, and Debugging (Carlson), 77
PPP Link Group, 217–220
PPP (Point-to-Point Protocol)
  authentication, 251
  configuration options, 112
  encapsulation, 183
  encapsulation and framing, 83
  events defined by standard, 88–91
  IP addresses and, 113
  logical link states, 78–83
  MIBs for, 217–225
  NCP and IPCP, 109–112
  NLPID assignments, 120
  problems, 146
  protocol used by T1 lines, 63
  standards for, 247–252
  transparency, 84
pppDiscardTest (object), 223
pppEchoTest (object), 223
pppIpConfigAdminStatus (object), 225
pppIpConfigCompression (object), 225
pppIpLocalMaxSlotId (object), 225
pppIpLocalToRemoteCompressionProtocol (object), 224
pppIpOperStatus (object), 224
pppIpRemoteMaxSlotId (object), 224
pppIpRemoteToLocalCompressionProtocol (object), 224
pppLinkConfigFcsSize (object), 220
pppLinkConfigInitialMRU (object), 220
pppLinkConfigMagicNumber (object), 220
pppLinkConfigReceiveACCMap (object), 220
pppLinkConfigTransmitACCMap (object), 220
pppLinkStatusBadAddresses (counter), 217
pppLinkStatusBadControls (counter), 218
pppLinkStatusBadFCSs (counter), 218
pppLinkStatusLocalMRU (object), 219
pppLinkStatusLocalToPeerACCMap (object), 219
pppLinkStatusLocalToRemoteACCompression (object), 219
pppLinkStatusLocalToRemoteProtocolCompression (object), 219
pppLinkStatusPacketTooLongs (counter), 218
pppLinkStatusPeerToLocalACCMap (object), 219
pppLinkStatusPhysicalIndex (object), 217
pppLinkStatusReceiveFcsSize (object), 219
pppLinkStatusRemoteMRU (object), 219
pppLinkStatusRemoteToLocalACCompression (object), 219
pppLinkStatusRemoteToLocalProtocolCompression (object), 219
pppLinkStatusTransmitFcsSize (object), 219
pppLqrConfigPeriod (object), 222
pppLqrConfigStatus (object), 222
pppLqrExtnsLastReceivedLqrPacket (object), 222
pppLqrInGoodOctets (object), 221
pppLqrInLQRs (object), 222
pppLqrLocalPeriod (object), 221
pppLqrOutLQRs (object), 222
pppLqrQuality (object), 221
pppLqrRemotePeriod (object), 221
PRI (see ISDN PRI)
primary DNS (IPCP), 110
primary NBNS (IPCP), 110
Primary Rate Interface (see ISDN PRI)
primary reference source (PRS), 32, 36
printing ASCII termination requests, 100
Private Branch Exchange (PBX), 47, 156, 243
PRM (performance report message), 193, 194–196
problems
  bipolar violations, 143
  cables, 134
  common V.35 cabling, 232
  frame relay, 146
  glare in E&M signaling, 155
  HDLC, 147
  in carrier's network, 134
  incorrect clock settings, 145
  PPP, 146

slip events, 195
timing, 143
timing in serial circuits, 134
using loopback testing, 134
(see also troubleshooting), 143
proprietary, 71
protection switching, 8
Protective Ground Shield (PGND SHD), 231
Protocol Field Compression (see PFC)
Protocol Identifier (PID), 121
protocol numbers (web site), 248
Protocol-Reject
address negotiation and, 113
bundling and, 168
LCP rejection message, 101
LCP type code, 98
protocols
bundling and, 168
cHDLC table codes, 71
configuring for cT1, 161
IANA standards, 239
LAPD, 30
major vendors and MMP, 186
NCPS and, 83
network control standards, 249
not publicly documented, 250
used by T1 lines, 63
PRS (primary reference source), 32, 36
Pulse Code Modulation (PCM), 18
pulse density (ones density), 24
pulses
alternating polarity and error detection, 19
clock inference from, 35
DSX-1 interface and, 47
tendencies of, 18
timing/synchronization, 20
PVC (permanent virtual circuit), 116, 117, 127
PVC status information element, 126

Q

Q.921, 163
Q.922, 118
Q.930, 163
Q.931, 163
Q.933, 129
Q.933 Annex A, 123
Q.933 protocol, 120
QRS (quasi random signal) tests, 25
Quality protocol (LCD option code), 104, 105
queries, bundle heads and, 189

R

R1 signaling in ITU recommendations, 159
radio services, AT&T experiments with, 3
RADIUS servers, 162
RAI (remote alarm indication)
failure, 198
troubleshooting, 141
yellow alarm signal, 48, 49
(see also yellow alarm)
RBOCs (Regional Bell Operating Companies), 6, 240
RC (Receive Clock), 229
RCA (Receive Configure-Ack) event
RCLK (Receive Clock), 229
RCN (Receive Configure-Nak), 89
RCR+ event (Reception of Configure-Request), 89
RCR- event (Reception of Configure-Request), 89
RD A (Receive Data A), 232
RD A (Receive Data B), 232
RD (Receive Data), 229
RD (receive data), 48
RD+ (Receive Data A), 232
RD- (Receive Data B), 232
RD (Request Disconnect) command, 68
reassembly, 171–177
receive alarm (condition), 26
Receive Clock (RC, RXC, RCLK), 229
Receive Configure-Ack (RCA), 89
Receive Configure-Nak (RCN), 89
Receive Data A (RD+), 232
Receive Data A (RD A), 232
Receive Data A (RX+), 232
Receive Data B (RD-), 232
Receive Data B (RD A), 232
Receive Data B (RX-), 232
Receive Data (RD, RX), 48, 229
receive data (RXD), 48
Receive Echo-Request (RXR) event, 90
Receive Not Ready (RNR), 67
Receive Protocol (RXJ+) event, 90
Receive Protocol (RXJ-) event, 90

Receive Terminate-Ack (RTA) event, 89
Receive Terminate-Request (RTR) event, 89
Receive Timing (RT), 229
Receive Timing (RT+), 232
Receive Timing (RT-), 232
Receive Timing (RxClk+), 232
Receive Timing (RxClk-), 232
Receive Unknown Code (RUC) event, 90
Received Line Signal Detector (RLSD), 231
Receiver Ready (RR), 67
Reception of Configure-Request (RCR+) event, 89
Reception of Configure-Request (RCR-) event, 89
record keeping, 160, 165
red alarm, 26, 48, 139
redundancy and SONET switching, 10
reframing buffers and, 42
Regional Bell Operating Companies (RBOCs), 6, 240
Registered Jack (RJ), 12, 233
REJ (Reject), 67
rejection messages (code and protocol), 100
remote alarm conditions, 26
remote alarm indication (see RAI)
Remote Loopback (RL), 133, 134, 231
repeaters, 8, 14, 18
Report Type information element, 126
Request Disconnect (RD), 68
Request for Comments (RFC), 239
Request Initialization Mode (RIM)
request to send (RS), 48
Request to Send (RTS), 48, 229, 231
Request-Sent (Req-Sent), 87
Reset (RSET) command, 68
Reset-Ack, 99, 103
Reset-Req, 103
Reset-Request, 99
resistive loss, 15
Reverse ARP Ethertype, 71
RFC 1055 (PPP standards), 247
RFC 1144 (PPP standards), 247
RFC 1149 (April 1), 239
RFC 1155 (SNMP standards), 252
RFC 1157 (SNMP standards), 252
RFC 1172 (style address assignment), 110
RFC 1213 (SNMP standards), 253
RFC 1332 (style address negotiation), 109, 110, 111, 249
RFC 1334 (PPP Authentication Protocols), 251
RFC 1376 (DNCP), 249
RFC 1377 (OSINLCP), 249
RFC 1378 (ATCP), 249
RFC 1471 (Link Control Protocol), 217–220, 220–222, 255
RFC 1472 (PPP Security Group), 255
RFC 1473 (PPP IP Group MIB), 223, 255
RFC 1490 (InARP), 120–122, 131, 245
RFC 1547 (PPP standards), 248
RFC 1552 (IPXCP), 249
RFC 1570 (PPP LCP Extensions), 102, 103, 251
RFC 1661 (PPP), 77, 106, 184, 248
RFC 1662
  PPP framing information, 77
  PPP in HDLC-like Framing, 248
  stuffing procedures, 84
  unnumbered mode control and, 108
RFC 1663 (Numbered Mode), 104, 107, 248
RFC 1717 (multilink specification), 179
RFC 1812 (requirements for IPv4 routers), 77
RFC 1877 (IPCP extensions), 111, 249
RFC 1905 (SNMPv2 standards), 253
RFC 1934 (MP+), 250
RFC 1962 (CCP), 251
RFC 1968 (ECP), 251
RFC 1989 (PPP Link Quality Monitoring), 105, 251
RFC 1990 (PPP Multilink Protocol (MP)), 108, 167, 183, 249
RFC 1994 (CHAP), 252
RFC 2043 (SNACP), 249
RFC 2097 (NBFCP), 249
RFC 2115 (SNMP MIB standard), 213–216, 255
RFC 2125 (PPP standards), 250
RFC 2153 (PPP Vendor Extensions), 103, 104, 251
RFC 2284 (EAP), 252
RFC 2290 (Mobile-IPv4 Configuration Option for PPP IPCP), 249
RFC 2341 (L2F), 250
RFC 2390 (Inverse ARP), 246

RFC 2427
  Multiprotocol Interconnect over Frame Relay, 245
  NLPID format preferred, 121
  update of RFC 1490, 120
RFC 2433 (CHAP extensions), 252
RFC 2495 (SNMP MIB standard), 205–213, 254
RFC 2570 (SNMPv3 standards), 253
RFC 2571 (SNMP Management), 254
RFC 2572 (SNMPv3 standards), 254
RFC 2574 (SNMPv3 standards), 254
RFC 2575 (SNMPv3 standards), 254
RFC 2576 (SNMPv3 standards), 254
RFC 2578 (SMIv2), 253
RFC 2579 (SMIv2), 253
RFC 2661 (L2TP), 250
RFC 2664 (IPv4 routers), 77
RFC 2701 (Bundle Discovery Protocol), 186, 250
RFC 2759 (CHAP Extensions, Version 2), 252
RFC 2878 (BCP), 249
RFC (Request For Comments), 239
RIM (Request Initialization Mode) command, 68
ringing, 26
RJ (Registered Jack), 12, 233
RJ-11 telephone jack, 233
RJ-45, 140, 233
RJ-48, 60, 140, 235
RJ-48X, 12, 233
RL (Remote Loopback), 133, 134, 231
RLSD (Received Line Signal Detector), 231
RNR (Receive Not Ready), 67
robbedBit (dsx1SignalMode), 208
robbed-bit signaling
  channels and, 31
  cT1 and, 153
  line provisions for, 150
  on T1 circuits, 30
  T-carrier systems and, 7
routers
  as offload MMP servers, 191
  as PPP configuration option, 113
  as serial DTE, 226
  bringing up circuits, 61
  frame relay configurations, 131
  installation considerations, 59
  SLARP protocol, 71–75
  to CSU/DSU with V.35, 10
  troubleshooting transmission problems, 40
  V.35 and, 227
RR (Receiver Ready), 67
RS (request to send), 48
RS-232 interface, 227
RS-422 interface, 227
RSET (Reset) command, 68
RT (Receive Timing), 229
RT+ (Receive Timing), 232
RT- (Receive Timing), 232
RTA (Receive Terminate-Ack) event, 89
RTR (Receive Terminate-Request) event, 89
RTS (Request to Send), 48, 229, 231
RUC (Receive Unknown Code) event, 90
rules, bundle establishment, 183
RX (receive data), 48
RX+ (Receive Data A), 232
RX- (Receive Data B), 232
RXC (Receive Clock), 229
RxClk+ (Receive Timing), 232
RxClk- (Receive Timing), 232
RXD (receive data), 48
RXJ+ (Receive Protocol or Code-Reject) event, 90
RXJ- (Receive Protocol or Code-Reject) event, 90
RXR (Receive Echo-Request) event, 90

S

SABM (Set Asynchronous Balanced Mode) command, 67, 68
SABME (Set Asynchronous Balanced Mode Extended) command, 68
SAIC (Bellcore/Telcordia), 240
Salo, Tim (quote attribution), 236
sampling theorem, 18
Sandburg, Carl (quote), 149
SAPI 15, path and signal testing with, 193
SAPI (Service Access Point Identifier), 193
SARM (Set Asynchronous Response Mode) command, 68
sca (Send Configure-Ack) action, 90
scj (Send Code-Reject) action, 91
scn (Send Configure-Nak or Configure-Rej) action, 90
SCR A (Serial Clock Receive A), 232

SCR B (Serial Clock Receive A), 232
scr (Send Configure-Request) action, 90
SCT A (Serial Clock Transmit A), 232
SCT B (Serial Clock Transmit B), 232
SCTE A (Serial Clock Transmit External A), 232
SCTE B (Serial Clock Transmit External B), 232
SCTE (serial clock timing external) label, 40
SCTE (Serial Clock Transmit External) serial signal, 229
SD (Send Data), 48, 229
SD+ (Send Data A), 232
SD- (Send Data B), 232
SDLC (Synchronous Data Link Control), 64, 66
secondary DNS (IPCP), 110
secondary NBNS (IPCP), 110
SEF (severely errored framing), 200
SEF/AIS second (SAS-P) performance data, 203, 204
SEFS (Severely Errored Framing Second), 210
Selective Reject (SREJ), 67
self-initiated link initialization, 91
self-initiated link termination, 95
Send Code-Reject (scj) action, 91
Send Configure-Ack (sca) action, 90
Send Configure-Nak (scn) action, 90
Send Configure-Request (scr) action, 90
Send Data A (SD+), 232
Send Data A (TD A), 232
Send Data A (TX+), 232
Send Data B (SD-), 232
Send Data B (TD B), 232
Send Data B (TX-), 232
Send Data (SD), 48, 229
send data (TX), 48
Send Data (TXD), 48
Send Echo-Reply (ser) action, 91
Send Terminate-Ack (sta) action, 91
Send Terminate-Request (str) action, 91
Send Timing (ST+), 232
Send Timing (ST-), 232
Send Timing (TxClk+), 232
Send Timing (TxClk-), 232
sequence numbers, SLARP protocol and, 73
ser (Send Echo-Reply) action, 91
Serial Clock Receive A (SCR A), 232
Serial Clock Receive A (SCR B), 232
serial clock timing external (SCTE) label, 40
Serial Clock Transmit A (SCT A), 232
Serial Clock Transmit B (SCT B), 232
Serial Clock Transmit External A (SCTE A), 232
Serial Clock Transmit External B (SCTE B), 232
Serial Clock Transmit External (SCTE) serial signal, 229
serial communications
  data ports as, 37
  overview, 226–229
  PPP and, 77
  routers to T1s, 10
Service Access Point Identifier (SAPI), 193
Service Profile Identifier (SPID), 164
SES (severely errored seconds), 203, 204, 210
Set Asynchronous Balanced Mode Extended (SABME) command, 68
Set Asynchronous Balanced Mode (SABM) command, 67, 68
Set Asynchronous Response Mode (SARM) command, 68
Set Initialization Mode (SIM), 68
Set Normal Response Mode Extended (SNRME) command, 68
Set Normal Response Mode (SNRM) command, 68
Severely Errored Framing Second (SEFS), 210
severely errored framing (SEF), 200
severely errored seconds (SES), 203, 204, 210
SF (superframe)
  A and B channels, 31
  CRC errors, 194
  CSU/DSU configuration options, 53
  drawbacks, 27
  FE and, 201
  multiple frames as, 23
  OOF defect and, 199
  PCV and, 201
  SES and, 203
  standard, 26

S-frame (supervisory frame), 67
SGBP (Stack Group Bidding Protocol), 191, 250
SGND (Signal Ground), 231
shielding
  pre-connection tasks for wiring, 60
  recommended distances for cables, 58
  signal leakage and, 58
short sequence number (SSN), 108, 178, 182
shorting bars, 12, 233, 234
Signal Ground (GND or SGND), 231
signal loss
  allowable levels, 14
  basic failure, 197
  deliberate with long-haul LBO, 45
signaling
  cT1 and, 153–159, 160, 161, 243
  degradation, 8
  E&M, 153–156
  failures and alarms, 197–200
  framing bits, 26
  ground start, 156
  ISDN, 163, 164
  leakage, 58
  loop start, 158
  methods to separate channels, 29
  robbed bits and channels, 31
  serial lines, 227–228
  strength, 2
  testing, 193
SIM (Set Initialization Mode) command, 68
Simple Network Management Protocol (see SNMP)
SLARP (Serial Line ARP)
  echo messages compared with, 102
  router-to-router, 71–75
sliding windows, 66
slips
  controlled, 41, 202
  timing problems and, 41–43
  timing problems and events, 195
  uncontrolled, 41
smart jacks
  AIS and, 198
  locations for, 13
  triggering remote loopback, 133
SNA Control Protocol (SNACP), 249
SNA protocol, 78, 249
SNMP (Simple Network Management Protocol)
  monitoring, 196
  standards, 252–255
  T1-related MIBs, 205–225
SNMPv standards, 253
SNRM (Set Normal Response Mode) command, 68
SNRME (Set Normal Response Mode Extended) command, 68
SONET (synchronous optical network)
  benefits of, 8
  data streams and rings, 16
  PPP use with, 76
  protection switching, 10
spans, 8, 26
spare capacity, 10
specific access station (SAS) signaling, 159
specifications
  differences between LMI, 129
  HDLC example, 238
  IEC, 238
  Q.922, 118
  Registered Jacks, 233
  RFC 1661, 77
  switches, 163
  (see also entries beginning with RFC)
speed (T-carrier comparison table), 22
SPID (Service Profile Identifier), 164
SREJ (Selective Reject), 67
SSN (short sequence number), 108, 178, 182
ST+ (Send Timing), 232
ST- (Send Timing), 232
sta (Send Terminate-Ack) action, 91
Stack Group Bidding Protocol (SGBP), 191, 250
Standard Ethernet, troubleshooting, 140
standards
  A-law, 19
  ANSI T1.403, 12
  dial-up and PPP, 77
  for PPP, 247–252
  frame relay, 244
  HDLC, 246
  ITU and, 11, 237
  JTC, 238
  kbps and, 18
  mu-law, 19

standards (*continued*)
ones density, 24
physical layer, 240–244
RJ numbers, 12
SNMP, 252–255
timing and synchronization, 20
standards bodies, 236–240
Starting (LCP engine state), 85
state engines, 78, 184
statistics
MIB and, 209
reported in PRM, 194–196
Status Polling Counter (N391), 124
Stopped (LCP engine state), 85
Stopping (LCP engine state), 86
str (Send Terminate-Request) action, 91
StrataCom (Gang of Four), 123
Stratum 1-4 Bell system timing sources, 33
subchannels (B and D channels), 162
subnet-mask assignment, 112
Subnetwork Access Protocol (SNAP), 120, 121
supervisory frame (S-frame), 67
SVC (switched virtual circuit), 116
switch types, configuring ISDN PRI, 165
switches, DCS hardware as, 15
synchronization
AIS and, 26
AMI and, 18
bit stuffing and, 70
CSU/DSU process, 24
framing bits and, 28
ones density and, 24
timing and, 20
Synchronous Data Link Control (SDLC), 64, 66
synchronous optical network (see SONET)

T

T1 circuits
architectural overview, 8–16
as T-carrier, 5
configuring ISDN PRI, 165
costs and alternatives, 7
getting connectivity, 56–62
interfaces, 133, 165
Internet chronology, 17
parameters, 161
performance monitoring, 192–204
pricing, 15
protocols for lines, 63
RJ-48X jacks and, 233
roots, 1
serial connections to routers, 10
specifications, 14
timing in, 35–41
top trouble spots, 137
T1.231, 195, 196, 202, 205
T1.403, 29
T3 circuits, 17
T391 (Link Integrity Verification Polling Timer), 123, 124
T392 (Polling Verification Timer), 123, 124
T5 circuits, 5
Tannenbaum, Andrew (quote), 236
TC (Transmit Clock) serial signal, 229
T-carriers
1950 through 1970, 4–5
multiplexing and, 22–31
problems with, 7
timing, clocking, and synchronization, 32–43
(see also T1 circuits)
TCLK (Transmit Clock) serial signal, 229
tcpdump, troubleshooting with, 137
TCP/IP, 66, 108, 247
TD A (Send Data A), 232
TD B (Send Data B), 232
TD (Transmit Data) serial signal, 229
TDM (time division multiplexing), 5, 22
technology, cyclic nature of, 3
TEI (Terminal Endpoint Identifier), 193
telco, 12, 21, 135
Telcordia, 163, 240
telecommunications, 8–16
telegraphs, 2, 11
telephone networks
basic digital transmission, 17–21
history of U.S., 1–7
signaling on T1 links, 30
U.S. as plesiochronous, 34
telephone numbers
ordering, 159, 160
ordering for ISDN, 164
transmission (MF and DTMF), 156
Terminal Endpoint Identifier (TEI), 193
terminal framing bits, 26
Terminate-Ack, 98, 100, 183

Terminate-Request
  address negotiation and, 113
  bundling maintenance and, 183
  LCP termination messages and, 100
  LCP type code, 98
Termination (PPP link state), 83
Test Mode (TM), 232
Test (TEST) command, 68
testing
  all-ones test, 26
  all-zeros test, 26
  loopback, 50, 132
  use of loopback retention signal, 30
  1-in-8 test, 25
  path and signal, 193
  3-in-24 test, 25
this layer down (tld) action, 90
this layer finished (tlf) action, 90
this layer started (tls) action, 90
this layer up (tlu) action, 90
3-in-24 test, 25
throughTiming
    (dsx1TransmitClockSource), 209
time division multiplexing (TDM), 5, 22
time slots, 24, 35, 52
Time-Remaining (LCP type code), 99
timing
  external timing, 52
  in T1 circuits, 35–41
  internal timing, 37
  loop timing, 36, 43, 52
  network timing, 36
  physical layer standards, 241
  problems, 143, 195
  synchronization and, 20
  taxonomy, 32–34
  variation and slips in signals, 41–43
  (see also clocking)
tip and ring scheme (RJ-48X), 234
tld (this layer down) action, 90
tlf (this layer finished) action, 90
tls (this layer started) action, 90
tlu (this layer up) action, 90
TM (Test Mode), 232
TO+ event, 89
TO- event, 89
total table (MIB), 212
TR 9577, 120
traffic, 118, 176
transistors, computerization and, 4
transmissions
  steps for IP packets, 80
  timing and synchronization, 20
  tip and ring scheme, 234
  unbalanced methods, 227
transmit clock source (CSU/DSU
    configuration options), 53
Transmit Clock (TC, TCLK, TXC), 229
Transmit Data (TD, TX), 229
Transmit Timing (see TT)
transparency, 70, 84
troubleshooting
  bringing up circuit, 62
  calling for help, 135–138
  common V.35 cabling problems, 232
  determining alarm declaration, 198
  error conditions, 143
  identification message usefulness in, 102
  link layer problems, 145–148
  near-end and far-end statistics for, 196
  outline, 138
  physical layer problems, 139–145
  rules of fixing things, 136
  T1 performance monitoring, 192–204
  tools and techniques, 132–138
  top T1 trouble spots, 137
  with debug trace, 137
  yellow alarm diagnostics table, 142
  (see also error checking/detection)
trunk systems (see carriers)
TT (Transmit Timing) serial signal, 229
tunneling techniques, 184
two-way calling, ordering, 159
TX (send data), 48
TX+ (Send Data A), 232
TX- (Send Data B), 232
TX (Transmit Data) serial signal, 229
TXC (Transmit Clock), 229
TxClk+ (Send Timing), 232
TxClk- (Send Timing), 232
TXD (send data), 48

U

UA (Unnumbered Acknowledgment)
    command, 68
UAS (Unavailable Seconds), 204, 210
U-frames (unnumbered frames), 67
UI (Unnumbered Information), 68

uncontrolled slips, 41
Unix, tcpdump for troubleshooting, 137
Unnumbered Poll (UP) command, 68
Up event
  as trigger, 109
  LCP and, 81
  Lower layer up, 88
  PPP initializing links, 78, 82
UP (Unnumbered Poll) command, 68
USOC (Universal Service Order Codes), 233
UTP (unshielded twisted pair) cable, 58

V

V.24 circuit, name definitions, 229
V.28 circuit, electrical characteristics, 229
V.35 circuit
  as serial interface, 227
  common cabling problems, 232
  explicit timing with, 35
  high-speed serial, 230–233
  pre-connection tasks, 60
  serial standard, 10, 11
  signal compared with, 82
V.42 specification, 66
Vail, Theodore, 3
Van Jacobson compression, 84, 224, 247
vendor, calling for help, 136
vendor extensions, 103, 104
Vendor-Extension (LCP type code), 98
virtual connections, 116
virtual private networks (VPNs), 116
voice transmissions
  channels, 2, 22
  digitalization, 18
  errors in, 19
voltage, 18, 37
VPNs (virtual private networks), 116

W

WAN, 63
wander, 35
web sites
  ANSI Committee T1, 237
  April 1 RFC, 239
  DISA Circular 300-175-9, 213, 242
  Ethernet type codes, 247
  Frame Relay Forum, 238
  IANA, 239
  IETF, 239
  ISO, 238
  ITU, 237
  ITU URL, 11
  PPP protocol numbers, 248
  quote attribution, 236
  RFCs, 239
  Telcordia, 240
  Van Jacobson Usenet post, 248
Western Electric Manufacturing Company, 2, 7, 26
Western Union, 2
Westfield, Bill, 246
Winchester block connector, 10
windows in HDLC, 66
wink start, 155
WINS servers, 111
wire pairs in T1 circuits, 8
wiring, 13, 133, 151

X

X.25, 115, 244
XCLK label, 40
XID (Exchange Information) command, 68
XOFF control characters, 70
XON control characters, 70

Y

yellow alarm
  CSU/DSU configuration options, 53
  RAI, 48, 49
  superframe alarm conditions, 26
  troubleshooting, 141

Z

Zeno (quote), 32
zero (0)
  as space, 19
  all-zeros test, 26
  B8ZS encoding, 21
  lack of voltage pulse, 18
Zeroth Rule, 136
zrc (zero restart counter) action, 90

About the Author

Matthew S. Gast is a former aspiring research physicist who found computer networking too addictive for his own good. He was initially hooked during a summer research program, when he found the Ethernet drop in his dorm room far more interesting than his assigned task of developing monitoring software for a particle accelerator. Upon returning to academic life in the fall, he suffered severe withdrawal symptoms, which he conquered only with the help of a new religion—Unix. He has been connected and sober since that fateful summer.

Matthew is currently a Research Fellow at Nokia. His work deals with the intersection of mobility and security, with frequent excursions into routing and cryptography. In the course of daydreaming about changing the world, Matthew has been known to imagine networks that are much less complicated and repetitive to manage. Frequent travels take him away from his Silicon Valley home to a variety of locations throughout the U.S. and Europe, including the occasional visit to the corporate headquarters in Finland. On his last trip to Finland, Matthew crossed the Arctic Circle on a snowmobile in February.

Matthew is the coauthor of *Network Printing* (also published by O'Reilly) as well as several other articles and white papers.

Colophon

Our look is the result of reader comments, our own experimentation, and feedback from distribution channels. Distinctive covers complement our distinctive approach to technical topics, breathing personality and life into potentially dry subjects.

The animal on the cover of *T1: A Survival Guide* is a caribou. Caribou, or reindeer, can be found in the arctic tundra, the mountain tundra, and the northern forests of North America, Russia, and Scandinavia. There are about 5 million caribou in the world, divided into three types: woodland, barren-ground, and Peary. A fourth type native to Canada, called the Queen Charlotte Island caribou, is extinct. One of the reasons the caribou is able to survive the northern climate is because its primary food source is lichen. The caribou's keen sense of smell enables it to find lichen buried under the snow.

Caribou are the only members of the deer family in which both sexes grow antlers. Adult bulls shed their antlers around November or December after mating, while cows and young caribou often carry their antlers through the entire winter. During growth, the antlers have a fuzzy covering, or *velvet*, which contains blood vessels that carry nutrients.

In addition to their antlers, caribou have lateral hooves that allow their feet to spread on snow or soft ground. The hooves also act as paddles, making the caribou an excellent swimmer.

Linley Dolby was the production editor and copyeditor for *T1: A Survival Guide*. Rachel Wheeler proofread the book. Nicole Arigo and Claire Cloutier provided quality control. Kimo Carter, Sarah Sherman, Mary Brady, and Sada Preisch provided production support. Lucie Haskins wrote the index.

Ellie Volckhausen designed the cover of this book, based on a series design by Edie Freedman. The cover image is from the *Illustrated Natural History: Mammalia*. Emma Colby produced the cover layout with QuarkXPress 4.1 using Adobe's ITC Garamond font.

Melanie Wang designed the interior layout based on a series design by Nancy Priest. Neil Walls converted the files from Microsoft Word to FrameMaker 5.5.6 using tools created by Mike Sierra. The text and heading fonts are ITC Garamond Light and Garamond Book; the code font is Constant Willison. The illustrations that appear in the book were produced by Robert Romano and Jessamyn Read using Macromedia FreeHand 9 and Adobe Photoshop 6. This colophon was written by Linley Dolby.

Whenever possible, our books use a durable and flexible lay-flat binding. If the page count exceeds this binding's limit, perfect binding is used.